AF411336

COVID-19 and Other Pleiotropic Actions of Vitamin D: Proceedings from the 5th International Conference "Vitamin D—Minimum, Maximum, Optimum"

COVID-19 and Other Pleiotropic Actions of Vitamin D: Proceedings from the 5th International Conference "Vitamin D—Minimum, Maximum, Optimum"

Editor

Pawel Pludowski

MDPI • Basel • Beijing • Wuhan • Barcelona • Belgrade • Manchester • Tokyo • Cluj • Tianjin

Editor
Pawel Pludowski
The Children's Memorial Health Institute
Poland

Editorial Office
MDPI
St. Alban-Anlage 66
4052 Basel, Switzerland

This is a reprint of articles from the Special Issue published online in the open access journal *Nutrients* (ISSN 2072-6643) (available at: https://www.mdpi.com/journal/nutrients/special_issues/COVID_Vitamin_D_Conference).

For citation purposes, cite each article independently as indicated on the article page online and as indicated below:

LastName, A.A.; LastName, B.B.; LastName, C.C. Article Title. *Journal Name* **Year**, *Volume Number*, Page Range.

ISBN 978-3-0365-6934-5 (Hbk)
ISBN 978-3-0365-6935-2 (PDF)

Contents

Review

Vitamin D Dosing: Basic Principles and a Brief Algorithm (2021 Update)

Andrius Bleizgys

Clinic of Internal Diseases, Family Medicine and Oncology, Faculty of Medicine, Vilnius University Santariškių 2, LT-08661 Vilnius, Lithuania; andrius.bleizgys@gmail.com

Abstract: Nowadays, in modern societies, many people can be at high risk to have low vitamin D levels. Therefore, testing of serum 25-hydroxy-vitamin D (25OH-D) levels should be performed before prescribing them vitamin D supplementation. However, in some cases the 25OH-D level assessment is not available at the right moment, e.g., due to mandatory quarantine of COVID-19 outpatients. Therefore, such patients could be advised to start taking moderate vitamin D doses (e.g., 4000 IU/day for adults), and their 25-OH-D levels could be checked later. The proposed algorithm also comprises vitamin D dosing principles when baseline 25OH-D levels are known.

Keywords: vitamin D; calcidiol; supplementation; COVID-19

Citation: Bleizgys, A. Vitamin D Dosing: Basic Principles and a Brief Algorithm (2021 Update). *Nutrients* **2021**, *13*, 4415. https://doi.org/10.3390/nu13124415

Academic Editor: Pawel Pludowski

Received: 24 November 2021
Accepted: 8 December 2021
Published: 10 December 2021

Publisher's Note: MDPI stays neutral with regard to jurisdictional claims in published maps and institutional affiliations.

1. Introduction

The numbers of new COVID-19 cases and deaths from COVID-19 are increasing in many countries, despite the availability of different vaccines, more or less strict lockdowns or other state-level infection control measures, and various treatment options. It seems that there is a need for other effective tools for combating the COVID-19 disaster. Vitamin D (Vit. D) was suggested as one such tool [1]. Is it well known that Vit. D, in the form of calcitriol, has a pleotropic activity in human organism [2]. There is some evidence from clinical trials regarding the benefits of Vit. D for COVID-19 patients [3–7], as it was also stated in two recent reviews [8,9]. In the previous paper, the main mechanisms of Vit. D action in regard to COVID-19 infection were already discussed, and some suggestions for Vit. D dosing in the COVID-19 era were proposed [10]. Recently it was suggested that Vit. D might also act as an important cofactor of strengthening the activity of COVID-19 vaccines [11].

It is still debatable if checking the 25-hydroxyvitamin D levels (25OH-D-a marker of Vit. D status) and supplementing with Vit. D should be included in the COVID-19 prevention and treatment guidelines. Nevertheless, diagnosing, preventing, and treating low Vit. D status is still an issue in regard to both COVID-19 patients and the whole population, particularly keeping in mind the reduced accessibility to health care services during the COVID-19 pandemic. The current paper discusses the main Vit. D dosing principles, outlines the most important low Vit. D risk groups, and suggests a brief Vit. dose selection algorithm for clinical practice.

2. Is There a Need for New Guidelines/Algorithms?

During the past 10–15 years, different international and regional guidelines for low Vit. D status prevention and treatment were published (e.g., [12–16]). However, for several reasons, physicians might currently need some new kinds of recommendations for clinical practice regarding Vit. D status evaluation and Vit. D dosing.

1. Despite the available evidence of vitamin's D important role for the human organism, including extra-skeletal health and the high prevalence of low Vit. D status in different regions of the world [17–21], many countries still do not have national, up-to-date,

approved Vit. D guidelines. The same applies also to Lithuania, which has only the Rickets' diagnosis and treatment guidelines approved in 2015. Moreover, in most countries, the potential beneficial role of Vit. D for COVID-19 prevention and treatment (i.e., Vit. D as an adjuvant) is still not accepted; consequently, no specialized relevant recommendations are developed. Paradoxically, it is the COVID-19 pandemic that inspired the author of the present article to start developing national Vit. D guidelines for Lithuania. Hopefully, the basic principles of those guidelines presented in the current paper could be an additional source for more specialized future recommendations both for Lithuania and for other countries.

2. Traditionally, any well-prepared Vit. D guidelines should reflect clinical practice and therefore must include the following domains: definition of risk groups for low vitamin D; principles of evaluation of Vit. D status by using laboratory measurements; and Vit. D dosing for prevention and treatment. However, the COVID-19 pandemic brought some challenges that aggravated our routine clinical practice. Firstly, due to reduced accessibility to health care facilities, mandatory isolation of some patients (due to diagnosed COVID-19 disease or due to close contact with a confirmed COVID-19 case), or a patient's fear of getting SARS-CoV-2 during visits to a clinic or laboratory, it is not possible to perform the measurements of serum 25OH-D levels at the desired time. Therefore, the recent Vit. D status of many outpatients could remain unknown. Secondly, with the absence of data on recent 25OH-D level measurements, it might be difficult for physicians to make decisions regarding Vit. D dosing, particularly for low Vit. D risk group patients. We need an extended list of risk factors that might suggest the clinician to presume that certain patients could be put into a Vit. D risk group and, consequently, to suggest him/her higher Vit. D doses for supplementation. Finally, even disregarding the potentially beneficial direct Vit. D role on COVID-19 prevention and treatment, it is wise to remember that the problem of low Vit. D in society has not disappeared during the pandemic. Moreover, some people, due to various reasons during lockdowns, may have even higher risk to newly develop Vit. D insufficiency, leading to poorer skeletal and extra-skeletal health [10]. Patients having low 25OH-D levels might be considered as high-risk group for getting severe illness from COVID-19 [22].

3. In older Vit. D guidelines, there is almost no talk about the causes that could result in failure to achieve the desired levels of 25OH-D by supplementing Vit. D, and the suggested actions for physicians. In the present article, the author also tried, in part, to fulfil those gaps.

4. It is the Vit. D supplementation that modern guidelines should be mostly oriented to. Production of vitamin D3 in the skin is not a reliable source for repletion of low Vit. D status. Firstly, human skin is able to produce only limited amount of vitamin D3 that can enter the circulation [23,24]. Secondly, it is difficult to predict the effect of solar radiation in regard to vitamin D3 production and its influence on 25OH-D levels, since a large number of factors might affect vitamin D3 synthesis in the skin, e.g., skin type, patient age, time of the day, altitude, etc. [23,25,26]. Finally, in some countries, e.g., Lithuania, that are located at the middle latitudes, the intensity of solar radiation decreases significantly during the cold season, and the synthesis of vitamin D3 in the skin is almost absent during the period from October till March [27,28]. Food, unless fortified with Vit. D, usually cannot serve as a valuable source of this vitamin, too [27,29,30]. Therefore, this paper does not discuss recommendations on exposure to sunlight or influence of certain types of food for prevention or treatment of low Vit. D status.

3. Risk Factors for Low Vitamin D Status

There are a number of diseases and conditions associated with low Vit. D status. Illnesses definitely caused by inadequate Vit. D status comprise only a small part of the group of all risk factors for low Vit. D status. Many diseases, conditions, or drugs *per se* can impair vitamin's D metabolism and/or increase the needs for this vitamin, thus contributing to development of low Vit. D status.

In addition, there are a lot of diseases and conditions where low Vit. D status can be considered only as an epiphenomenon. In other words, low Vit. D status itself does not necessarily have cause−consequence relationships with certain diseases or conditions, but it frequently accompanies them and could have common causes. In many cases, an unhealthy lifestyle can act as such a common cause, and low Vit. D status might serve as an indicator of that lifestyle [31–34].

It is worth to try to identify those risk factors, since some of them can be corrected (or prevented) and this may help to prevent and treat low Vit. D. In addition, COVID-19 disease and low Vit. D status share many risk factors, e.g., older age and obesity [10]. Therefore, considering together those risk factors (including low Vit. D status) can help to correctly evaluate the risk of severe COVID-19 and, in some cases, also advocate vaccination. On the other hand, confirmed symptomatic COVID-19 disease might be considered as risk factor for suspecting low Vit. D status.

For simplicity, risk factors for low Vit. D status can be divided into several groups (Table 1) [12,13,20,24,35–41]. Patients having one or several risk factors should be tested for their serum 25OH-D levels, since the analysis results helps in making better decisions regarding Vit. D supplementation [12,42,43].

Table 1. Risk factors for low Vit. D status.

Groups of Risk Factors	Examples: Diseases, Conditions, Lifestyle Features
Musculoskeletal disorders	Rickets, osteoporosis, osteopenia, "bone pains", muscle pain, myopathy, myodystrophy, recurrent ("low energy") bone fractures, recurrent falls, bone deformities
Endocrine and metabolic diseases/conditions	Diabetes mellitus (type I and II), metabolic syndrome, obesity, overweight, hypo- and hyperparathyroidism, hypo- and hyperthyroidism, hypocalcemia, calciuria, phosphatemia, hypo- and hyperphosphatasia, phosphaturia, dyslipidemias
Increased demand for physiological reasons	Childhood, adolescence, pregnancy, breastfeeding
Malabsorption syndromes	Pancreatic exocrine insufficiency (old age, pancreatitis, type II diabetes, etc.), inflammatory bowel disease (Crohn's disease, ulcerative colitis), cystic fibrosis, lactose intolerance, celiac disease, bariatric surgery
Diseases of the liver and bile ducts	Hepatic insufficiency, cirrhosis of the liver, cholestasis, hepatosteatosis
Kidney diseases	Renal insufficiency, chronic kidney disease (especially stages III–V), nephrotic syndrome
Respiratory diseases	Bronchial asthma, chronic obstructive pulmonary disease
Infectious diseases	Tuberculosis, recurrent respiratory infections
Systemic connective tissue diseases	Rheumatoid arthritis, systemic lupus erythematosus, dermatomyositis, fibromyalgia
Skin diseases	Atopic dermatitis, psoriasis
Diseases of the nervous system	Multiple sclerosis, Parkinson's disease, dementia, cerebral palsy, autism
Decreased production of vitamin D3 in the skin	Older age (especially >70 years) Active protection against sun exposure (sunscreens, etc.) Cultural features (usual full-body clothing) Rare outdoor activities (work and leisure predominantly indoors; living in a care home) Increased air pollution (living in a city) Winter season (at medium latitudes) Dark-skinned (especially Africans)
Nutritional features	Veganism and other types of vegetarianism Allergy to cow's milk Low-fat diet Insufficient magnesium intake Insufficient calcium intake

Table 1. *Cont.*

Groups of Risk Factors	Examples: Diseases, Conditions, Lifestyle Features
Long-term use of drugs	Antiepileptic drugs (e.g., valproate, phenytoin); antiretroviral drugs; glucocorticoids; systemic antifungal drugs; rifampin; bile acid sequestrants (cholestyramine); lipase inhibitors (orlistat)
Malignant neoplasms	Colon cancer, lymphatic system and blood cancers, breast cancer, ovarian cancer, prostate cancer
Granulomatous diseases	Sarcoidosis, histoplasmosis, coccidiomycosis, berylliosis
Mental illnesses	Depression, schizophrenia, anorexia nervosa
Cardiovascular diseases	Arterial hypertension, ischemic heart disease, heart failure
Others	Chronic fatigue syndrome Inpatient treatment (especially in the resuscitation and intensive care unit) Awaiting organ transplantation and post-transplant

4. Evaluation of Vitamin D Status

Vit. D status can be categorized by evaluating serum 25OH-D levels (Table 2) [10,12,24,35,40,44]. For many years, it has been argued that levels of 25OH-D should be at least 50 nmol/L, since this is sufficient to maintain good skeletal health in almost all individuals [20]. However, many experts claim that levels of 75 nmol/L and above are those sufficient to ensure normal skeletal and muscular structure and function [12,24,45,46]. There is growing evidence that minimum 100 nmol/L of 25OH-D levels are needed to reduce the risk of some cancers (e.g., colorectal), cardiovascular disease, infectious diseases, pathological pregnancies (e.g., preeclampsia, gestational diabetes, preterm birth), systemic connective tissue diseases, diabetes, and also COVID-19 [9,14,16,22,41,44,47–50]. An optimal (at least 100 nmol/L) levels of 25OH-D mean that Vit. D is sufficient for all systems in the human body, not only for bones [16]. Some authors speculated that a laboratory-determined concentration of 100 nmol/L indicates that the true serum 25OH-D levels of the individual are greater than 75 nmol/L [12]. In summary, 25OH-D levels of 75–150 nmol/L should be considered as "normal". The term "low vitamin D status" used in the present paper comprises both Vit. D deficiency and Vit. D insufficiency, as defined in Table 2.

Table 2. Vit. D status categories by 25OH-D levels.

Category	25OH-D Levels, nmol/L
Severe deficiency	<25
Moderate deficiency	25–<50
Insufficiency	50–<75
Sufficiency	75–<100
Optimal levels (optimal levels in tissues/cells)	100–<150
Increased levels	150–<250
Overdose	≥250
Intoxication *	≥375

* Intoxication category also includes lower 25OH-D levels, if hypercalcemia is caused by vitamin D supplements. 25OH-D–serum 25-hydroxy-vitamin D levels.

5. Vitamin D Dosing Principles

In order to simplify and not to overload the final brief algorithm of Vit. D dose selection, it is reasonable to present both prophylactic Vit. D doses and Vit. D doses for treatment separately. The aim of this article is not to discuss various advices on Vit. D dosing from different guidelines in depth; therefore, only summarized recommendations are presented.

Table 3 presents recommended Vit. D doses for prevention of low Vit. D status in different age groups [12,13,24,51]. In cases of Vit. D deficiency or insufficiency, therapeutic Vit. D doses should be prescribed according to both baseline 25OH-D levels and patient age (Table 4) [12,13,16,24,45,52–55].

For Vit. D risk group patients, particularly obese individuals or those having weight >90 kg as well as persons with malabsorption syndromes, Vit. D dose should be increased two-fold or sometimes even three-fold [12,16,24,35,40,46,48,56,57]. It is noteworthy that Vit. D doses up to 10,000 IU/d are considered safe for the vast majority of patients [44].

In patients with or at high risk of hypercalcemia, such as those with granulomatous disease, Vit. D dose should be adjusted individually by routine monitoring of calcemia, calciuria, 25OH-D, parathormone, and 1,25-dihydroxy-vitamin D levels [24]. For those patients, small Vit. D doses are suggested in order to maintain serum 25OH-D levels below 75 nmol/L [12].

In case of Vit. D overdosing, Vit. D supplementation should be stopped or at least halved and, if indicated, serum calcium levels should be measured. In case of Vit. D intoxication, which is an extremely rare condition when hypercalcemia occurs due to use of Vit. D supplements, Vit. D supplementation should be stopped, and the treatment of hypercalcemia should be applied (see [51,58,59] for details).

Table 3. Vitamin D prophylactic doses.

Patient Age	Recommended Daily Dose (IU/d)	Recommended Intermittent Dose	Upper Tolerable Daily Dose (IU)
Infants < 6 months	400–600	–	1000
Infants 6–<12 months	600–800	–	1000
Children 1–10 yrs.	600–1000	–	2000
Teens 11–<18 yrs.	800–2000	25,000 IU in 5–2 weeks	4000
Adults 18–<75 yrs.	1000–2000	25,000 IU in 4–2 weeks	4000
Adults ≥ 75 yrs.	2000–4000	25,000 IU in 2–1 weeks	4000

IU–international units.

Table 4. Vitamin D therapeutic doses.

Patient Age	Recommended Daily Dose and Duration	Recommended Intermittent Dose and Duration
25OH-D Levels < 25 nmol/L		
Infants < 1 month	1000 IU/d 3 months	–
Infants 1–<12 months	2000 IU/d 3 months	–
Children 1–<11 yrs.	3000–6000 IU/d 3 months	–
Children 11–<18 yrs.	6000 IU/d 3 months	50,000 IU/week 1.5–2 months
Adults	6000 IU/d 3 months	50,000 IU/week 2 months

Table 4. *Cont.*

Patient Age	Recommended Daily Dose and Duration	Recommended Intermittent Dose and Duration
25OH-D Levels 25–<75 nmol/L		
Infants < 1 month	• Previously taking vit. D supplements: increase dose 1.5–2-fold	–
Children 1–10 yrs.		–
Children 11–<18 yrs.	• No previous vit. D supplementation: largest prophylactic dose for age group (Table 3) • Duration 2 months	25,000 IU/week 2 months
Adults	• Previously taking vit. D supplements: increase dose 1.5–2-fold • No previous vit. D supplementation: largest prophylactic dose for age group (Table 3) • Duration 2–3 months	50,000 IU/week 2 months

IU–international units; 25OH-D–serum 25-hydroxy-vitamin D levels.

6. A Brief Algorithm for Vitamin D Dosing

As mentioned above, even for Vit. D risk group patients, recent measurements of serum 25-OH-D are not always available. Therefore, presuming that the true levels of 25OH-D for many individuals could be below 75 nmol/L, it is reasonable to suggest starting vit. D supplementation with 4000 IU/d or an equivalent weekly dose. For patients definitely belonging to Vit. D risk group, except those with or at high risk of hypercalcemia (sarcoidosis, etc.), the initial Vit. D dose might be doubled (Figure 1). For patients that are already taking Vit. D supplements without having performed 25OH-D measurements and without physician's advice before beginning of supplementation, it could be presumed that they might have low Vit. D status, i.e., they decided to start taking Vit. D on the grounds of their symptoms that potentially could have been caused by Vit. D insufficiency/deficiency. Therefore, it is reasonable to suggest doubling the Vit. D dose that they are currently using, but not exceeding the upper "safe" dose limits (10,000 IU/d).

In all those cases, 25OH-D levels should be checked at 1–1.5 months after the initiation of Vit. D supplementation or the enlargement of Vit. D dose, respectively. The span of 1–1.5 months was chosen for three reasons: (i) after that period, the mandatory isolation for the majority of patients is ended or other obstacles precluding their visit to the laboratory are solved; (ii) in some cases, it might help to evaluate the efficacy of supplementation (and to suspect, e.g., malabsorption), and (iii) to detect Vit. D overdose early enough [10,12,24,36,60].

The suggested original, brief or "working" algorithm presented in Figure 1 also comprises the main principles of Vit. D dose selection discussed earlier in this paper when baseline 25OH-D levels are known ([10,13,24,49]).

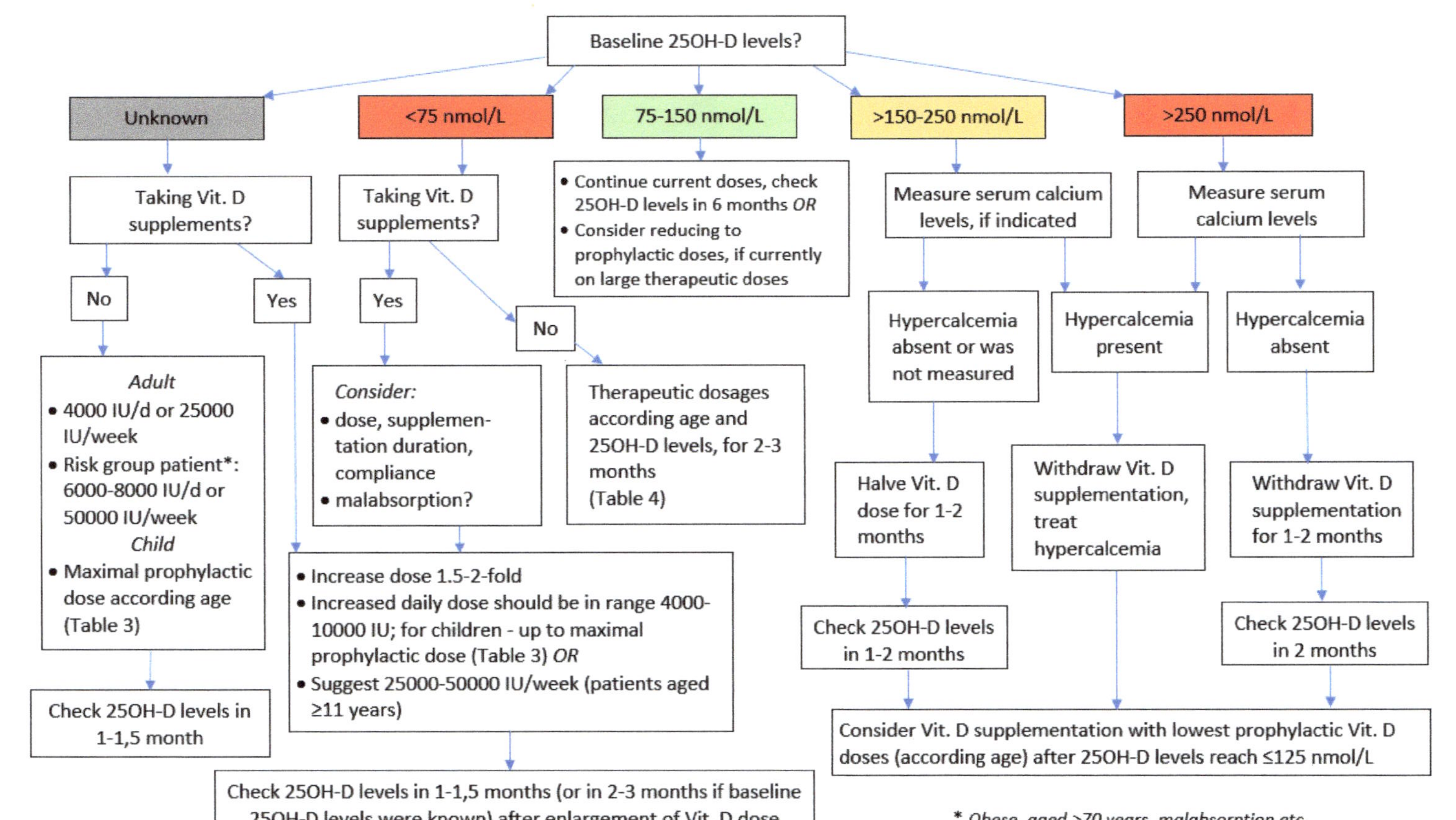

Figure 1. Brief algorithm for vitamin D dosing.

7. Dealing with Failure to Increase 25-Hydroxy-Vitamin D Levels

In cases where "adequate" Vit. D supplementation fails to improve Vit. D status, it is worth thinking over several things, such as:

1. The dose that was prescribed and the duration of supplementation. If the Vit. D dose could have been too small, it can be increased two-fold, and the next check of 25OH-D levels can be performed at 1.5–2 months after dose correction.
2. The compliance. Some patients prefer not to take large Vit. D doses even by physician prescription and, in fact, consume only small doses, for the fear of Vit. D overdose.
3. Possibility of unreported chronic diseases or use of certain drugs that could impair Vit. D metabolism. Some patients might be candidates to be examined for possible malabsorption syndrome, particularly in cases when 25OH-D levels did not increase significantly even after the supplementation with doubled dose. In some cases, e.g., for those with celiac disease, severe liver disease, or after bariatric surgery, calcidiol might be suggested, since it has better intestinal absorption than Vit. D and appears to be two to three times more effective in increasing serum 25OH-D levels than vitamin D3 [20], and this feature of calcidiol might be very important also in early stages of the COVID-19 disease, when low serum 25OH-D levels need to be increased as soon as possible [8,61].
4. Adequacy of calcium (Ca) and/or magnesium (Mg) intake. During the treatment of low Vit. D, supplementation with Mg (daily dose in the range 250–500 mg/d) is recommended, since Mg acts as a cofactor in many enzymes involved in Vit. D metabolism [44]. In addition, it is worth understanding that long-term decreased intake of Ca with food can, in turn, aggravate low Vit. D status because of compensatory hyperparathyroidism that increases the production of calcitriol in the kidney from 25OH-D, and this consequently contributes to diminishing serum 25OH-D levels. Therefore, with adequate Ca intake (including Ca from supplementation, if necessary), a better response to Vit. D preparations can be expected [39,62]. The recommended daily intake of Ca for adults is ~1000–1200 mg; more data about Ca inadequacy can be found elsewhere [63].

8. Conclusions

Due to various risk factors, many COVID-19 and other patients are at high risk to develop low vitamin D status. If possible, it is reasonable to check their serum 25-hydroxy-vitamin D levels, and only after that, an appropriate dose of vitamin D supplements should be suggested. In case 25-hydroxy-vitamin D measurements are not available, taking moderate vitamin D doses (e.g., at least 4000 IU/d) could be advised for at least 1–1.5 months, presuming that this supplementation can contribute to reaching adequate vitamin D status and can help to maintain better overall health status—both skeletal and non-skeletal—until serum 25-hydroxy-vitamin D testing will be accessible for the current patient. If low vitamin D status is confirmed, an appropriate, large enough vitamin D dose must be suggested for supplementation, and the next check of serum 25-hydroxy-vitamin D levels should be advised after the treatment in order to evaluate treatment results and choose the right tactic regarding further supplementation.

Funding: This work received no external funding.

Institutional Review Board Statement: Not applicable.

Informed Consent Statement: Not applicable.

Data Availability Statement: Not applicable.

Conflicts of Interest: The author declares no conflict of interest.

References

1. Over 200 Scientists & Doctors Call for Increased Vitamin D Use to Combat COVID-19. Available online: https://vitamind4all.org/letter.html (accessed on 2 February 2021).
2. Holick, M.F. Vitamin D: Extraskeletal health. *Endocrinol. Metab. Clin. Am.* **2010**, *39*, 381–400. [CrossRef]
3. Entrenas Castillo, M.; Entrenas Costa, L.M.; Vaquero Barrios, J.M.; Alcala Diaz, J.F.; Lopez Miranda, J.; Bouillon, R.; Quesada Gomez, J.M. Effect of calcifediol treatment and best available therapy versus best available therapy on intensive care unit admission and mortality among patients hospitalized for COVID-19: A pilot randomized clinical study. *J. Steroid. Biochem. Mol. Biol.* **2020**, *203*, 105751. [CrossRef]
4. Rastogi, A.; Bhansali, A.; Khare, N.; Suri, V.; Yaddanapudi, N.; Sachdeva, N.; Puri, G.D.; Malhotra, P. Short term, high-dose vitamin D supplementation for COVID-19 disease: A randomised, placebo-controlled, study (SHADE study). *Postgrad. Med. J.* **2020**. [CrossRef]
5. Tan, C.W.; Ho, L.P.; Kalimuddin, S.; Cherng, B.P.Z.; Teh, Y.E.; Thien, S.Y.; Wong, H.M.; Tern, P.J.W.; Chandran, M.; Chay, J.W.M.; et al. Cohort study to evaluate the effect of vitamin D, magnesium, and vitamin B12 in combination on progression to severe outcomes in older patients with coronavirus (COVID-19). *Nutrition* **2020**, *79*, 111017. [CrossRef]
6. Annweiler, G.; Corvaisier, M.; Gautier, J.; Dubee, V.; Legrand, E.; Sacco, G.; Annweiler, C. Vitamin D Supplementation Associated to Better Survival in Hospitalized Frail Elderly COVID-19 Patients: The GERIA-COVID Quasi-Experimental Study. *Nutrients* **2020**, *12*, 3377. [CrossRef] [PubMed]
7. Annweiler, C.; Hanotte, B.; Grandin de l'Eprevier, C.; Sabatier, J.M.; Lafaie, L.; Celarier, T. Vitamin D and survival in COVID-19 patients: A quasi-experimental study. *J. Steroid. Biochem. Mol. Biol.* **2020**, *204*, 105771. [CrossRef]
8. Grant, W.B.; Lordan, R. Vitamin D for COVID-19 on trial: An update on prevention and therapeutic application. *Endocr. Pract.* **2021**, *27*, 1266–1268. [CrossRef]
9. Yisak, H.; Ewunetei, A.; Kefale, B.; Mamuye, M.; Teshome, F.; Ambaw, B.; Yideg Yitbarek, G. Effects of Vitamin D on COVID-19 Infection and Prognosis: A Systematic Review. *Risk Manag. Health Policy* **2021**, *14*, 31–38. [CrossRef] [PubMed]
10. Bleizgys, A. Vitamin D and COVID-19: It is time to act. *Int. J. Clin. Pract.* **2020**, *75*, e13748. [CrossRef]
11. Chiu, S.K.; Tsai, K.W.; Wu, C.C.; Zheng, C.M.; Yang, C.H.; Hu, W.C.; Hou, Y.C.; Lu, K.C.; Chao, Y.C. Putative Role of Vitamin D for COVID-19 Vaccination. *Int. J. Mol. Sci.* **2021**, *22*, 8988. [CrossRef] [PubMed]
12. Holick, M.F.; Binkley, N.C.; Bischoff-Ferrari, H.A.; Gordon, C.M.; Hanley, D.A.; Heaney, R.P.; Murad, M.H.; Weaver, C.M.; Endocrine, S. Evaluation, treatment, and prevention of vitamin D deficiency: An Endocrine Society clinical practice guideline. *J. Clin. Endocrinol. Metab.* **2011**, *96*, 1911–1930. [CrossRef] [PubMed]
13. Pludowski, P.; Karczmarewicz, E.; Bayer, M.; Carter, G.; Chlebna-Sokol, D.; Czech-Kowalska, J.; Debski, R.; Decsi, T.; Dobrzanska, A.; Franek, E.; et al. Practical guidelines for the supplementation of vitamin D and the treatment of deficits in Central Europe—Recommended vitamin D intakes in the general population and groups at risk of vitamin D deficiency. *Endokrynol. Pol.* **2013**, *64*, 319–327. [CrossRef] [PubMed]
14. Pludowski, P.; Holick, M.F.; Grant, W.B.; Konstantynowicz, J.; Mascarenhas, M.R.; Haq, A.; Povoroznyuk, V.; Balatska, N.; Barbosa, A.P.; Karonova, T.; et al. Vitamin D supplementation guidelines. *J. Steroid Biochem. Mol. Biol.* **2018**, *175*, 125–135. [CrossRef] [PubMed]
15. Ross, A.C.; Manson, J.E.; Abrams, S.A.; Aloia, J.F.; Brannon, P.M.; Clinton, S.K.; Durazo-Arvizu, R.A.; Gallagher, J.C.; Gallo, R.L.; Jones, G.; et al. The 2011 report on dietary reference intakes for calcium and vitamin D from the Institute of Medicine: What clinicians need to know. *J. Clin. Endocrinol. Metab.* **2011**, *96*, 53–58. [CrossRef]
16. Kimball, S.M.; Holick, M.F. Official recommendations for vitamin D through the life stages in developed countries. *Eur. J. Clin. Nutr.* **2020**, *74*, 1514–1518. [CrossRef] [PubMed]
17. van Schoor, N.M.; Lips, P. Worldwide vitamin D status. *Best Pract. Res. Clin. Endocrinol. Metab.* **2011**, *25*, 671–680. [CrossRef]
18. Lips, P. Vitamin D status and nutrition in Europe and Asia. *J. Steroid Biochem. Mol. Biol.* **2007**, *103*, 620–625. [CrossRef]
19. Pludowski, P.; Grant, W.B.; Bhattoa, H.P.; Bayer, M.; Povoroznyuk, V.; Rudenka, E.; Ramanau, H.; Varbiro, S.; Rudenka, A.; Karczmarewicz, E.; et al. Vitamin D status in central europe. *Int. J. Endocrinol.* **2014**, *2014*, 589587. [CrossRef]
20. Lips, P.; Cashman, K.D.; Lamberg-Allardt, C.; Bischoff-Ferrari, H.A.; Obermayer-Pietsch, B.; Bianchi, M.L.; Stepan, J.; El-Hajj Fuleihan, G.; Bouillon, R. Current vitamin D status in European and Middle East countries and strategies to prevent vitamin D deficiency: A position statement of the European Calcified Tissue Society. *Eur. J. Endocrinol. Eur. Fed. Endocr. Soc.* **2019**, *180*, P23–P54. [CrossRef]
21. Kara, M.; Ekiz, T.; Ricci, V.; Kara, O.; Chang, K.V.; Ozcakar, L. "Scientific Strabismus" or two related pandemics: Coronavirus disease and vitamin D deficiency. *Br. J. Nutr.* **2020**, *124*, 736–741. [CrossRef]
22. Ebadi, M.; Montano-Loza, A.J. Perspective: Improving vitamin D status in the management of COVID-19. *Eur. J. Clin. Nutr.* **2020**, *74*, 856–859. [CrossRef] [PubMed]
23. Holick, M.F. Vitamin D deficiency. *N. Engl. J. Med.* **2007**, *357*, 266–281. [CrossRef]
24. Rusinska, A.; Pludowski, P.; Walczak, M.; Borszewska-Kornacka, M.K.; Bossowski, A.; Chlebna-Sokol, D.; Czech-Kowalska, J.; Dobrzanska, A.; Franek, E.; Helwich, E.; et al. Vitamin D Supplementation Guidelines for General Population and Groups at Risk of Vitamin D Deficiency in Poland-Recommendations of the Polish Society of Pediatric Endocrinology and Diabetes and the Expert Panel with Participation of National Specialist Consultants and Representatives of Scientific Societies-2018 Update. *Front. Endocrinol.* **2018**, *9*, 246. [CrossRef]

25. Giustina, A.; Adler, R.A.; Binkley, N.; Bollerslev, J.; Bouillon, R.; Dawson-Hughes, B.; Ebeling, P.R.; Feldman, D.; Formenti, A.M.; Lazaretti-Castro, M.; et al. Consensus statement from 2(nd) International Conference on Controversies in Vitamin D. *Rev. Endocr. Metab. Disord.* **2020**, *21*, 89–116. [CrossRef]

26. Gueli, N.; Verrusio, W.; Linguanti, A.; Di Maio, F.; Martinez, A.; Marigliano, B.; Cacciafesta, M. Vitamin D: Drug of the future. A new therapeutic approach. *Arch. Gerontol. Geriatr.* **2012**, *54*, 222–227. [CrossRef] [PubMed]

27. Foss, Y.J. Vitamin D deficiency is the cause of common obesity. *Med. Hypotheses* **2009**, *72*, 314–321. [CrossRef]

28. Diffey, B.L. Is casual exposure to summer sunlight effective at maintaining adequate vitamin D status? *Photodermatol. Photoimmunol. Photomed.* **2010**, *26*, 172–176. [CrossRef]

29. Grant, W.B.; Boucher, B.J. Requirements for Vitamin D across the life span. *Biol. Res. Nurs.* **2011**, *13*, 120–133. [CrossRef]

30. Strucinska, M.; Rowicka, G.; Dylag, H.; Riahi, A.; Bzikowska, A. Dietary intake of vitamin D in obese children aged 1–3 years. *Rocz. Panstw. Zakl. Hig.* **2015**, *66*, 353–360.

31. Guessous, I. Role of Vitamin D deficiency in extraskeletal complications: Predictor of health outcome or marker of health status? *Biomed. Res. Int.* **2015**, *2015*, 563403. [CrossRef] [PubMed]

32. Black, L.J.; Anderson, D.; Clarke, M.W.; Ponsonby, A.L.; Lucas, R.M.; Ausimmune Investigator Group. Analytical Bias in the Measurement of Serum 25-Hydroxyvitamin D Concentrations Impairs Assessment of Vitamin D Status in Clinical and Research Settings. *PLoS ONE* **2015**, *10*, e0135478. [CrossRef] [PubMed]

33. Madsen, K.H.; Rasmussen, L.B.; Mejborn, H.; Andersen, E.W.; Molgaard, C.; Nissen, J.; Tetens, I.; Andersen, R. Vitamin D status and its determinants in children and adults among families in late summer in Denmark. *Br. J. Nutr.* **2014**, *112*, 776–784. [CrossRef] [PubMed]

34. Autier, P.; Boniol, M.; Pizot, C.; Mullie, P. Vitamin D status and ill health: A systematic review. *Lancet Diabetes Endocrinol.* **2014**, *2*, 76–89. [CrossRef]

35. Holick, M.F. The vitamin D deficiency pandemic: Approaches for diagnosis, treatment and prevention. *Rev. Endocr. Metab. Disord.* **2017**, *18*, 153–165. [CrossRef]

36. Collins, A. Practice implications for preventing population vulnerability related to vitamin D status. *J. Am. Assoc. Nurse Pract.* **2013**, *25*, 109–118. [CrossRef]

37. LeFevre, M.L. Screening for vitamin D deficiency in adults: U.S. Preventive Services Task Force recommendation statement. *Ann. Intern. Med.* **2015**, *162*, 133–140. [CrossRef]

38. Cooper, I.D.; Crofts, C.A.P.; DiNicolantonio, J.J.; Malhotra, A.; Elliott, B.; Kyriakidou, Y.; Brookler, K.H. Relationships between hyperinsulinaemia, magnesium, vitamin D, thrombosis and COVID-19: Rationale for clinical management. *Open Heart* **2020**, *7*, e001356. [CrossRef]

39. Mazahery, H.; von Hurst, P.R. Factors Affecting 25-Hydroxyvitamin D Concentration in Response to Vitamin D Supplementation. *Nutrients* **2015**, *7*, 5111–5142. [CrossRef] [PubMed]

40. Pilz, S.; Zittermann, A.; Trummer, C.; Theiler-Schwetz, V.; Lerchbaum, E.; Keppel, M.H.; Grubler, M.R.; Marz, W.; Pandis, M. Vitamin D testing and treatment: A narrative review of current evidence. *Endocr. Connect.* **2019**, *8*, R27–R43. [CrossRef]

41. Amrein, K.; Scherkl, M.; Hoffmann, M.; Neuwersch-Sommeregger, S.; Kostenberger, M.; Tmava Berisha, A.; Martucci, G.; Pilz, S.; Malle, O. Vitamin D deficiency 2.0: An update on the current status worldwide. *Eur. J. Clin. Nutr.* **2020**, *74*, 1498–1513. [CrossRef]

42. Bonham, M.P.; Lamberg-Allardt, C. Vitamin D in public health nutrition. *Public Health Nutr.* **2014**, *17*, 717–720. [CrossRef] [PubMed]

43. Boyages, S.C. Vitamin D testing: New targeted guidelines stem the overtesting tide. *Med. J. Aust.* **2016**, *204*, 18. [CrossRef]

44. Grant, W.B.; Lahore, H.; McDonnell, S.L.; Baggerly, C.A.; French, C.B.; Aliano, J.L.; Bhattoa, H.P. Evidence that Vitamin D Supplementation Could Reduce Risk of Influenza and COVID-19 Infections and Deaths. *Nutrients* **2020**, *12*, 988. [CrossRef] [PubMed]

45. Lopez, A.G.; Kerlan, V.; Desailloud, R. Non-classical effects of vitamin D: Non-bone effects of vitamin D. *Ann. Endocrinol.* **2021**, *82*, 43–51. [CrossRef]

46. American Geriatrics Society Workgroup on Vitamin D Supplementation for Older Adults. Recommendations abstracted from the American Geriatrics Society Consensus Statement on vitamin D for Prevention of Falls and Their Consequences. *J. Am. Geriatr. Soc.* **2014**, *62*, 147–152. [CrossRef]

47. Wimalawansa, S.J. Global epidemic of coronavirus—COVID-19: What can we do to minimize risks. *Eur. J. Biomed. Pharm. Sci.* **2020**, *7*, 432–438.

48. Charoenngam, N.; Holick, M.F. Immunologic Effects of Vitamin D on Human Health and Disease. *Nutrients* **2020**, *12*, 2097. [CrossRef]

49. Grober, U.; Holick, M.F. The coronavirus disease (COVID-19)—A supportive approach with selected micronutrients. *Int. J. Vitam. Nutr. Res.* **2021**, 1–22. [CrossRef] [PubMed]

50. Charoenngam, N.; Shirvani, A.; Holick, M.F. Vitamin D and Its Potential Benefit for the COVID-19 Pandemic. *Endocr. Pract.* **2021**, *27*, 484–493. [CrossRef]

51. Rachito Diagnostika ir Gydymas (Protokolas). Available online: https://sam.lrv.lt/uploads/sam/documents/files/Veiklos_sritys/Programos_ir_projektai/Asmens_sveikatos_prieziuros_kokybes_grinimas/Vaiku%20ligu%20protokolai%20(papildomi%2023).zip (accessed on 7 February 2021).

52. Munns, C.F.; Shaw, N.; Kiely, M.; Specker, B.L.; Thacher, T.D.; Ozono, K.; Michigami, T.; Tiosano, D.; Mughal, M.Z.; Makitie, O.; et al. Global Consensus Recommendations on Prevention and Management of Nutritional Rickets. *J. Clin. Endocrinol. Metab.* **2016**, *101*, 394–415. [CrossRef]

53. Misra, M. *Vitamin D Insufficiency and Deficiency in Children and Adolescents.* 2020. Available online: https://www.uptodate.cn/contents/zh-Hans/vitamin-d-insufficiency-and-deficiency-in-children-and-adolescents?search (accessed on 21 November 2021).

54. Carpenter, T. *Etiology and Treatment of Calcipenic Rickets in Children.* 2020. Available online: https://www.uptodate.cn/contents/zh-Hans/etiology-and-treatment-of-calcipenic-rickets-in-children?search (accessed on 21 November 2021).

55. Dawson-Hughes, B. *Vitamin D Deficiency in Adults: Definition, Clinical Manifestations, and Treatment.* 2020. Available online: https://www.uptodate.cn/contents/zh-Hans/vitamin-d-deficiency-in-adults-definition-clinical-manifestations-and-treatment?search (accessed on 21 November 2021).

56. Rhodes, J.M.; Subramanian, S.; Laird, E.; Griffin, G.; Kenny, R.A. Perspective: Vitamin D deficiency and COVID-19 severity—Plausibly linked by latitude, ethnicity, impacts on cytokines, ACE2 and thrombosis. *J. Intern. Med.* **2021**, *289*, 97–115. [CrossRef] [PubMed]

57. Ferder, L.; Martin Gimenez, V.M.; Inserra, F.; Tajer, C.; Antonietti, L.; Mariani, J.; Manucha, W. Vitamin D Supplementation as a Rational Pharmacological Approach in the Covid-19 Pandemic. *Am. J. Physiol. Lung Cell. Mol. Physiol.* **2020**, *319*, L941–L948. [CrossRef]

58. Marcinowska-Suchowierska, E.; Kupisz-Urbanska, M.; Lukaszkiewicz, J.; Pludowski, P.; Jones, G. Vitamin D Toxicity-A Clinical Perspective. *Front. Endocrinol.* **2018**, *9*, 550. [CrossRef] [PubMed]

59. Girdžiūtė, M. Hipokalcemija ir hiperkalcemija: Diagnostikos ir gydymo aspektai. *Internistas* **2017**, *2*, 10–13.

60. Ali, N. Role of vitamin D in preventing of COVID-19 infection, progression and severity. *J. Infect. Public Health* **2020**, *13*, 1373–1380. [CrossRef]

61. Maghbooli, Z.; Sahraian, M.A.; Jamalimoghadamsiahkali, S.; Asadi, A.; Zarei, A.; Zendehdel, A.; Varzandi, T.; Mohammadnabi, S.; Alijani, N.; Karimi, M.; et al. Treatment With 25-Hydroxyvitamin D3 (Calcifediol) Is Associated with a Reduction in the Blood Neutrophil-to-Lymphocyte Ratio Marker of Disease Severity in Hospitalized Patients With COVID-19: A Pilot Multicenter, Randomized, Placebo-Controlled, Double-Blinded Clinical Trial. *Endocr. Pract.* **2021**, *27*, 1242–1251. [CrossRef]

62. Christakos, S.; Li, S.; De La Cruz, J.; Bikle, D.D. New developments in our understanding of vitamin metabolism, action and treatment. *Metabolism* **2019**, *98*, 112–120. [CrossRef]

63. Calcium. Fact Sheet for Health Professionals. Available online: https://ods.od.nih.gov/factsheets/Calcium-HealthProfessional (accessed on 21 November 2021).

Review

Critical Appraisal of Large Vitamin D Randomized Controlled Trials

Stefan Pilz [1,*], Christian Trummer [1], Verena Theiler-Schwetz [1], Martin R. Grübler [1], Nicolas D. Verheyen [2], Balazs Odler [3], Spyridon N. Karras [4], Armin Zittermann [5] and Winfried März [6,7,8]

1 Division of Endocrinology and Diabetology, Department of Internal Medicine, Medical University of Graz, 8036 Graz, Austria; christian.trummer@medunigraz.at (C.T.); verena.schwetz@medunigraz.at (V.T.-S.); martin.gruebler@gmx.net (M.R.G.)
2 Division of Cardiology, Department of Internal Medicine, Medical University of Graz, 8036 Graz, Austria; nicolas.verheyen@medunigraz.at
3 Division of Nephrology, Department of Internal Medicine, Medical University of Graz, 8036 Graz, Austria; balazs.odler@medunigraz.at
4 National Scholarship Foundation, 55535 Thessaloniki, Greece; karraspiros@yahoo.gr
5 Clinic for Thoracic and Cardiovascular Surgery, Herz- und Diabeteszentrum Nordrhein-Westfalen (NRW), Ruhr University Bochum, 32545 Bad Oeynhausen, Germany; azittermann@hdz-nrw.de
6 Clinical Institute of Medical and Chemical Laboratory Diagnostics, Medical University of Graz, 8036 Graz, Austria; winfried.maerz@synlab.de
7 SYNLAB Academy, Synlab Holding Deutschland GmbH, 68159 Mannheim, Germany
8 V[th] Department of Medicine (Nephrology, Hypertensiology, Rheumatology, Endocrinology, Diabetology, Lipidology), Medical Faculty Mannheim, University of Heidelberg, 68167 Mannheim, Germany
* Correspondence: stefan.pilz@medunigraz.at; Tel.: +43-316-385-81143

Citation: Pilz, S.; Trummer, C.; Theiler-Schwetz, V.; Grübler, M.R.; Verheyen, N.D.; Odler, B.; Karras, S.N.; Zittermann, A.; März, W. Critical Appraisal of Large Vitamin D Randomized Controlled Trials. *Nutrients* **2022**, *14*, 303. https://doi.org/10.3390/nu14020303

Academic Editor: Connie Weaver

Received: 22 December 2021
Accepted: 10 January 2022
Published: 12 January 2022

Publisher's Note: MDPI stays neutral with regard to jurisdictional claims in published maps and institutional affiliations.

Abstract: As a consequence of epidemiological studies showing significant associations of vitamin D deficiency with a variety of adverse extra-skeletal clinical outcomes including cardiovascular diseases, cancer, and mortality, large vitamin D randomized controlled trials (RCTs) have been designed and conducted over the last few years. The vast majority of these trials did not restrict their study populations to individuals with vitamin D deficiency, and some even allowed moderate vitamin D supplementation in the placebo groups. In these RCTs, there were no significant effects on the primary outcomes, including cancer, cardiovascular events, and mortality, but explorative outcome analyses and meta-analyses revealed indications for potential benefits such as reductions in cancer mortality or acute respiratory infections. Importantly, data from RCTs with relatively high doses of vitamin D supplementation did, by the vast majority, not show significant safety issues, except for trials in critically or severely ill patients or in those using very high intermittent vitamin D doses. The recent large vitamin D RCTs did not challenge the beneficial effects of vitamin D regarding rickets and osteomalacia, that therefore continue to provide the scientific basis for nutritional vitamin D guidelines and recommendations. There remains a great need to evaluate the effects of vitamin D treatment in populations with vitamin D deficiency or certain characteristics suggesting a high sensitivity to treatment. Outcomes and limitations of recently published large vitamin D RCTs must inform the design of future vitamin D or nutrition trials that should use more personalized approaches.

Keywords: vitamin D; RCT; clinical trial; cholecalciferol; randomized controlled trial; epidemiology; supplementation; mortality; infections

1. Introduction

Vitamin D deficiency is a public health problem because it has a relatively high worldwide prevalence and is associated with poor skeletal health, i.e., with rickets and osteomalacia [1–3]. Whether vitamin D exerts additional health benefits, in particular with reference to extra-skeletal diseases, is subject to an intense and ongoing debate within the scientific community.

Vitamin D has a unique metabolism, with its synthesis from endogenous precursors in the skin, a process induced by ultraviolet-B exposure, whereas dietary intake usually plays only a minor role [4]. Vitamin D is mainly metabolized to 25-hydroxyvitamin D (25(OH)D) in the liver and finally to 1,25-dihydroxyvitamin D (1,25(OH)2D, also termed calcitriol) in the kidneys. Importantly, 1,25(OH)2D is a steroid hormone. Its target receptors, i.e., vitamin D receptors (VDRs), are expressed in almost all human tissues [4]. In general, vitamin D can exert endocrine, paracrine, and autocrine effects as part of a complex regulation and interactions of vitamin D metabolism [4,5].

In mechanistic studies, VDR signaling has been implicated in the pathogenesis of many skeletal and extra-skeletal diseases [6]. In line with this, epidemiological studies have shown that serum concentrations of 25(OH)D, the accepted marker of vitamin D status, were inversely associated with adverse health outcomes including, e.g., cancer, cardiovascular diseases, and mortality, thus suggesting a widespread role of vitamin D for human health [6,7]. Based on this background, large randomized controlled trials (RCTs) were designed and have been recently published to evaluate the role of vitamin D supplementation for clinical outcomes with relevance for public health [8,9]. Cancer, cardiovascular and musculoskeletal health, infections, and mortality were the main outcomes of these trials and are thus the subject of this review. Given that the current opinion on vitamin D seems to be significantly, if not mainly, determined by these large RCTs, there exists a great need for a structured summary and guidance through all of these study results. While we acknowledge and appreciate previous work on this topic, we aim to extend these publications by considering important additional and modified aspects [2,10–14].

In this brief narrative review, we will summarize and discuss findings from major vitamin D RCTs and their potential impact on the appreciation of vitamin D regarding public health and science. We will start with a critical appraisal of clinical vitamin D research and then summarize major findings (primary outcomes) from large vitamin D RCTs that will then be discussed concerning their interpretations and implications. We will also outline current meta-analyses on vitamin D and major health outcomes before we provide an outlook on potential future research and clinical applications for vitamin D.

2. Critical Appraisal of Clinical Vitamin D Research

Scientific vitamin D publications have substantially increased over the last few decades, providing a wealth of data from experimental and observational studies. Overall, they support the hypothesis that vitamin D supplementation might be a sort of panacea for human health [15]. We agree that accumulating evidence points in the direction of a beneficial role of vitamin D in many diseases, thus justifying the call for large vitamin D RCTs. Nevertheless, before discussing the findings from these trials, a critical appraisal of vitamin D research leading to the initiation of these trials is warranted, as well as some basic considerations regarding general aspects of the trial designs.

Beforehand, we wish to stress the rather provocative but scientifically well supported statement that most claimed research findings are false, as pointed out by the landmark essay by Prof. John Ioannidis [16]. In summary, research findings are less likely to be true, the smaller the studies and the effect sizes are, the greater the number and the lesser the selection of tested relationships are, the greater the flexibility in designs, definitions, outcomes, and analytical models are, the greater the financial and other interests and prejudices are, and the hotter a scientific field (with more scientific teams involved) is [16]. While these considerations apply to clinical research in general, they definitely apply to the recent hype in vitamin D research and thus have to be taken into account when aiming for a balanced judgement of the scientific vitamin D literature.

As observational studies linking low 25(OH)D concentrations to poor health were one of the main drivers for the high public health interest in vitamin D, we want to underline that epidemiological studies on vitamin D status are particularly prone to confounding [17].

Low serum 25(OH)D concentrations are a consequence of, e.g., an unhealthy lifestyle with less outdoor activity (and thus less sun exposure), obesity, and poor nutrition [17]. In addition, reverse causation needs to be considered, which means that underlying diseases, in particular those related to inflammatory processes or limiting physical activity, may themselves decrease 25(OH)D concentrations [17]. Therefore, it has been postulated that a poor vitamin D status may merely be a marker of ill health [17]. These factors have been addressed in many well-conducted epidemiological vitamin D studies, but limitations inherent to observational study designs such as residual confounding or reverse causation cannot be completely excluded.

When designing large RCTs with the intention to prove or refute the hypothesis of a beneficial clinical value of vitamin D supplementation, it should be logical that findings from past nutrient RCTs are considered. The scientific literature contains several examples of vitamins (e.g., vitamin E) that showed promising results in experimental and association studies but that have failed to show benefits or that were even harmful in large RCTs [18–20]. When interpreting the results of these "disappointing" trials, a common conclusion was that large RCTs should primarily target populations who are particularly sensitive to the beneficial effects of the intervention, e.g., populations who are deficient for the respective vitamin. With reference to vitamin D, observational studies have largely shown that there is no meaningful association of 25(OH)D and health outcomes such as mortality over a wide range of the 25(OH)D distribution, whereas there was a steep increase in risk at very low 25(OH)D concentrations [21–23]. Unfortunately, large vitamin D RCTs were mainly designed to evaluate effects of a vitamin D supplementation in the general population with fairly "normal" 25(OH)D concentrations, thus not considering the findings from past nutrient RCTs mentioned above [8,9].

3. Results of Large Vitamin D RCTs

The main results, i.e., the primary outcomes, of three major vitamin D-randomized placebo-controlled trials in the general older population have been published recently: the Vitamin D and Omega-3 Trial (VITAL) from the United States, the Vitamin D Assessment (ViDA) study from New Zealand, and the Vitamin D3-Omega3-Home Exercise-Healthy Ageing and Longevity Trial (DO-HEALTH) from five different European countries (Switzerland, France, Germany, Portugal, and Austria) [24–28]. The characteristics and results of these trials are summarized in this chapter along with findings from a selection of recently published large vitamin D RCTs in specific populations that are considered relevant to current knowledge on vitamin D [29–36]. The main results of the D-Health Trial, an RCT with 21,315 participants aged 60 to 79 years old from Australia receiving either monthly doses of 60,000 IU of vitamin D or placebo are still pending [37]. The primary outcomes of this trial are all-cause mortality and total as well colorectal cancer incidence. Publications of a few other large vitamin D RCTs are also still pending [8,38]. All of the trials described below were randomized for vitamin D versus placebo in a 1:1 ratio if not otherwise specified, and effect sizes (e.g., hazard ratios (HRs)) are shown for the vitamin D group compared to the placebo group. When referring to vitamin D, we actually mean vitamin D3 (cholecalciferol) if not otherwise indicated.

3.1. VITAL

The VITAL study is a multicentre RCT from the United States among men aged 50 years or older and women aged 55 years or older with no history of cancer or cardiovascular diseases at baseline and who were (amongst other inclusion/exclusion criteria) required to agree to limit the use of vitamin D from all supplemental sources to 800 international units (IU) per day [27]. This trial was conducted using a two-by-two factorial design with 2000 international units (IU) of vitamin D per day and 1 g of marine n-3 fatty acids per day. Primary endpoints were invasive cancer of any type and major cardiovascular events (composite of myocardial infarction, stroke, or death from cardiovascular causes). A total of 25,871 participants were randomized (12,927 to vitamin D and 12,944 to placebo) and followed-up for a median of 5.3 years. Invasive cancer was diagnosed in 793 participants in the vitamin D group and in 824 participants in the placebo group, corresponding to a HR (with 95% confidence interval (CI)) of 0.96 (0.88 to 1.06; $p = 0.47$). Major cardiovascular events occurred in 396 participants from the vitamin D group and in 409 participants from the placebo group, translating into an HR (95% CI) of 0.97 (0.85 to 1.12; $p = 0.69$). At baseline, the current intake of vitamin D supplements ($\leq$800 IU per day) was reported by 42.5% of the participants in the vitamin D group and by 42.7% of the participants in the placebo group. At 2 years, the prevalence of outside use of vitamin D supplements (>800 IU daily) was 3.8% in the vitamin D group and 5.6% in the placebo group, with an increase to 6.4% and 10.8%, respectively, after 5 years [27].

3.2. VIDA

The ViDA study is an RCT from New Zealand among adults aged 50 to 84 years old, with one of the exclusion criteria being the current use of vitamin D supplements, including cod liver oil at a dose of >600 IU per day if aged 50–70 years old and of >800 IU per day if aged 71–84 years old [25]. A total of 5108 participants received either vitamin D (n = 2558) at an initial dose of 200,000 IU followed by monthly doses of 100,000 IU one month later or a placebo (n = 2550) for a median of 3.3 years (range 2.5 to 4.2 years). Incident cardiovascular disease events, the primary outcome measure, were recorded in 303 participants from the vitamin D group and in 293 participants from the placebo group, resulting in an HR (95% CI) of 1.02 (0.87 to 1.29). At baseline, 8% of the study population were taking vitamin D supplements.

3.3. Do-Health

The DO-HEALTH study is a multicentre study from Europe in 2157 community dwelling adults aged 70 years or older with a 2 × 2 × 2 factorial design with three treatment comparisons, i.e., 2000 IU vitamin D per day, 1 g omega-3 fatty acids, and a strength-training exercise program [28]. The mean changes (with 99% CI) in the vitamin D versus the placebo group for the six primary outcomes were, -0.8 (-2.1 to 0.5; $p = 0.13$) mm Hg for systolic blood pressure, 0 (-0.7 to 0.8; $p = 0.88$) mm Hg for diastolic blood pressure, -0.1 (-0.3 to 0.1; $p = 0.26$) points for Short Physical Performance Battery, -0.1 (-0.4 to 0.1; $p = 0.11$) for the Montreal Cognitive Assessment (MoCA), 1.03 (0.75 to 1.43; $p = 0.79$) for nonvertebral fractures, and 0.95 (0.84 to 1.08; $p = 0.33$) for infections. At baseline, 10.8% of all of the study participants reported a supplement intake of 800 IU of vitamin D per day or more.

3.4. Vitamin D RCTs in Specific Populations

In the D2d study from the United States, 2423 participants with pre-diabetes were randomized to receive either vitamin D3 4000 IU per day or placebo [29]. After a median follow-up of 2.5 years, the HR (95% CI) for new onset diabetes was 0.88 (0.75 to 1.04; $p = 0.12$). In a RCT conducted in Bangladesh, 1300 pregnant women were randomly allocated into five groups to receive vitamin D supplementation either from 17 to 24 weeks of gestation until birth at a dose of 4200, 16,800, or 28,000 IU, respectively, per week, or from 17 to 24 weeks of gestation until 26 weeks postpartum at a dose of 28,000 IU peer week or placebo (equal group sizes) [30]. Among 1164 infants, at 1 year of age, there was no significant group difference in terms of the primary outcome, i.e., length-for-age z scores. In the Vitamin D to Improve Outcomes by Leveraging Early Treatment (VIOLET) trial, 1078 critically ill patients with 25(OH)D concentrations below 20 ng/mL (multiply by 2.496 to convert ng/mL to nmol/L) were randomized to receive a single enteral dose of 540,000 IU vitamin D or a placebo [31]. The 90-day mortality was 23.5% in the vitamin D and 20.6% in the placebo group, respectively, with a 2.9% (95% CI: -2.1 to 7.9%; $p = 0.26$) difference. Importantly, after the first interim analysis, the VIOLET trial was stopped for futility, i.e., the inability of this trial to achieve its objectives. An RCT from Nebraska (United States) randomized 2303 postmenopausal women aged 55 years old or older to either 2000 IU of vitamin D plus 1500 mg calcium per day or a placebo [32]. After four years, the HR (95% CI) for all-type cancer (excluding nonmelanoma skin cancer) was 0.70 (0.47 to 1.02). An RCT performed in 18 public schools in Mongolia randomized 8851 schoolchildren aged 6 to 13 years to either 14,000 IU vitamin D weekly or a placebo for 3 years [34]. The primary outcome, i.e., a positive QuantiFERON-TB Gold In-Tube assay test result, was recorded in 3.6% of the children in the vitamin D group and in 3.3% of the children in the placebo group, with a respective HR (95% CI) of 1.10 (0.87 to 1.38; $p = 0.42$). In a trial from Germany, 400 patients with advanced heart failure and 25(OH)D concentrations below 30 ng/mL were randomized to receive either 4000 IU of vitamin D per day or a placebo for 3 years [33]. All-cause mortality was 19.6% in the vitamin D group and 17.9% in the placebo group, with an HR (95% CI) of 1.09 (0.69 to 1.71; $p = 0.73$). In a so-called safety trial from Canada, 311 healthy participants aged 55 to 70 years with a 25(OH)D concentration from 12 to 40 ng/mL were randomized to either 400, 4000, or 10,000 IU of vitamin D per day for 3 years [36]. Primary outcomes were total volumetric bone mineral density (BMD), which was measured by high resolution peripheral quantitative computed tomography, and bone strength (failure load) at the radius and tibia. Compared to the 400 IU group, radial volumetric BMD was significantly lower for the 4000 and 10,000 IU group, and tibial volumetric BMD was significantly lower for the 10,000 IU group, with no significant differences being observed for other primary outcome measures.

3.5. Secondary Outcome and Subgroup Analyses

Publications of the above described primary outcomes of the large vitamin D RCTs were accompanied and followed (or in the case of, e.g., the D-Health Trial even preceeded) by a total number of several hundreds of secondary outcome and subgroup analyses (as well as meta-analyses) that should only be interpreted with great caution [35,39–48]. In general, the risks of type 1 errors ("false positive results") and type 2 errors ("false negative results") should be taken into account when interpreting such trial results. Major problems with secondary and subgroup analyses are, e.g., the common lack of pre-specified power analyses, no adjustments for multiple testing (e.g., setting the p value for statistical significance as 0.05 divided by the number of tests according to Bonferroni or using the Benjamini–Hochberg procedure), or missing assumptions of minimal important difference (MID) effect estimates [40]. These limitations, inherent to many RCTs including the large vitamin D RCTs, suggest that the published secondary outcome and subgroup analyses can only be interpreted as so called "explorative outcome" analyses. Therefore, we only briefly discuss some of the findings from such analyses. In general, the vast majority of these explorative outcome analyses did not reveal findings in favour of rejecting the null hypothesis of no vitamin D effect. The indication of no vitamin D effect applied to cardiovascular events, fractures, or falls [26,27]. Of interest were some analyses suggesting a potential beneficial effect of vitamin D on cancer mortality and advanced cancer (metastatic or fatal) [27,38,45]. In this context, vitamin D supplementation significantly reduced cancer mortality in the VITAL study when excluding the first year or the first two years of follow-up [27]. Moreover, in analyes restricted to participants with a body mass index below 25 kg/m^2, cancer incidence was significantly reduced in the vitamin D when compared to the placebo group [27]. In addition, some explorative analyses suggest that participants with low 25(OH)D concentrations may benefit regarding some surrogate parameters such as BMD, arterial, or lung function [38]. In secondary analyses of the D2d trial, participants who maintained intra-trial 25(OH)D concentrations of at least 40 ng/mL had a significantly reduced risk of progression from prediabetes to diabetes mellitus [49]. By contrast, subgroup analyses from the VIOLET trial in ICU patients with sepsis, infection, or respiratory distress syndrome even suggest increased mortality in patients receiving vitamin D [31]. Similarly, an increased need for mechanical circulatory support (MCS) was observed in the vitamin D group of an explorative analyses of 400 heart failure patients from the EVITA trial [33]. Apart from these two latter studies, there were no major signs of concern regarding adverse clinical events (such as kidney stones) of vitamin D supplementation [46]. With regard to the safety of vitamin D supplementation, it must be emphasized that some previous RCTs using high intermittent doses of vitamin D have partially shown an increased risk of fractures and/or falls, including one RCT documenting an increased risk of falls at a dose of 60,000 IU of vitamin D per month [50,51]. Other RCTs, such as the ViDA study using vitamin D doses of 100,000 IU per month, could, however, not confirm these potential adverse effects of intermittent bolus doses, but uncertainty and concerns remain regarding this issue due to the relatively short half-life of 25(OH)D and other vitamin D metabolites, so it appears prudent to prefer daily or weekly dosing intervals.

4. Interpretations and Implications of Large Vitamin D RCTs

As none of the above listed vitamin D RCTs reported any significant effect on the primary outcomes, it has to be concluded that there are no obvious overall major health benefits of vitamin D supplementation in the setting and in the populations included. It can also be concluded that any potential health benefits in the subgroups are likely to be relatively small if present at all. Despite these clear signs of lacking beneficial vitamin D effects, we wish to critically discuss the study designs of these RCTs before drawing final conclusions that may have a huge impact on the current use of vitamin D treatment and future vitamin D research.

It has to be stressed that vitamin D RCTs or nutrient RCTs in general have fundamental differences compared to drug RCTs [52]. With reference to vitamin D, it is not biologically possible that there is no vitamin D exposure in the placebo group, so any group comparison within vitamin D RCTs is always based on a placebo group with a certain vitamin D exposure compared to the intervention group with a higher vitamin D exposure. Therefore, any conclusion that vitamin D supplementation at a given dose is of no health benefit has to refer to the baseline 25(OH)D concentration of the study population and to the achieved vitamin D status after treatment. We therefore outline the 25(OH)D concentrations at baseline and follow-up of important vitamin D RCTs in Table 1 along with the vitamin D dosages and dosing schedules.

The baseline and follow-up vitamin D status of the above-mentioned RCTs has to be viewed in light of observational data on 25(OH)D and hard clinical outcomes such as mortality [21–23]. Epidemiological studies from Europe and the US observed the lowest risk of mortality at 25(OH)D concentrations of 31.3 and 32.5 ng/mL, respectively [21,22]. Importantly, data from the Third National Health and Nutrition Survey (NHANES III) showed no association of 25(OH)D with total mortality for values ranging from 16 to 48 ng/mL, but reported a significant increase in mortality at 25(OH)D concentrations below 16 ng/mL [22]. Similar results were observed for associations between 25(OH)D concentrations and incidence rates of cardiovascular disease or cancer [53–55]. It is thus in line with (or one may argue a confirmation of) previous observational studies that there was no vitamin D effect on the primary outcomes when studying populations with baseline 25(OH)D concentrations that were mostly above, e.g., 16 ng/mL [56,57]. Bolland and colleagues therefore concluded that these large vitamin D RCTs could be considered research waste because they enrolled participants who were not vitamin D deficient [57]. We are not that critical, as we greatly appreciate the efforts to conduct these large vitamin D RCTs, but we stress that most large vitamin D RCTs enrolled participants in whom, based on their baseline 25(OH)D concentration, no significant effect on the primary outcome could realistically be expected.

Only the VIOLET, EVITA, and MDIG trial enrolled participants with very low 25(OH)D concentrations, but these trials also failed to document significant effects on the primary outcomes [30,31,33]. One conclusion from the VIOLET and EVITA trial in critically ill or severely ill (advanced heart failure) patients is that for such patients, there may be safety concerns when using high dose vitamin D treatment, so we should refrain from this in clinical routine [31,33]. With reference to pregnancy outcomes in the MDIG trial, it has to be acknowledged vitamin D supplementation started relatively late during pregnancy and that the overall rate of pregnancy complications was relatively low for Bangladesh [30].

Moreover, other RCTs and meta-analyses do suggest potential benefits of vitamin D supplementation during pregnancy for e.g. gestational diabetes mellitus or pre-eclampsia [58,59]. In view of the totality of evidence we continue to strongly recommend a sufficient vitamin D status with a 25(OH)D concentration of at least 20 ng/mL in pregnant women. This can be safely and effectively achieved by vitamin D supplementation with a dose of 800 to 1000 IU per day [59,60].

Table 1. Baseline and follow-up 25(OH)D concentrations and vitamin D dosing regimens of selected recent large vitamin D RCTs.

Study Acronym or First Author	Study Population	Baseline 25(OH)D in the Entire Cohort (ng/mL)	Baseline 25(OH)D in the Placebo Group (ng/mL)	Follow-Up 25(OH)D in the Placebo Group (ng/mL)	Baseline 25(OH)D in the Vitamin D Group (ng/mL)	Follow-Up 25(OH)D in the Vitamin D Group (ng/mL)	Vitamin D Supplement Dose	Study Duration
VITAL	Older general population	30.8 ± 10.0	30.8 ± 10.0	minus 0.7 from baseline	30.9 ± 10.0	41.8 (mean)	2000 IU per day	5.3 years (median)
ViDA	Older general population	25.3 ± 9.5	24.4 ± 9.6	26.4 ± 11.6	24.4 ± 9.6	54.1 ± 16.0	Initial 200,000 IU, followed by 100,000 IU per month	3.3 years (median)
DO-HEALTH	Older general population	22.4 ± 8.4	22.4 ± 8.5	24.4 (mean)	22.4 ± 8.4	37.6 (mean)	2000 IU per day	2.99 years (median)
D2d	Patients with prediabetes	28.0 ± 10.2	28.2 ± 10.1	28.8 (mean)	27.7 ± 10.2	54.3 (mean)	4000 IU per day	2.5 years (median)
MDIG	Pregnant women	11.0 ± 5.7	11.1 ± 5.5	9.5 ± 5.6	11.0 ± 5.7, 11.5 ± 5.6, 10.8 ± 5.9	27.9 ± 7.8, 40.4 ± 9.4, 44.3 ± 11.2	4200 IU per week, 16,000 IU per week, or 28,000 IU per week	From 17 to 24 weeks of gestation until birth
VIOLET	Critically ill patients	Not reported	11.0 ± 4.7	11.4 ± 5.6	11.2 ± 4.8	46.9 ± 23.2	Single enteral dose of 540,000 IU	90 days
CAPS	Postmenopausal women	32.8 ± 10.5	32.7 (95% CI: 32.1 to 33.3)	30.9 (95% CI: 30.2 to 31.6)	33.0 (95% CI: 32.3 to 33.6)	42.5 (95% CI: 41.7 to 43.3)	2000 IU plus 1500 mg calcium per day	4 years
Ganmaa	School children	11.9 ± 4.2	11.9 ± 4.2	10.7 ± 5.3	11.9 ± 4.2	31.0 ± 9.1	14,000 IU per week	3 years (median)
EVITA	Patients with heart failure	14.6 ± 6.7	14.1 (10.3 to 19.7)	16.3 (12.5 to 23.2)	12.5 (8.6 to 17.9)	37.2 (25.0 to 51.4)	4000 IU per day	3 years
Burt	Older general population	31.3 ± 7.8	No placebo group	No placebo group	30.6 ± 8.4, 32.5 ± 8.0, 31.3 ± 7.4	31.0 (mean), 52.9 (mean), 57.8 (mean)	400 IU per day, 4000 IU per day, 10,000 IU per day	3 years

Data are shown as mean ± standard deviation (SD) or as medians with 25th to 75th percentile, if not otherwise indicated; for the MDIG trial, only groups with no postpartum intervention are shown.

5. Meta-Analyses of Vitamin D RCTs

It is also crucial to outline the knowledge provided by meta-analyses of RCTs that have partially considered the evidence provided by the above-mentioned trials [10,11,61–67].

Regarding cancer outcomes, meta-analyses of RCTs conclude that vitamin D supplementation does not have an effect on cancer incidence, but vitamin D supplementation reduces cancer mortality up to 16% with a respective HR (95% CI) of 0.84 (0.74 to 0.95) based on 12 RCTs in 45,578 participants with 939 cancer deaths [10,11,14,68]. These data on vitamin D and cancer mortality are in line with experimental and observational studies on this topic.

Meta-analyses of RCTs reported that vitamin D supplementation does not reduce overall or individual (such as stroke or myocardial infarction) cardiovascular events [14,26,66]. Similarly, there were also no overall major effects on vascular function or cardiovascular risk factors, although some explorative subgroup analyses of individual RCTs and meta-analyses suggest potential (minor) benefits in subgroups such as those with low 25(OH)D and prediabetes [10,24,29,38,67,69].

With reference to bone health or musculoskeletal health in general, it has to be stressed that the effects of vitamin D in terms of the prevention and treatment of nutritional rickets and osteomalacia are historically established and provide the basis for the conclusion of nutritional vitamin D guidelines that serum 25(OH)D concentrations below 10 or 12 ng/mL have to be prevented and treated [1]. This consensus has not been challenged by recent large vitamin D trials or their meta-analyses. Of note, explorative data from the MDIG trial reported three radiographically confirmed cases of rickets in the placebo group and just one case in the intervention groups that included four times more patients than the placebo group did overall [30]. Regarding other musculoskeletal health outcomes, i.e., fractures, falls, and BMD, the conclusions from meta-analyses of the RCTs are inconsistent and puzzling [10,14,63–65]. While some meta-analyses conclude that vitamin D supplementation per se does not prevent fractures and falls or has meaningful effects on BMD, there are other meta-analyses documenting fracture prevention by daily combined calcium plus vitamin D supplementation in older adults [10,14,63–65]. Data interpretation has to consider evidence for a U-shaped effect, as high vitamin D bolus doses increase the risk of fractures and falls, whereas any beneficial effects, if present, are particularly observed with moderate vitamin D doses of about 800 IU per day in older individuals with a poor vitamin D status [10,14,63–65].

A recently published updated meta-analysis of RCTs reported that vitamin D supplementation significantly prevents acute respiratory infections [62]. By using data from 48,488 participants from 43 RCTs, the odds ratio (95% CI) for having one or more acute respiratory infections in the vitamin D versus the placebo group was 0.92 (0.86 to 0.99). Explorative subgroup analyses indicated a protective effect in groups with daily vitamin D doses, a dose equivalent of 400 to 1000 IU per day, a trial duration of up to 12 months and in those aged 1 to 16 years old. Accordingly, there are also findings from meta-analyses of RCTs suggesting that vitamin D may prevent exacerbations of COPD and asthma [70–72].

Coronavirus Disease 2019 (COVID-19), a disease caused by the severe acute respiratory syndrome coronavirus 2 (SARS-CoV-2), has not been addressed in the above listed large vitamin D RCTs or meta-analyses of RCTs [73–75]. Although the data on vitamin D and acute respiratory infections are promising and support recommendations for a sufficient vitamin D status during this pandemic, it is unclear whether this also applies to COVID-19 [76,77]. In view of the exploding publication output on vitamin D and COVID-19, we should keep in mind that research findings are less likely to be true in a hot scientific research field (with more scientific teams involved) [16]. This can be partially attributed to the fact that many groups have started to work on this topic, and those with significant findings are more likely to publish their results (publication bias), do not consider the multiple testing problem of many similar investigations around the world and/or are less critical, careful, and balanced when following a publication hype. We should also consider that vitamin D is effective as a preventive measure and not as a therapy for acute respiratory infections using high doses in patients already suffering from severe infection [61,62]. In this context, the subgroup analyses of the VIOLET trial reported increased mortality in those receiving high doses of vitamin D with sepsis, infection, and respiratory distress syndrome [31]. In general, public health strategies to fight the COVID-19 pandemic should primarily consider well established and relevant data, such as the effectiveness of vaccination and protection after natural infection, as well as the enormous age-dependent association of the infection fatality rate [73–75].

Older meta-analyses of vitamin D supplementation and all-cause mortality have reported that vitamin D may moderately but significantly reduce all-cause mortality, whereas updated meta-analyses on this topic failed to document a significant effect [66,78,79]. In detail, the relative risks (95% CI) for the effect of vitamin D in two meta-analyses of RCTs were 0.97 (0.93 to 1.02) and 0.98 (0.95 to 1.02), respectively [66,79].

In addition to classic RCTs, we wish to point out the clinical value of Mendelian Randomization (MR) studies that can be considered as a sort of randomized trials of human nature [6,7,14,61,80–82]. MR studies use genetically determined 25(OH)D concentrations as an instrumental variable to evaluate associations with clinical outcomes [82]. This strategy, although observational by definition with all inherent limitations, is useful to address the question of causality, as it can be assumed that a certain genetic variant that predisposes people to higher or lower 25(OH)D concentrations should not be associated with common confounding factors [82]. It is beyond the scope of this article to summarize all MR studies on vitamin D, but it bears mentioning that a recent large MR study including more than half a million participants reported an increased all-cause mortality risk for 25(OH)D concentrations below 10 ng/mL [81]. When using a finer stratification of participants in the same study, there was an inverse association between genetically determined 25(OH)D concentrations and mortality up to 16 ng/mL [81]. Findings of another non-linear MR study also suggest that the correction of vitamin D deficiency could reduce cardiovascular disease incidence and blood pressure [83].

Epidemiological studies including data on the use of vitamin D supplementation and baseline as well as follow-up 25(OH)D concentrations may also be of value, but they require careful interpretation in light of their limitations due to their observational study design [84–87].

6. Future Outlook

Results of a few other large vitamin D RCTs will be published in the near future, but given that, e.g., in the D-HEALTH trial, the 25(OH)D concentration in the placebo group is even above 30 ng/mL, we do not expect results that significantly differ from previous studies with similar designs [44].

Considering that even one year after finishing a trial with 20,000 IU of vitamin D per week for 3 to 5 years, there was still a significant difference in serum 25(OH)D concentrations between the vitamin D and the placebo group (i.e., 33.8 versus 29.2 ng/mL), suggesting a much longer half-life of 25(OH)D than the frequently quoted approximately 3 weeks, we recommend evaluating health outcomes in large vitamin D RCTs after finishing the active treatment periods to capture the potential latency or legacy effects of vitamin D [88]. Such an approach has already been used for the EVITA trial, confirming the null effect of vitamin D on total mortality in this cohort of heart failure patients [89].

When designing and interpreting vitamin D RCTs, it has to be considered that there are significant differences regarding the individual molecular responses to vitamin D [90,91]. A variety of nutrients interact with vitamin D and its metabolism, such as, e.g., magnesium, calcium, vitamin K and A, etc. and may thus modulate the individual sensitivity to effects of vitamin D supplementation [92,93]. Similarly, genetic polymorphisms related to VDR signaling or vitamin D metabolism may also contribute to individual differences in the response to vitamin D supplementation [94]. Overall, future vitamin D research should put more emphasis on a personalized approach. The "fire and forget" concept of recent large vitamin D RCTs with a single vitamin D dose that must fit everyone should be modified for future trial designs by enrolling participants who are most likely to benefit from vitamin D treatment and in whom individual differences are accounted for. It should, for example, be recognized that there are higher vitamin D requirements in obese individuals, and pre-specified optimal 25(OH)D concentrations should be targeted by vitamin D dosing adaptations during the trial. Individual participant baseline and achieved 25(OH)D concentrations should therefore be considered for the design and analysis of RCTs. Potential ethnic differences should also be accounted for [95]. Accurate and standardized measurements of vitamin D status are also crucial for future vitamin D trials, and additional measurements of vitamin D metabolites and the consideration of bioavailable fractions are also worth considering [96]. Finally, seasonal variations in vitamin D status and the various different sources of vitamin D should also be considered in the design of future vitamin D trials as well as in the optimal follow-up time for the given outcomes of interest.

7. Conclusions

In conclusion, recent RCTs have failed to document significant extra-skeletal benefits of vitamin D supplementation in individuals with largely normal 25(OH)D concentrations. While the limitations of trial designs increased the likelihood of not achieving significant effects on the primary outcomes, we aimed to interpret these findings regarding their potential impact on the current view of vitamin D in terms of clinical practice and science.

Current nutritional vitamin D guidelines are mainly based on the beneficial effects of vitamin D with reference to rickets and osteomalacia [97–100]. As noted above, this knowledge has not been challenged by recent large vitamin D RCTs, so their null findings do not have an impact on current vitamin D intake recommendations. With reference to other musculoskeletal health outcomes such as fractures and falls, we conclude that increasing vitamin D doses beyond intakes of about 800 to 1000 IU per day does not confer additional benefits but may even be harmful when very high doses are used, particularly intermittent bolus doses [35,36,50,51].

Regarding non-skeletal health outcomes, there is evidence from RCTs suggesting that vitamin D supplementation may prevent acute respiratory infections and cancer mortality. In addition, limited evidence suggests the potential benefits of vitamin D for some other health outcomes such as diabetes mellitus, but this still requires further evaluation. Moreover, vitamin D RCTs indicate that "more is not always better". Several lines of evidence suggest that moderate daily vitamin D doses are superior to very high daily doses or intermittent high doses [33,35,36,50,51,62]. Nevertheless, vitamin D requirements and the "optimal vitamin D status" may be different according to the outcomes studied, and we do have firm evidence for some kind of threshold effect in the sense that those with very low 25(OH)D concentrations are most likely to benefit from vitamin D treatment [12,21,22,81].

Finally, we want to point out that further large trials using more personalized approaches are needed to evaluate potential effects of vitamin D treatment in populations with severe vitamin D deficiency or certain characteristics that suggest a high sensitivity to vitamin D treatment. From the public health perspective, there is still an urgent need to erase the worldwide burden of vitamin D deficiency with 25(OH)D concentrations below 10 to 12 ng/mL. This requires, beyond a healthy lifestyle that considers nutrition and physical activity, general actions that can be taken to improve vitamin D status such as vitamin D food fortification and vitamin D supplementation in those with vitamin D deficiency [14,101].

Author Contributions: S.P. Writing—original draft preparation; S.P., C.T., V.T.-S., M.R.G., N.D.V., B.O., S.N.K., A.Z. and W.M. contributed to writing—review and editing. S.P., C.T., V.T.-S., M.R.G., N.D.V., B.O., S.N.K., A.Z. and W.M. have read and agreed to the published version of the manuscript. All authors have read and agreed to the published version of the manuscript.

Funding: This research received no external funding.

Conflicts of Interest: S.P. received speaker's honoraria from Diasorin, Synlab, Merk, Procter & Gamble, and Wörwag Pharma. The other authors declare no conflict of interest.

References

1. Pilz, S.; Zittermann, A.; Trummer, C.; Theiler-Schwetz, V.; Lerchbaum, E.; Keppel, M.H.; Grubler, M.R.; Marz, W.; Pandis, M. Vitamin D testing and treatment: A narrative review of current evidence. *Endocr. Connect.* **2019**, *8*, R27–R43. [CrossRef]
2. Giustina, A.; Bouillon, R.; Binkley, N.; Sempos, C.; Adler, R.A.; Bollerslev, J.; Dawson-Hughes, B.; Ebeling, P.R.; Feldman, D.; Heijboer, A.; et al. Controversies in Vitamin D: A Statement From the Third International Conference. *JBMR Plus* **2020**, *4*, e10417. [CrossRef] [PubMed]
3. Cashman, K.D.; Dowling, K.G.; Skrabakova, Z.; Gonzalez-Gross, M.; Valtuena, J.; De Henauw, S.; Moreno, L.; Damsgaard, C.T.; Michaelsen, K.F.; Molgaard, C.; et al. Vitamin D deficiency in Europe: Pandemic? *Am. J. Clin. Nutr.* **2016**, *103*, 1033–1044. [CrossRef] [PubMed]
4. Saponaro, F.; Saba, A.; Zucchi, R. An Update on Vitamin D Metabolism. *Int. J. Mol. Sci.* **2020**, *21*, 6573. [CrossRef]
5. Hanel, A.; Carlberg, C. Vitamin D and evolution: Pharmacologic implications. *Biochem. Pharmacol.* **2020**, *173*, 113595. [CrossRef]

6. Bouillon, R.; Marcocci, C.; Carmeliet, G.; Bikle, D.; White, J.H.; Dawson-Hughes, B.; Lips, P.; Munns, C.F.; Lazaretti-Castro, M.; Giustina, A.; et al. Skeletal and Extraskeletal Actions of Vitamin D: Current Evidence and Outstanding Questions. *Endocr. Rev.* **2019**, *40*, 1109–1151. [CrossRef]

7. Zittermann, A.; Trummer, C.; Theiler-Schwetz, V.; Lerchbaum, E.; Marz, W.; Pilz, S. Vitamin D and Cardiovascular Disease: An Updated Narrative Review. *Int. J. Mol. Sci.* **2021**, *22*, 2896. [CrossRef] [PubMed]

8. Kupferschmidt, K. Uncertain verdict as vitamin D goes on trial. *Science* **2012**, *337*, 1476–1478. [CrossRef]

9. Pilz, S.; Rutters, F.; Dekker, J.M. Disease prevention: Vitamin D trials. *Science* **2012**, *338*, 883. [CrossRef]

10. Ganmaa, D.; Enkhmaa, D.; Nasantogtokh, E.; Sukhbaatar, S.; Tumur-Ochir, K.E.; Manson, J.E. Vitamin D, respiratory infections, and chronic disease: Review of meta-analyses and randomized clinical trials. *J. Intern. Med.* **2021**. [CrossRef]

11. Sluyter, J.D.; Manson, J.E.; Scragg, R. Vitamin D and Clinical Cancer Outcomes: A Review of Meta-Analyses. *JBMR Plus* **2021**, *5*, e10420. [CrossRef] [PubMed]

12. Scragg, R. Emerging Evidence of Thresholds for Beneficial Effects from Vitamin D Supplementation. *Nutrients* **2018**, *10*, 561. [CrossRef]

13. Pilz, S. Marine n-3 Fatty Acids and Vitamin D Supplementation and Primary Prevention. *N. Engl. J. Med.* **2019**, *380*, 1878–1879. [CrossRef]

14. Bouillon, R.; Manousaki, D.; Rosen, C.; Trajanoska, K.; Rivadeneira, F.; Richards, J.B. The health effects of vitamin D supplementation: Evidence from human studies. *Nat. Rev. Endocrinol.* **2021**, *11*, 1–15. [CrossRef] [PubMed]

15. Cashman, K.D.; Kiely, M. Contribution of nutrition science to the vitamin D field-Clarity or confusion? *J. Steroid Biochem. Mol. Biol.* **2019**, *187*, 34–41. [CrossRef] [PubMed]

16. Ioannidis, J.P. Why most published research findings are false. *PLoS Med.* **2005**, *2*, e124. [CrossRef]

17. Autier, P.; Boniol, M.; Pizot, C.; Mullie, P. Vitamin D status and ill health: A systematic review. *Lancet Diabetes Endocrinol.* **2014**, *2*, 76–89. [CrossRef]

18. Yang, C.S.; Chen, J.X.; Wang, H.; Lim, J. Lessons learned from cancer prevention studies with nutrients and non-nutritive dietary constituents. *Mol. Nutr. Food Res.* **2016**, *60*, 1239–1250. [CrossRef]

19. Vastag, B. Nutrients for prevention: Negative trials send researchers back to drawing board. *J. Natl. Cancer Inst.* **2009**, *101*, 446–448. [CrossRef] [PubMed]

20. Robinson, I.; de Serna, D.G.; Gutierrez, A.; Schade, D.S. Vitamin E in humans: An explanation of clinical trial failure. *Endocr. Pract.* **2006**, *12*, 576–582. [CrossRef]

21. Gaksch, M.; Jorde, R.; Grimnes, G.; Joakimsen, R.; Schirmer, H.; Wilsgaard, T.; Mathiesen, E.B.; Njolstad, I.; Lochen, M.L.; Marz, W.; et al. Vitamin D and mortality: Individual participant data meta-analysis of standardized 25-hydroxyvitamin D in 26916 individuals from a European consortium. *PLoS ONE* **2017**, *12*, e0170791. [CrossRef]

22. Durazo-Arvizu, R.A.; Dawson-Hughes, B.; Kramer, H.; Cao, G.; Merkel, J.; Coates, P.M.; Sempos, C.T. The Reverse J-Shaped Association Between Serum Total 25-Hydroxyvitamin D Concentration and All-Cause Mortality: The Impact of Assay Standardization. *Am. J. Epidemiol.* **2017**, *185*, 720–726. [CrossRef] [PubMed]

23. Schottker, B.; Jorde, R.; Peasey, A.; Thorand, B.; Jansen, E.H.; Groot, L.; Streppel, M.; Gardiner, J.; Ordonez-Mena, J.M.; Perna, L.; et al. Vitamin D and mortality: Meta-analysis of individual participant data from a large consortium of cohort studies from Europe and the United States. *BMJ* **2014**, *348*, g3656. [CrossRef] [PubMed]

24. Scragg, R. The Vitamin D Assessment (ViDA) study—Design and main findings. *J. Steroid Biochem. Mol. Biol.* **2020**, *198*, 105562. [CrossRef] [PubMed]

25. Scragg, R.; Stewart, A.W.; Waayer, D.; Lawes, C.M.M.; Toop, L.; Sluyter, J.; Murphy, J.; Khaw, K.T.; Camargo, C.A., Jr. Effect of Monthly High-Dose Vitamin D Supplementation on Cardiovascular Disease in the Vitamin D Assessment Study: A Randomized Clinical Trial. *JAMA Cardiol.* **2017**, *2*, 608–616. [CrossRef]

26. Manson, J.E.; Bassuk, S.S.; Buring, J.E.; Group, V.R. Principal results of the VITamin D and OmegA-3 TriaL (VITAL) and updated meta-analyses of relevant vitamin D trials. *J. Steroid Biochem. Mol. Biol.* **2020**, *198*, 105522. [CrossRef]

27. Manson, J.E.; Cook, N.R.; Lee, I.M.; Christen, W.; Bassuk, S.S.; Mora, S.; Gibson, H.; Gordon, D.; Copeland, T.; D'Agostino, D.; et al. Vitamin D Supplements and Prevention of Cancer and Cardiovascular Disease. *N. Engl. J. Med.* **2019**, *380*, 33–44. [CrossRef]

28. Bischoff-Ferrari, H.A.; Vellas, B.; Rizzoli, R.; Kressig, R.W.; da Silva, J.A.P.; Blauth, M.; Felson, D.T.; McCloskey, E.V.; Watzl, B.; Hofbauer, L.C.; et al. Effect of Vitamin D Supplementation, Omega-3 Fatty Acid Supplementation, or a Strength-Training Exercise Program on Clinical Outcomes in Older Adults: The DO-HEALTH Randomized Clinical Trial. *JAMA* **2020**, *324*, 1855–1868. [CrossRef]

29. Pittas, A.G.; Dawson-Hughes, B.; Sheehan, P.; Ware, J.H.; Knowler, W.C.; Aroda, V.R.; Brodsky, I.; Ceglia, L.; Chadha, C.; Chatterjee, R.; et al. Vitamin D Supplementation and Prevention of Type 2 Diabetes. *N. Engl. J. Med.* **2019**, *381*, 520–530. [CrossRef]

30. Roth, D.E.; Morris, S.K.; Zlotkin, S.; Gernand, A.D.; Ahmed, T.; Shanta, S.S.; Papp, E.; Korsiak, J.; Shi, J.; Islam, M.M.; et al. Vitamin D Supplementation in Pregnancy and Lactation and Infant Growth. *N. Engl. J. Med.* **2018**, *379*, 535–546. [CrossRef]

31. National Heart, L.; Blood Institute, P.C.T.N.; Ginde, A.A.; Brower, R.G.; Caterino, J.M.; Finck, L.; Banner-Goodspeed, V.M.; Grissom, C.K.; Hayden, D.; Hough, C.L.; et al. Early High-Dose Vitamin D3 for Critically Ill, Vitamin D-Deficient Patients. *N. Engl. J. Med.* **2019**, *381*, 2529–2540. [CrossRef]

32. Lappe, J.; Watson, P.; Travers-Gustafson, D.; Recker, R.; Garland, C.; Gorham, E.; Baggerly, K.; McDonnell, S.L. Effect of Vitamin D and Calcium Supplementation on Cancer Incidence in Older Women: A Randomized Clinical Trial. *JAMA* **2017**, *317*, 1234–1243. [CrossRef] [PubMed]

33. Zittermann, A.; Ernst, J.B.; Prokop, S.; Fuchs, U.; Dreier, J.; Kuhn, J.; Knabbe, C.; Birschmann, I.; Schulz, U.; Berthold, H.K.; et al. Effect of vitamin D on all-cause mortality in heart failure (EVITA): A 3-year randomized clinical trial with 4000 IU vitamin D daily. *Eur. Heart J.* **2017**, *38*, 2279–2286. [CrossRef]

34. Ganmaa, D.; Uyanga, B.; Zhou, X.; Gantsetseg, G.; Delgerekh, B.; Enkhmaa, D.; Khulan, D.; Ariunzaya, S.; Sumiya, E.; Bolortuya, B.; et al. Vitamin D Supplements for Prevention of Tuberculosis Infection and Disease. *N. Engl. J. Med.* **2020**, *383*, 359–368. [CrossRef] [PubMed]

35. Billington, E.O.; Burt, L.A.; Rose, M.S.; Davison, E.M.; Gaudet, S.; Kan, M.; Boyd, S.K.; Hanley, D.A. Safety of High-Dose Vitamin D Supplementation: Secondary Analysis of a Randomized Controlled Trial. *J. Clin. Endocrinol. Metab.* **2020**, *105*, 1261–1273. [CrossRef] [PubMed]

36. Burt, L.A.; Billington, E.O.; Rose, M.S.; Raymond, D.A.; Hanley, D.A.; Boyd, S.K. Effect of High-Dose Vitamin D Supplementation on Volumetric Bone Density and Bone Strength: A Randomized Clinical Trial. *JAMA* **2019**, *322*, 736–745. [CrossRef]

37. Waterhouse, M.; English, D.R.; Armstrong, B.K.; Baxter, C.; Duarte Romero, B.; Ebeling, P.R.; Hartel, G.; Kimlin, M.G.; McLeod, D.S.A.; O'Connell, R.L.; et al. A randomized placebo-controlled trial of vitamin D supplementation for reduction of mortality and cancer: Statistical analysis plan for the D-Health Trial. *Contemp. Clin. Trials Commun.* **2019**, *14*, 100333. [CrossRef]

38. Scragg, R.; Sluyter, J.D. Is There Proof of Extraskeletal Benefits From Vitamin D Supplementation From Recent Mega Trials of Vitamin D? *JBMR Plus* **2021**, *5*, e10459. [CrossRef]

39. Bauchner, H.; Golub, R.M.; Fontanarosa, P.B. Reporting and Interpretation of Randomized Clinical Trials. *JAMA* **2019**, *322*, 732–735. [CrossRef]

40. Jakobsen, J.C.; Ovesen, C.; Winkel, P.; Hilden, J.; Gluud, C.; Wetterslev, J. Power estimations for non-primary outcomes in randomised clinical trials. *BMJ Open* **2019**, *9*, e027092. [CrossRef]

41. VanderWeele, T.J.; Knol, M.J. Interpretation of subgroup analyses in randomized trials: Heterogeneity versus secondary interventions. *Ann. Intern. Med.* **2011**, *154*, 680–683. [CrossRef] [PubMed]

42. Freemantle, N. Interpreting the results of secondary end points and subgroup analyses in clinical trials: Should we lock the crazy aunt in the attic? *BMJ* **2001**, *322*, 989–991. [CrossRef] [PubMed]

43. Sluyter, J.D.; Camargo, C.A., Jr.; Stewart, A.W.; Waayer, D.; Lawes, C.M.M.; Toop, L.; Khaw, K.T.; Thom, S.A.M.; Hametner, B.; Wassertheurer, S.; et al. Effect of Monthly, High-Dose, Long-Term Vitamin D Supplementation on Central Blood Pressure Parameters: A Randomized Controlled Trial Substudy. *J. Am. Heart Assoc.* **2017**, *6*, e006802. [CrossRef]

44. Pham, H.; Waterhouse, M.; Baxter, C.; Duarte Romero, B.; McLeod, D.S.A.; Armstrong, B.K.; Ebeling, P.R.; English, D.R.; Hartel, G.; Kimlin, M.G.; et al. The effect of vitamin D supplementation on acute respiratory tract infection in older Australian adults: An analysis of data from the D-Health Trial. *Lancet Diabetes Endocrinol.* **2021**, *9*, 69–81. [CrossRef]

45. Chandler, P.D.; Chen, W.Y.; Ajala, O.N.; Hazra, A.; Cook, N.; Bubes, V.; Lee, I.M.; Giovannucci, E.L.; Willett, W.; Buring, J.E.; et al. Effect of Vitamin D3 Supplements on Development of Advanced Cancer: A Secondary Analysis of the VITAL Randomized Clinical Trial. *JAMA Netw. Open* **2020**, *3*, e2025850. [CrossRef]

46. Malihi, Z.; Lawes, C.M.M.; Wu, Z.; Huang, Y.; Waayer, D.; Toop, L.; Khaw, K.T.; Camargo, C.A.; Scragg, R. Monthly high-dose vitamin D supplementation does not increase kidney stone risk or serum calcium: Results from a randomized controlled trial. *Am. J. Clin. Nutr.* **2019**, *109*, 1578–1587. [CrossRef] [PubMed]

47. Jakobsen, J.C.; Wetterslev, J.; Winkel, P.; Lange, T.; Gluud, C. Thresholds for statistical and clinical significance in systematic reviews with meta-analytic methods. *BMC Med. Res. Methodol.* **2014**, *14*, 120. [CrossRef]

48. Khaw, K.T.; Stewart, A.W.; Waayer, D.; Lawes, C.M.M.; Toop, L.; Camargo, C.A., Jr.; Scragg, R. Effect of monthly high-dose vitamin D supplementation on falls and non-vertebral fractures: Secondary and post-hoc outcomes from the randomised, double-blind, placebo-controlled ViDA trial. *Lancet Diabetes Endocrinol.* **2017**, *5*, 438–447. [CrossRef]

49. Dawson-Hughes, B.; Staten, M.A.; Knowler, W.C.; Nelson, J.; Vickery, E.M.; LeBlanc, E.S.; Neff, L.M.; Park, J.; Pittas, A.G.; Group, D.d.R. Intratrial Exposure to Vitamin D and New-Onset Diabetes Among Adults With Prediabetes: A Secondary Analysis From the Vitamin D and Type 2 Diabetes (D2d) Study. *Diabetes Care* **2020**, *43*, 2916–2922. [CrossRef]

50. Bischoff-Ferrari, H.A.; Dawson-Hughes, B.; Orav, E.J.; Staehelin, H.B.; Meyer, O.W.; Theiler, R.; Dick, W.; Willett, W.C.; Egli, A. Monthly High-Dose Vitamin D Treatment for the Prevention of Functional Decline: A Randomized Clinical Trial. *JAMA Intern. Med.* **2016**, *176*, 175–183. [CrossRef]

51. Sanders, K.M.; Stuart, A.L.; Williamson, E.J.; Simpson, J.A.; Kotowicz, M.A.; Young, D.; Nicholson, G.C. Annual high-dose oral vitamin D and falls and fractures in older women: A randomized controlled trial. *JAMA* **2010**, *303*, 1815–1822. [CrossRef]

52. Heaney, R.P. Guidelines for optimizing design and analysis of clinical studies of nutrient effects. *Nutr. Rev.* **2014**, *72*, 48–54. [CrossRef] [PubMed]

53. Wang, L.; Song, Y.; Manson, J.E.; Pilz, S.; Marz, W.; Michaelsson, K.; Lundqvist, A.; Jassal, S.K.; Barrett-Connor, E.; Zhang, C.; et al. Circulating 25-hydroxy-vitamin D and risk of cardiovascular disease: A meta-analysis of prospective studies. *Circ. Cardiovasc. Qual. Outcomes* **2012**, *5*, 819–829. [CrossRef]

54. Zhang, R.; Li, B.; Gao, X.; Tian, R.; Pan, Y.; Jiang, Y.; Gu, H.; Wang, Y.; Wang, Y.; Liu, G. Serum 25-hydroxyvitamin D and the risk of cardiovascular disease: Dose-response meta-analysis of prospective studies. *Am. J. Clin. Nutr.* **2017**, *105*, 810–819. [CrossRef]

55. Han, J.; Guo, X.; Yu, X.; Liu, S.; Cui, X.; Zhang, B.; Liang, H. 25-Hydroxyvitamin D and Total Cancer Incidence and Mortality: A Meta-Analysis of Prospective Cohort Studies. *Nutrients* **2019**, *11*, 2295. [CrossRef] [PubMed]
56. Boucher, B.J. Why do so many trials of vitamin D supplementation fail? *Endocr. Connect.* **2020**, *9*, R195–R206. [CrossRef]
57. Bolland, M.J.; Grey, A.; Avenell, A. Assessment of research waste part 2: Wrong study populations- an exemplar of baseline vitamin D status of participants in trials of vitamin D supplementation. *BMC Med. Res. Methodol.* **2018**, *18*, 101. [CrossRef] [PubMed]
58. Palacios, C.; Kostiuk, L.K.; Pena-Rosas, J.P. Vitamin D supplementation for women during pregnancy. *Cochrane Database Syst. Rev.* **2019**, *7*, CD008873. [CrossRef]
59. Pilz, S.; Zittermann, A.; Obeid, R.; Hahn, A.; Pludowski, P.; Trummer, C.; Lerchbaum, E.; Perez-Lopez, F.R.; Karras, S.N.; Marz, W. The Role of Vitamin D in Fertility and during Pregnancy and Lactation: A Review of Clinical Data. *Int. J. Environ. Res. Public Health* **2018**, *15*, 2241. [CrossRef]
60. O'Callaghan, K.M.; Hennessy, A.; Hull, G.L.J.; Healy, K.; Ritz, C.; Kenny, L.C.; Cashman, K.D.; Kiely, M.E. Estimation of the maternal vitamin D intake that maintains circulating 25-hydroxyvitamin D in late gestation at a concentration sufficient to keep umbilical cord sera >/=25–30 nmol/L: A dose-response, double-blind, randomized placebo-controlled trial in pregnant women at northern latitude. *Am. J. Clin. Nutr.* **2018**, *108*, 77–91. [CrossRef]
61. Maretzke, F.; Bechthold, A.; Egert, S.; Ernst, J.B.; Melo van Lent, D.; Pilz, S.; Reichrath, J.; Stangl, G.I.; Stehle, P.; Volkert, D.; et al. Role of Vitamin D in Preventing and Treating Selected Extraskeletal Diseases-An Umbrella Review. *Nutrients* **2020**, *12*, 969. [CrossRef]
62. Jolliffe, D.A.; Camargo, C.A., Jr.; Sluyter, J.D.; Aglipay, M.; Aloia, J.F.; Ganmaa, D.; Bergman, P.; Bischoff-Ferrari, H.A.; Borzutzky, A.; Damsgaard, C.T.; et al. Vitamin D supplementation to prevent acute respiratory infections: A systematic review and meta-analysis of aggregate data from randomised controlled trials. *Lancet Diabetes Endocrinol.* **2021**, *9*, 276–292. [CrossRef]
63. Bolland, M.J.; Grey, A.; Avenell, A. Effects of vitamin D supplementation on musculoskeletal health: A systematic review, meta-analysis, and trial sequential analysis. *Lancet Diabetes Endocrinol.* **2018**, *6*, 847–858. [CrossRef]
64. Yao, P.; Bennett, D.; Mafham, M.; Lin, X.; Chen, Z.; Armitage, J.; Clarke, R. Vitamin D and Calcium for the Prevention of Fracture: A Systematic Review and Meta-analysis. *JAMA Netw. Open* **2019**, *2*, e1917789. [CrossRef]
65. Li, S.; Xi, C.; Li, L.; Long, Z.; Zhang, N.; Yin, H.; Xie, K.; Wu, Z.; Tian, J.; Wang, F.; et al. Comparisons of different vitamin D supplementation for prevention of osteoporotic fractures: A Bayesian network meta-analysis and meta-regression of randomised controlled trials. *Int. J. Food Sci. Nutr.* **2021**, *72*, 518–528. [CrossRef]
66. Barbarawi, M.; Kheiri, B.; Zayed, Y.; Barbarawi, O.; Dhillon, H.; Swaid, B.; Yelangi, A.; Sundus, S.; Bachuwa, G.; Alkotob, M.L.; et al. Vitamin D Supplementation and Cardiovascular Disease Risks in More Than 83000 Individuals in 21 Randomized Clinical Trials: A Meta-analysis. *JAMA Cardiol.* **2019**, *4*, 765–776. [CrossRef]
67. Beveridge, L.A.; Khan, F.; Struthers, A.D.; Armitage, J.; Barchetta, I.; Bressendorff, I.; Cavallo, M.G.; Clarke, R.; Dalan, R.; Dreyer, G.; et al. Effect of Vitamin D Supplementation on Markers of Vascular Function: A Systematic Review and Individual Participant Meta-Analysis. *J. Am. Heart Assoc.* **2018**, *7*, e008273. [CrossRef]
68. Gnagnarella, P.; Muzio, V.; Caini, S.; Raimondi, S.; Martinoli, C.; Chiocca, S.; Miccolo, C.; Bossi, P.; Cortinovis, D.; Chiaradonna, F.; et al. Vitamin D Supplementation and Cancer Mortality: Narrative Review of Observational Studies and Clinical Trials. *Nutrients* **2021**, *13*, 3285. [CrossRef] [PubMed]
69. Pittas, A.G.; Jorde, R.; Kawahara, T.; Dawson-Hughes, B. Vitamin D Supplementation for Prevention of Type 2 Diabetes Mellitus: To D or Not to D? *J. Clin. Endocrinol. Metab.* **2020**, *105*, 3721–3733. [CrossRef]
70. Jolliffe, D.A.; Greenberg, L.; Hooper, R.L.; Griffiths, C.J.; Camargo, C.A., Jr.; Kerley, C.P.; Jensen, M.E.; Mauger, D.; Stelmach, I.; Urashima, M.; et al. Vitamin D supplementation to prevent asthma exacerbations: A systematic review and meta-analysis of individual participant data. *Lancet Respir. Med.* **2017**, *5*, 881–890. [CrossRef]
71. Jolliffe, D.A.; Greenberg, L.; Hooper, R.L.; Mathyssen, C.; Rafiq, R.; de Jongh, R.T.; Camargo, C.A.; Griffiths, C.J.; Janssens, W.; Martineau, A.R. Vitamin D to prevent exacerbations of COPD: Systematic review and meta-analysis of individual participant data from randomised controlled trials. *Thorax* **2019**, *74*, 337–345. [CrossRef] [PubMed]
72. Li, X.; He, J.; Yu, M.; Sun, J. The efficacy of vitamin D therapy for patients with COPD: A meta-analysis of randomized controlled trials. *Ann. Palliat. Med.* **2020**, *9*, 286–297. [CrossRef] [PubMed]
73. Ioannidis, J.P.A.; Axfors, C.; Contopoulos-Ioannidis, D.G. Population-level COVID-19 mortality risk for non-elderly individuals overall and for non-elderly individuals without underlying diseases in pandemic epicenters. *Environ. Res.* **2020**, *188*, 109890. [CrossRef] [PubMed]
74. Polack, F.P.; Thomas, S.J.; Kitchin, N.; Absalon, J.; Gurtman, A.; Lockhart, S.; Perez, J.L.; Perez Marc, G.; Moreira, E.D.; Zerbini, C.; et al. Safety and Efficacy of the BNT162b2 mRNA Covid-19 Vaccine. *N. Engl. J. Med.* **2020**, *383*, 2603–2615. [CrossRef]
75. Pilz, S.; Chakeri, A.; Ioannidis, J.P.; Richter, L.; Theiler-Schwetz, V.; Trummer, C.; Krause, R.; Allerberger, F. SARS-CoV-2 re-infection risk in Austria. *Eur. J. Clin. Investig.* **2021**, *51*, e13520. [CrossRef]
76. Zemb, P.; Bergman, P.; Camargo, C.A., Jr.; Cavalier, E.; Cormier, C.; Courbebaisse, M.; Hollis, B.; Joulia, F.; Minisola, S.; Pilz, S.; et al. Vitamin D deficiency and the COVID-19 pandemic. *J. Glob. Antimicrob. Resist.* **2020**, *22*, 133–134. [CrossRef]
77. Stroehlein, J.K.; Wallqvist, J.; Iannizzi, C.; Mikolajewska, A.; Metzendorf, M.I.; Benstoem, C.; Meybohm, P.; Becker, M.; Skoetz, N.; Stegemann, M.; et al. Vitamin D supplementation for the treatment of COVID-19: A living systematic review. *Cochrane Database Syst. Rev.* **2021**, *5*, CD015043. [CrossRef]

78. Bjelakovic, G.; Gluud, L.L.; Nikolova, D.; Whitfield, K.; Wetterslev, J.; Simonetti, R.G.; Bjelakovic, M.; Gluud, C. Vitamin D supplementation for prevention of mortality in adults. *Cochrane Database Syst. Rev.* **2014**, CD007470. [CrossRef]

79. Zhang, Y.; Fang, F.; Tang, J.; Jia, L.; Feng, Y.; Xu, P.; Faramand, A. Association between vitamin D supplementation and mortality: Systematic review and meta-analysis. *BMJ* **2019**, *366*, l4673. [CrossRef]

80. Aspelund, T.; Grubler, M.R.; Smith, A.V.; Gudmundsson, E.F.; Keppel, M.; Cotch, M.F.; Harris, T.B.; Jorde, R.; Grimnes, G.; Joakimsen, R.; et al. Effect of Genetically Low 25-Hydroxyvitamin D on Mortality Risk: Mendelian Randomization Analysis in 3 Large European Cohorts. *Nutrients* **2019**, *11*, 74. [CrossRef]

81. Emerging Risk Factors Collaboration/EPIC-CVD/Vitamin D Studies Collaboration. Estimating dose-response relationships for vitamin D with coronary heart disease, stroke, and all-cause mortality: Observational and Mendelian randomisation analyses. *Lancet Diabetes Endocrinol.* **2021**, *9*, 837–846. [CrossRef]

82. Mukamal, K.J.; Stampfer, M.J.; Rimm, E.B. Genetic instrumental variable analysis: Time to call mendelian randomization what it is. The example of alcohol and cardiovascular disease. *Eur. J. Epidemiol.* **2020**, *35*, 93–97. [CrossRef]

83. Zhou, A.; Selvanayagam, J.B.; Hypponen, E. Non-linear Mendelian randomization analyses support a role for vitamin D deficiency in cardiovascular disease risk. *Eur. Heart J.* **2021**, *11*, 809. [CrossRef] [PubMed]

84. McDonnell, S.L.; Baggerly, C.A.; French, C.B.; Baggerly, L.L.; Garland, C.F.; Gorham, E.D.; Hollis, B.W.; Trump, D.L.; Lappe, J.M. Breast cancer risk markedly lower with serum 25-hydroxyvitamin D concentrations >/=60 vs <20 ng/ml (150 vs 50 nmol/L): Pooled analysis of two randomized trials and a prospective cohort. *PLoS ONE* **2018**, *13*, e0199265. [CrossRef] [PubMed]

85. McDonnell, S.L.; Baggerly, K.A.; Baggerly, C.A.; Aliano, J.L.; French, C.B.; Baggerly, L.L.; Ebeling, M.D.; Rittenberg, C.S.; Goodier, C.G.; Mateus Nino, J.F.; et al. Maternal 25(OH)D concentrations >/=40 ng/mL associated with 60% lower preterm birth risk among general obstetrical patients at an urban medical center. *PLoS ONE* **2017**, *12*, e0180483. [CrossRef] [PubMed]

86. Acharya, P.; Dalia, T.; Ranka, S.; Sethi, P.; Oni, O.A.; Safarova, M.S.; Parashara, D.; Gupta, K.; Barua, R.S. The Effects of Vitamin D Supplementation and 25-Hydroxyvitamin D Levels on the Risk of Myocardial Infarction and Mortality. *J. Endocr. Soc.* **2021**, *5*, bvab124. [CrossRef] [PubMed]

87. Mirhosseini, N.; Vatanparast, H.; Kimball, S.M. The Association between Serum 25(OH)D Status and Blood Pressure in Participants of a Community-Based Program Taking Vitamin D Supplements. *Nutrients* **2017**, *9*, 1244. [CrossRef]

88. Martinaityte, I.; Kamycheva, E.; Didriksen, A.; Jakobsen, J.; Jorde, R. Vitamin D Stored in Fat Tissue During a 5-Year Intervention Affects Serum 25-Hydroxyvitamin D Levels the Following Year. *J. Clin. Endocrinol. Metab.* **2017**, *102*, 3731–3738. [CrossRef] [PubMed]

89. Zittermann, A.; Ernst, J.B.; Prokop, S.; Fuchs, U.; Berthold, H.K.; Gouni-Berthold, I.; Gummert, J.F.; Pilz, S. A 3 year post-intervention follow-up on mortality in advanced heart failure (EVITA vitamin D supplementation trial). *ESC Heart Fail.* **2020**, *7*, 3754–3761. [CrossRef]

90. Carlberg, C. Nutrigenomics of Vitamin D. *Nutrients* **2019**, *11*, 676. [CrossRef] [PubMed]

91. Hanel, A.; Neme, A.; Malinen, M.; Hamalainen, E.; Malmberg, H.R.; Etheve, S.; Tuomainen, T.P.; Virtanen, J.K.; Bendik, I.; Carlberg, C. Common and personal target genes of the micronutrient vitamin D in primary immune cells from human peripheral blood. *Sci. Rep.* **2020**, *10*, 21051. [CrossRef] [PubMed]

92. Zittermann, A. Magnesium deficit? Overlooked cause of low vitamin D status? *BMC Med.* **2013**, *11*, 229. [CrossRef]

93. van Ballegooijen, A.J.; Pilz, S.; Tomaschitz, A.; Grubler, M.R.; Verheyen, N. The Synergistic Interplay between Vitamins D and K for Bone and Cardiovascular Health: A Narrative Review. *Int. J. Endocrinol.* **2017**, *2017*, 7454376. [CrossRef] [PubMed]

94. Krasniqi, E.; Boshnjaku, A.; Wagner, K.H.; Wessner, B. Association between Polymorphisms in Vitamin D Pathway-Related Genes, Vitamin D Status, Muscle Mass and Function: A Systematic Review. *Nutrients* **2021**, *13*, 3109. [CrossRef]

95. Cashman, K.D.; Kiely, M.E.; Andersen, R.; Gronborg, I.M.; Tetens, I.; Tripkovic, L.; Lanham-New, S.A.; Lamberg-Allardt, C.; Adebayo, F.A.; Gallagher, J.C.; et al. Individual participant data (IPD)-level meta-analysis of randomised controlled trials to estimate the vitamin D dietary requirements in dark-skinned individuals resident at high latitude. *Eur. J. Nutr.* **2021**, 1–20. [CrossRef] [PubMed]

96. Makris, K.; Bhattoa, H.P.; Cavalier, E.; Phinney, K.; Sempos, C.T.; Ulmer, C.Z.; Vasikaran, S.D.; Vesper, H.; Heijboer, A.C. Recommendations on the measurement and the clinical use of vitamin D metabolites and vitamin D binding protein - A position paper from the IFCC Committee on bone metabolism. *Clin. Chim. Acta* **2021**, *517*, 171–197. [CrossRef]

97. Bouillon, R. Comparative analysis of nutritional guidelines for vitamin D. *Nat. Rev. Endocrinol.* **2017**, *13*, 466–479. [CrossRef]

98. Pilz, S.; Trummer, C.; Pandis, M.; Schwetz, V.; Aberer, F.; Grubler, M.; Verheyen, N.; Tomaschitz, A.; Marz, W. Vitamin D: Current Guidelines and Future Outlook. *Anticancer. Res.* **2018**, *38*, 1145–1151. [CrossRef]

99. Cashman, K.D.; Ritz, C.; Kiely, M.; Odin, C. Improved Dietary Guidelines for Vitamin D: Application of Individual Participant Data (IPD)-Level Meta-Regression Analyses. *Nutrients* **2017**, *9*, 469. [CrossRef]

100. Ross, A.C.; Manson, J.E.; Abrams, S.A.; Aloia, J.F.; Brannon, P.M.; Clinton, S.K.; Durazo-Arvizu, R.A.; Gallagher, J.C.; Gallo, R.L.; Jones, G.; et al. The 2011 report on dietary reference intakes for calcium and vitamin D from the Institute of Medicine: What clinicians need to know. *J. Clin. Endocrinol. Metab.* **2011**, *96*, 53–58. [CrossRef]

101. Pilz, S.; Marz, W.; Cashman, K.D.; Kiely, M.E.; Whiting, S.J.; Holick, M.F.; Grant, W.B.; Pludowski, P.; Hiligsmann, M.; Trummer, C.; et al. Rationale and Plan for Vitamin D Food Fortification: A Review and Guidance Paper. *Front. Endocrinol.* **2018**, *9*, 373. [CrossRef] [PubMed]

Review

Evaluating the Evidence in Clinical Studies of Vitamin D in COVID-19

Tom D. Thacher

Department of Family Medicine, Mayo Clinic 200 First Street SW, Rochester, MN 55905, USA;
Thacher.Thomas@Mayo.edu

Abstract: Laboratory evidence provides a biological rationale for the benefits of vitamin D in COVID-19, and vitamin D supplementation is associated with reduced risk of respiratory infections. Most of the clinical studies of vitamin D in COVID-19 have been observational, and the most serious problem with observational study design is that of confounding. Observational studies typically assess the relationship of 25(OH)D values with COVID-19 outcomes. Many conditions associated with low vitamin D status are also associated with worse COVID-19 outcomes. Randomized controlled trials (RCTs) overcome the problem of confounding, typically comparing outcomes between groups receiving vitamin D supplementation or placebo. However, any benefit of vitamin D in COVID-19 may be related to the dose, duration, daily vs. bolus administration, interaction with other treatments, and timing of administration prior to or during the illness. Serum 25(OH)D values >50 nmol/L have been associated with reduced infection rates, severity of COVID-19, and mortality in observational studies. Few RCTs of vitamin D supplementation have been completed, and they have shown no benefit of vitamin D in hospitalized patients. Vitamin D may benefit those with mild or asymptomatic COVID-19, and those with greater 25(OH)D values may have lower risk of acquiring infection. Because those at greatest risk of COVID-19 are also at greatest risk of vitamin D deficiency, it is reasonable to recommend vitamin D supplementation 15–20 mcg (600–800 IU) daily for the general population during the COVID-19 pandemic. Vitamin D doses greater than 100 mcg (4000 IU) daily should not be used without monitoring serum 25(OH)D and calcium.

Keywords: nutrition; cholecalciferol; 25-hydroxyvitamin D; infection; study design; epidemiology; clinical trials

Citation: Thacher, T.D. Evaluating the Evidence in Clinical Studies of Vitamin D in COVID-19. *Nutrients* **2022**, *14*, 464. https://doi.org/10.3390/nu14030464

Academic Editor: Michael F. Holick

Received: 23 December 2021
Accepted: 19 January 2022
Published: 21 January 2022

Publisher's Note: MDPI stays neutral with regard to jurisdictional claims in published maps and institutional affiliations.

1. Theoretical Benefits of Vitamin D in COVID-19

Vitamin D deficiency has classically been associated with the bone diseases of rickets in growing children and osteomalacia in adults [1]. Increasing interest in the non-skeletal effects of vitamin D relates to the finding of vitamin D receptors (VDR) widely distributed in most human tissues, as is the 1α-hydroxylase enzyme (CYP27B1). Within these tissues, 1α-hydroxylase converts 25(OH)D to 1,25(OH)$_2$D in order to exert localized effects (paracrine effects), without altering serum 1,25(OH)$_2$D concentrations. Local 1,25(OH)$_2$D enters the cell nucleus to influence the expression of genes, unrelated to calcium absorption or bone metabolism.

Severe COVID-19 results from an exuberant and dysregulated immune response to the SARS-CoV-2 virus. The term COVID-associated acute respiratory distress syndrome (CARDS) has been used to describe the similar clinical manifestations and pathophysiology of severe COVID-19 and those of acute respiratory distress syndrome, including multisystemic effects from the release of proinflammatory cytokines.

The COVID-19 pandemic continues, and current evidence for the role of vitamin D in the treatment and prevention of COVID-19 deserves regular review. Studies of vitamin D in COVID-19 are based on a biological rationale for the benefits of vitamin D in COVID-19. There are three primary lung defenses against infection: airway epithelia, alveolar macrophages, and dendritic cells involved in cytokine production. All these cells express 1a-hydroxylase and are capable of locally producing $1,25(OH)_2D$, which acts as an important immune and inflammatory modulator. Vitamin D can alter the expression of genes involved in infection and inflammation and could theoretically decrease the severity of COVID-19 infection. Specifically, vitamin D could have benefit in COVID-19 in three different ways: (1) reducing the risk of acquiring SARS-CoV-2 infection, (2) enhancing viral neutralization and clearance, and (3) reducing the severity of the inflammatory response [2].

Laboratory evidence demonstrates that $1,25(OH)_2D$ promotes the expression of antimicrobial proteins cathelicidin and β-defensin2 by pulmonary macrophages and epithelium. It also suppresses antigen presentation by dendritic cells and activation of T cells, thereby inhibiting proinflammatory cytokine production. Additionally, $1,25(OH)2D$ regulates ACE2 to reduce vascular sensitivity and SARS-CoV-2 attachment to ACE2 receptors [2].

Daily or weekly doses of vitamin D, but not large bolus doses, were associated with a reduced risk of respiratory infection in a meta-analysis of randomized controlled trials of vitamin D supplementation, particularly in subjects with $25(OH)D$ values <25 nmol/L [3]. A single high dose of vitamin D can induce the 24-hydroxylase enzyme (CYP24A1), resulting in greater production of $24,25(OH)_2D$ relative to $25(OH)D$ than daily supplementation [4]. A daily dose of vitamin D may increase $25(OH)D$ with less diversion of $25(OH)D$ to $24,25(OH)_2D$ than bolus dosing.

2. Strengths and Limitations of Clinical Study Designs

The aim of this narrative review is to equip the reader with a framework to evaluate the evidence related to vitamin D in the treatment and prevention of COVID-19. Studies have been selected from a search of English language, peer-reviewed publications on PubMed, using the terms "vitamin D" and "COVID". Selected studies represent those with larger sample sizes and higher study quality, including meta-analyses. The GRADE scoring methodology can be used as a means of formally evaluating study quality and making recommendations [5], and the focus of this review is to enable the reader to understand the limitations of study design and appropriately assess study quality. Many trials are yet to be completed: clinicaltrials.gov on 25 January 2022 displayed 67 active trials of vitamin D in COVID-19.

Two broad categories of study design are used in most clinical studies: observational and experimental studies. Observational study designs include case-control, cross-sectional, and cohort studies. In observational studies of vitamin D, the serum $25(OH)D$ concentration, reflecting vitamin D status, is typically the independent variable. Observational studies can only demonstrate associations and not prove causation. Experimental studies are generally designed as randomized controlled trials (RCT). Vitamin D supplement intake is typically the independent variable. Results of RCTs are usually considered higher quality evidence than those of observational designs.

The most serious problem with observational studies is that of confounding. A confounder is a disease or behavior that is both associated with $25(OH)D$ and a risk factor for the outcome. It is possible to statistically adjust for confounding variables if they have been measured and are included in multivariable models. High risk conditions associated with severe COVID-19 include age >65 years, obesity, chronic kidney disease, diabetes, cerebrovascular disease, heart conditions, chronic lung diseases, chronic liver disease, cancer, mental health disorders, pregnancy, and smoking [6]. All increase the risk of adverse outcomes from COVID-19, but most of these conditions have been associated with lower vitamin D status, as measured by serum $25(OH)D$. In any study of COVID outcomes, these should be considered confounding variables, because they are related both to vitamin D status and to COVID-19 outcomes. Studies must report and adjust for these confounding variables in analysis of outcomes. Sample sizes should be large enough to allow for adjustment, i.e., generally at least 10 subjects for each variable in an adjusted multivariate model.

Multiple examples of confounding in observational studies of vitamin D are applicable in studies of vitamin D in COVID-19. Obesity is both associated with lower 25(OH)D and worse outcomes in COVID-19. Seasonal variation of 25(OH)D will be associated with diseases having seasonal variation, like respiratory illnesses. Additionally, seasonal variation in gene expression related to immune responses has been described [7], and the benefits of vitamin D on health outcomes have been proposed to be season dependent [8]. People with chronic illnesses have less outdoor activity and lower 25(OH)D levels. Chronic disease and critical illness can lower 25(OH)D, so low 25(OH)D can be the result rather than the cause of more severe illness. Serum 25(OH)D values may be inversely related to inflammatory and acute phase markers in severe illness [9,10]. Dietary intake of foods with vitamin D may improve vitamin D status, but other nutrients in those foods may be related to COVID-19 outcomes. People from racial groups with dark skin generally have lower serum 25(OH)D concentrations than white Caucasians and are at greater risk for severe COVID-19.

Besides vitamin D intake and sunlight exposure, additional factors influence 25(OH)D values. The methodology for measurement of 25(OH)D can be an important source of variation between studies, particularly when cut-point values are used, highlighting the need for standardized 25(OH)D measurements [11]. Polymorphisms of genes related to vitamin D transport and metabolism are predictive of 25(OH)D concentrations [12]. The 25(OH)D concentration indicative of adequate vitamin D status and the effect of $1,25(OH)_2D$ on genetic expression is likely to be highly variable between individuals [13]. Serum 25(OH)D is a marker of vitamin D supply but not a functional measure of the action of $1,25(OH)_2D$ on vitamin D receptor-mediated gene expression. The degree of vitamin D deficiency that has detrimental effects on inflammatory and immune regulation is unknown. Investigators identified two susceptibility loci with genome-wide significance for severe COVID-19 with respiratory failure in Spain [14], and low 25(OH)D levels could theoretically be linked to genetic loci also associated with an increased risk of severe COVID-19. It is the free 25(OH)D, unbound to vitamin D binding protein (DBP), that is available intracellularly. DBP affects the bioavailability of 25(OH)D to monocytes, and immune responses may be related to DBP polymorphisms [2,15].

Limitations of the current evidence base need to be recognized. Many of the studies assessing the relationship of vitamin D status with clinical outcomes are observational, retrospective studies. Low vitamin D status could be a cause or consequence of severe illness. Serum 25(OH)D values may be inversely related to inflammatory and acute phase markers in COVID-19 [16,17]. Low serum 25(OH)D levels may be a marker of chronic illness or mortality risk and may not necessarily indicate a therapeutic benefit of improving the vitamin D status. For example, patients with chronic illnesses may have low vitamin D status resulting from their disease, inadequate dietary intake, or limited sun exposure, but it is their chronic disease that puts them at greater risk of severe COVID-19.

There are additional limitations of observational studies. One is selection bias: patients who had 25(OH)D measurements available are those selected for study, and these subjects may differ from those who did not have 25(OH)D measured. Healthy user bias may result when healthier people are more likely to take vitamin D than less healthy individuals. One major issue with observational studies, given their retrospective nature, results from post-hoc analysis. Investigators do not always provide a pre-specified hypothesis or specify the primary outcome tested in observational studies. A pre-specified hypothesis is needed to correctly apply significance testing with a p value of 0.05. Multiple post-hoc comparisons may be performed examining multiple outcomes, varying subgroups of patients, or alternative cut-points for 25(OH)D. Each statistical comparison increases the risk of a type 1 statistical error, i.e., concluding that a relationship with 25(OH)D is significant when it is not. For example, if one makes 20 statistical comparisons, the probability of finding one comparison with $p < 0.05$ is not 5%, but 64%. For these reasons, a post-hoc analysis should be considered exploratory, to identify possible relationships that need confirmation.

Other limitations are related to reporting study outcomes, and these are more likely to occur with observational studies than with RCTs. The first of these is publication and reporting bias. Medical journals are more likely to publish studies that show potential benefit of an intervention than studies with negative results. This can skew the published literature toward positive results. Negative results are frequently due to inadequate study power, and authors may abandon efforts to publish negative studies. The second concern relates to preprint servers. These allow authors to post their manuscript on a public server prior to peer-review. This is intended to allow authors to quickly disseminate research findings, and the use of preprint servers has proliferated during the COVID-19 pandemic. Because these manuscripts are not peer-reviewed, it is difficult to be assured of the quality of the work and more likely for fraudulent material to be published than in peer-reviewed publications. Their results are intended to be considered preliminary.

Compared with observational studies, a much stronger level of evidence is provided by well-designed RCTs. The major advantage of RCTs is the control for confounding variables, even those that are not measured, because the randomization process generally balances confounding variables between vitamin D and placebo groups. RCTs have not consistently confirmed beneficial effects of vitamin D found in observational studies. Although RCTs generally represent a higher quality of evidence than observational studies, they also have their limitations. Subjects in the control group may also take vitamin D through diet or supplements, potentially attenuating any observed benefit. RCTs may have an inadequate number of persons with vitamin D deficiency, or subjects with vitamin D deficiency may have been excluded from the RCT. Additional vitamin D may be of no benefit for persons with an adequate vitamin D status. The dose, duration, or bolus vs. daily administration of vitamin D may be related to benefit [18]. Finally, the timing of vitamin D administration in relation to illness onset is essential to consider in COVID-19, because the beneficial effects may vary by the stage or severity of illness.

The temporal course of COVID-19 has been described in three phases. During the first week, there is viral replication of SARS-CoV-2. Severe COVID-19 typically develops in the second week, in up to 5% of infected patients. Severe COVID-19 results from exuberant and dysregulated immune responses to high viral loads. The term COVID-associated acute respiratory distress syndrome (CARDS) has been used to describe the similar clinical manifestations and pathophysiology of severe COVID-19 to acute respiratory distress syndrome, including multisystemic effects from the release of proinflammatory cytokines, also called a cytokine storm. Post-acute sequelae of SARS-CoV-2 (PASC or long COVID) refer to persistent symptoms after recovery, such as chronic fatigue, headache, brain fog, and dizziness. Any benefit of vitamin D in COVID-19 may vary with the phase or severity of illness. This is not a unique consideration for vitamin D. For example, dexamethasone is beneficial to those hospitalized with COVID-19 on oxygen but not in those who are not on oxygen. Some monoclonal antibody treatments for COVID-19 are beneficial in the first phase, but not in the second phase. Tocilizumab is beneficial in hospitalized patients on corticosteroids, but not when used alone. Vitamin D may be beneficial for a specific phase or severity of illness, or when used in combination with other treatments. Table 1 summarizes selected limitations to be considered related to study design in vitamin D and infection.

Table 1. Selected Limitations to be Considered Related to Study Design in Vitamin D and Infection.

Observational Studies
Independent variable: vitamin D status (serum 25(OH)D concentration)
Confounding variables: associated with both 25(OH)D and the outcome
Sample size must be adequate to adjust for known confounding variables
Seasonal variation of 25(OH)D and respiratory illnesses
Those with chronic illness have less sunlight exposure to produce 25(OH)D
Obesity is associated with both lower 25(OH)D and adverse outcomes
25(OH)D may be inversely related to inflammatory markers in severe illness
Racial groups with dark skin may have lower 25(OH)D and different outcomes than Caucasian whites
Vitamin D fortified foods increase 25(OH)D, but other nutrients in food may be related to outcomes
25(OH)D level is related to genes involved in vitamin D transport and metabolism, which could be linked to other genes affecting disease outcomes
Laboratory variation in 25(OH)D measurements and methodology requires standardization
Selection bias: Those with 25(OH)D measurements available were selected for study. They likely differ from those who did not have 25(OH)D measured.
Healthy user bias: Those who take vitamin D may be healthier than those who do not.
Post-hoc analysis: A pre-specified hypothesis is needed to correctly apply significance testing. Analyses of multiple outcomes, subgroups, and 25(OH)D cut points can lead to erroneous conclusions (statistical type 1 error).
Publication and reporting bias: Journals are more likely to publish studies that show potential benefit of an intervention than studies with negative results.
Preprint server publications are not peer-reviewed and results should be considered preliminary.
Randomized Controlled Trials
Independent variable: vitamin D supplementation (dose of vitamin D)
Control group may also take vitamin D, potentially attenuating any observed benefit
Inadequate number of persons with vitamin D deficiency (or vitamin D deficient subjects excluded)
Dose and duration of vitamin D may be related to benefit
Formulation of vitamin D may be related to benefit (e.g., cholecalciferol vs. calcifediol)
Daily vs. bolus dosing of vitamin D may have different metabolic effects
Timing of vitamin D administration in relation to illness onset, stage of disease, or illness severity
Interaction of vitamin D with other treatments for disease (e.g., corticosteroids)

3. Observational Studies

Studies of the association of vitamin D status with the severity of COVID-19 have been carried out in many countries. The majority of studies include patients with mean ages in the range of 50–65 years, and the age-related increase in risk begins at age 50 and rises continuously with advancing age. Among 216 hospitalized patients in Spain, 82% had 25(OH)D levels below 50 nmol/L, compared with 47% in population-based controls, matched only for sex [16]. However, there was no relationship between the severity of COVID-19 with 25(OH)D level. In the same study, lower levels of 25(OH)D were associated with higher levels of ferritin, D-dimer, and CRP, which are inflammatory markers that are commonly elevated in patients with COVID-19. This suggests that low vitamin D status may also be a marker of more severe inflammation.

Using a retrospective cohort design in racially diverse, hospitalized patients with COVID-19 in the U.S., investigators found that patients with serum 25(OH)D values <75 nmol/L within 6 months before or during hospitalization for COVID-19 had increased mortality and need for invasive mechanical ventilation [19]. Among hospitalized patients in the UAE with COVID-19 and 25(OH)D measured on admission, a 25(OH)D concentration <30 nmol/L was significantly associated with 1.8 times greater odds of severe illness and 2.6 times greater odds of death [20]. In a study of over 80,000 patients in the UK with COVID-19 and a 25(OH)D measured within 12 months prior to diagnosis, a value of 25(OH)D <50 nmol/L was associated with 2.4 times greater odds of hospitalization [21]. Among hospitalized patients with COVID-19 in Spain and 25(OH)D measured on admission, a 25(OH)D value <50 nmol/L was associated with 4.2 times greater odds of admission to ICU but not with mortality [22].

A study of hospitalized patients with COVID-19 in the US and 25(OH)D measured during the prior year explored multiple outcomes and included adjustment for multiple confounding variables [23]. The subgroup of patients 65 years-old or greater with 25(OH)D values >75 nmol/L had reduced odds of death, ARDS, and severe sepsis, compared with those having 25(OH)D ≤75 nmol/L, possibly indicating a higher inflammatory burden of COVID-19 in older patients.

In a very large population-based study in Spain, subjects who were on a vitamin D supplement were compared with propensity-matched controls that were not taking vitamin D [24]. Patients taking vitamin D had a slightly lower risk of SARS-CoV2 infection (4.0% vs. 4.2%). In a subgroup analysis, those taking vitamin D with 25(OH)D values >75 nmol/L had a lower risk of infection (3.3%) and mortality (0.6%) than those not taking vitamin D with 25(OH)D values <50 nmol/L (5.6% and 1.3%, respectively).

However, not all observational studies have found a relationship between vitamin D status and outcomes of COVID-19. Studies in the USA [25,26], Italy [27], the UK [28], and India [29] failed to confirm a beneficial effect of greater 25(OH)D levels with outcomes of length of stay, days on oxygen, ICU admission, need for assisted ventilation, or mortality, but some of these studies may have lacked sufficient power to detect a relationship between vitamin D status and less frequent outcomes, like mortality.

Vitamin D status may interact with other treatments for COVID-19. Investigators in the UK compared outcomes associated with vitamin D status before (March–April 2020) and after (September–December 2020) the use of dexamethasone in hospitalized patients [30]. Vitamin D deficiency was associated with elevated CRP and need for ventilation in hospitalized COVID-19 patients prior to use of dexamethasone but not during the interval of dexamethasone use. The primary outcome was mortality, and no mortality difference was evident during either interval. Differences in ventilation and mortality rates between these two time intervals suggested greater severity of illness in the dexamethasone-treated group. Dexamethasone may attenuate the adverse effects of vitamin D deficiency. Glucocorticoids increase 24-hydroxylase gene (*CYP24A1*) transcription [31] and are associated with lower 25(OH)D concentrations, but they also increase *VDR* transcription which can enhance 1,25(OH)$_2$D effects [32].

A meta-analysis combined the results of observational studies in 2020 [33]. Vitamin D deficiency was variably defined as total 25(OH)D) level less than 30 nmol/L (seven studies), less than 50 nmol/L (eight studies) and less than 62.5 nmol/L (one study). Vitamin D deficiency was associated with 2.5 times greater odds of mortality. Vitamin D deficiency was also associated with higher rates of hospital admission and longer hospital stays but no significant difference in ICU admissions. However, the authors found substantial heterogeneity and a high risk of bias in the included studies due to a lack of control for confounding variables.

A retrospective analysis of 191,779 individuals tested for SARS-CoV-2 at a national reference laboratory and matched with 25(OH)D results in the preceding 12 month found lower SARS-CoV-2 positivity rates among those with greater 25(OH)D concentrations [34]. This relationship was consistent irrespective of latitude, race, age, and sex. Similarly, in a retrospective population study of 7807 patients in Israel tested for SARS-CoV-2 with any previous 25(OH)D level, those with 25(OH)D <50 nmoL/L had a 60% greater odds of testing positive for COVID-19, while adjusting for the fact that those who had positive tests were significantly younger than those who tested negative (36 vs. 47 years) [35].

4. Randomized Controlled Trials

RCTs of vitamin D supplementation in COVID-19 are necessary to conclusively demonstrate benefit [36]. A RCT comparing a single bolus dose of vitamin D 540,000 IU with placebo in critically ill patients (unrelated to COVID-19) found no benefit in patients with baseline 25(OH)D levels less than 50 nmol/L [37]. Subjects in the vitamin D group had no survival benefit, even among those with severe vitamin D deficiency (25(OH)D <30 nmol/L) and, relevant to COVID-19, mortality was greater in subgroups with infection or ARDS who received vitamin D.

A methodologically sound RCT of vitamin D in COVID-19 was performed in 240 hospitalized patients in Brazil, randomized to a single oral dose of vitamin D 200,000 IU or placebo [38]. The study excluded those with severe illness, admitted to ICU, or requiring invasive ventilation. The primary outcome was the probability of hospital discharge over the course of the hospital stay. The investigators found no difference in the hospital length of stay between the vitamin D and placebo groups, even in a post hoc analysis of those with baseline 25D <50 nmol/L. They found no difference between groups in secondary outcomes of mortality, ICU admission, or mechanical ventilation. On limitation of this study was that the mean time from the onset of symptoms to randomization was 10.3 days and from hospitalization to randomization was 1.4 days. Giving vitamin D at the time of hospitalization may be too late to observe benefit against the severity of illness.

A meta-analysis combining the results of three RCTs and two quasi-experimental studies of vitamin D in COVID-19 found no conclusive evidence that vitamin D supplementation reduced mortality, invasive ventilation, or ICU admission [39]. A RCT in 76 patients in Spain that was included in the meta-analysis used oral calcifediol (25(OH)D) rather than vitamin D and found that calcifediol significantly reduced the risk of ICU admission (adjusted OR 0.03) [40].

Vitamin D may have greater benefit if given earlier in COVID-19 or prior to infection, similar to monoclonal antibodies that have benefit early in COVID-19 to prevent severe illness and hospitalization. In a RCT of vitamin D 60,000 IU/day for 7 days in 40 subjects in north India with baseline 25(OH)D <50 nmol/L and mild or asymptomatic COVID-19, viral clearance before day 21 occurred three times more frequently in those randomized to vitamin D (63% vs. 21%) [41].

5. Conclusions

To summarize the evidence, a serum 25(OH)D >50 nmol/L has been associated with reduced infection rates, reduced severity of COVID-19, and reduced mortality in observational studies, but observational studies have a high risk of bias and are limited by the relationship of vitamin D status with comorbidities. Few RCTs of vitamin D supplementation have been completed, and they have shown no benefit of vitamin D in hospitalized patients. A small study suggested benefit of vitamin D in mild or asymptomatic COVID-19. Those with greater 25(OH)D levels may have lower risk of acquiring infection. The timing of vitamin D administration and phase of illness may be critical to observe benefit in COVID-19. Randomized controlled trials are necessary to confirm beneficial effects of vitamin D suggested by observational studies. Further studies of the dose, timing, and interaction of vitamin D with other treatments are indicated.

Because those at greatest risk of COVID-19 are also at greatest risk of vitamin D deficiency, it is reasonable to recommend vitamin D supplementation for the general population during the COVID-19 pandemic. The recommended dietary allowances (RDA) for vitamin D in the US were set by the US National Academy of Medicine to achieve a concentration of 25(OH)D of 50 nmol/L [42]. The RDA was set at 15 mcg (600 IU) daily for persons aged 1–70 years and 20 mcg (800 IU) daily for those over age 70 years. An upper limit vitamin D intake of 100 mcg (4000 IU) daily does not require monitoring, but higher intakes should be monitored. No harms are associated with this dose range of vitamin D, and there is potential benefit in reducing the severity of COVID-19 and risk of infection. Doses of vitamin D greater than 100 mcg (4000 IU) daily have no established role in the treatment or prevention of COVID-19, and excessive vitamin D can lead to toxicity, manifested as hypercalcemia and nephro-calcinosis.

Funding: This research received no external funding.

Conflicts of Interest: The author declares no conflict of interest.

References

1. Munns, C.F.; Shaw, N.; Kiely, M.; Specker, B.L.; Thacher, T.; Ozono, K.; Michigami, T.; Tiosano, D.; Mughal, M.Z.; Mäkitie, O.; et al. Global Consensus Recommendations on Prevention and Management of Nutritional Rickets. *J. Clin. Endocrinol. Metab.* **2016**, *101*, 394–415. [CrossRef]
2. Bilezikian, J.P.; Bikle, D.; Hewison, M.; Lazaretti-Castro, M.; Formenti, A.M.; Gupta, A.; Madhavan, M.V.; Nair, N.; Babalyan, V.; Hutchings, N.; et al. Vitamin D and COVID-19. *Eur. J. Endocrinol.* **2020**, *183*, R133–R147. [CrossRef] [PubMed]
3. Martineau, A.R.; Jolliffe, D.A.; Hooper, R.L.; Greenberg, L.; Aloia, J.F.; Bergman, P.; Dubnov-Raz, G.; Esposito, S.; Ganmaa, D.; Ginde, A.A.; et al. Vitamin D supplementation to prevent acute respiratory tract infections: Systematic review and meta-analysis of individual participant data. *BMJ* **2017**, *356*, i6583. [CrossRef] [PubMed]
4. Ketha, H.; Thacher, T.D.; Oberhelman, S.S.; Fischer, P.R.; Singh, R.J.; Kumar, R. Comparison of the effect of daily versus bolus dose maternal vitamin D3 supplementation on the 24,25-dihydroxyvitamin D3 to 25-hydroxyvitamin D3 ratio. *Bone* **2018**, *110*, 321–325. [CrossRef]
5. Guyatt, G.H.; Oxman, A.D.; Kunz, R.; Vist, G.E.; Falck-Ytter, Y.; Schunemann, H.J.; Group, G.W. What is "quality of evidence" and why is it important to clinicians? *BMJ* **2008**, *336*, 995–998. [CrossRef] [PubMed]
6. Centers for Disease Control and Prevention (CDC). Underlying Medical Conditions Associated with Higher Risk for Severe COVID-19: Information for Healthcare Providers. Available online: https://www.cdc.gov/coronavirus/2019-ncov/hcp/clinical-care/underlyingconditions.html (accessed on 20 November 2021).
7. Dopico, X.C.; Evangelou, M.; Ferreira, R.C.; Guo, H.; Pekalski, M.L.; Smyth, D.; Cooper, N.; Burren, O.S.; Fulford, A.J.; Hennig, B.J.; et al. Widespread seasonal gene expression reveals annual differences in human immunity and physiology. *Nat. Commun.* **2015**, *6*, 7000. [CrossRef] [PubMed]
8. Milani, G.P.; Simonetti, G.D.; Edefonti, V.; Lava, S.A.G.; Agostoni, C.; Curti, M.; Stettbacher, A.; Bianchetti, M.G.; Muggli, F. Seasonal variability of the vitamin D effect on physical fitness in adolescents. *Sci. Rep.* **2021**, *11*, 182. [CrossRef]
9. Reid, D.; Toole, B.J.; Knox, S.; Talwar, D.; Harten, J.; O'reilly, D.S.J.; Blackwell, S.; Kinsella, J.; McMillan, D.C.; Wallace, A.M.; et al. The relation between acute changes in the systemic inflammatory response and plasma 25-hydroxyvitamin D concentrations after elective knee arthroplasty. *Am. J. Clin. Nutr.* **2011**, *93*, 1006–1011. [CrossRef] [PubMed]
10. Amer, M.; Qayyum, R. Relation between serum 25-hydroxyvitamin D and C-reactive protein in asymptomatic adults (from the continuous National Health and Nutrition Examination Survey 2001 to 2006). *Am. J. Cardiol.* **2012**, *109*, 226–230. [CrossRef] [PubMed]
11. Binkley, N.; Dawson-Hughes, B.; Durazo-Arvizu, R.; Thamm, M.; Tian, L.; Merkel, J.M.; Jones, J.C.; Carter, G.D.; Sempos, C.T. Vitamin D measurement standardization: The way out of the chaos. *J. Steroid Biochem. Mol. Biol.* **2017**, *173*, 117–121. [CrossRef]
12. Revez, J.A.; Lin, T.; Qiao, Z.; Xue, A.; Holtz, Y.; Zhu, Z.; Zeng, J.; Wang, H.; Sidorenko, J.; Kemper, K.E.; et al. Genome-wide association study identifies 143 loci associated with 25 hydroxyvitamin D concentration. *Nat. Commun.* **2020**, *11*, 1647. [CrossRef] [PubMed]
13. Thacher, T.D.; Clarke, B.L. Vitamin D insufficiency. *Mayo Clin. Proc.* **2011**, *86*, 50–60. [CrossRef]
14. Severe Covid GWAS Group. Genomewide Association Study of Severe Covid-19 with Respiratory Failure. *N. Engl. J. Med.* **2020**, *383*, 1522–1534. [CrossRef] [PubMed]
15. Chun, R.F.; Lauridsen, A.L.; Suon, L.; Zella, L.A.; Pike, J.W.; Modlin, R.L.; Martineau, A.R.; Wilkinson, R.; Adams, J.; Hewison, M. Vitamin D-binding protein directs monocyte responses to 25-hydroxy- and 1,25-dihydroxyvitamin D. *J. Clin. Endocrinol. Metab.* **2010**, *95*, 3368–3376. [CrossRef] [PubMed]
16. Hernández, J.L.; Nan, D.; Fernandez-Ayala, M.; García-Unzueta, M.; A Hernández-Hernández, M.; López-Hoyos, M.; Muñoz-Cacho, P.; Olmos, J.M.; Gutiérrez-Cuadra, M.; Ruiz-Cubillán, J.J.; et al. Vitamin D Status in Hospitalized Patients with SARS-CoV-2 Infection. *J. Clin. Endocrinol. Metab.* **2020**, *106*, e1343–e1353. [CrossRef]
17. Maghbooli, Z.; Sahraian, M.A.; Ebrahimi, M.; Pazoki, M.; Kafan, S.; Tabriz, H.M.; Hadadi, A.; Montazeri, M.; Nasiri, M.; Shirvani, A.; et al. Vitamin D sufficiency, a serum 25-hydroxyvitamin D at least 30 ng/mL reduced risk for adverse clinical outcomes in patients with COVID-19 infection. *PLoS ONE* **2020**, *15*, e0239799. [CrossRef] [PubMed]
18. Griffin, G.; Hewison, M.; Hopkin, J.; Kenny, R.A.; Quinton, R.; Rhodes, J.; Subramanian, S.; Thickett, D. Preventing vitamin D deficiency during the COVID-19 pandemic: UK definitions of vitamin D sufficiency and recommended supplement dose are set too low. *Clin. Med. (Lond.)* **2021**, *21*, e48–e51. [CrossRef]
19. Angelidi, A.M.; Belanger, M.J.; Lorinsky, M.K.; Karamanis, D.; Chamorro-Pareja, N.; Ognibene, J.; Palaiodimos, L.; Mantzoros, C.S. Vitamin D Status Is Associated With In-Hospital Mortality and Mechanical Ventilation: A Cohort of COVID-19 Hospitalized Patients. *Mayo Clin. Proc.* **2021**, *96*, 875–886. [CrossRef]

20. AlSafar, H.; Grant, W.B.; Hijazi, R.; Uddin, M.; Alkaabi, N.; Tay, G.; Mahboub, B.; Al Anouti, F. COVID-19 Disease Severity and Death in Relation to Vitamin D Status among SARS-CoV-2-Positive UAE Residents. *Nutrients* **2021**, *13*, 1714. [CrossRef]

21. Jude, E.B.; Ling, S.F.; Allcock, R.; Yeap, B.X.Y.; Pappachan, J.M. Vitamin D Deficiency Is Associated with Higher Hospitalization Risk from COVID-19: A Retrospective Case-Control Study. *J. Clin. Endocrinol. Metab.* **2021**, *106*, e4708–e4715. [CrossRef]

22. Diaz-Curiel, M.; Cabello, A.; Arboiro-Pinel, R.; Mansur, J.L.; Heili-Frades, S.; Mahillo-Fernandez, I.; Herrero-Gonzalez, A.; Andrade-Poveda, M. The relationship between 25(OH) vitamin D levels and COVID-19 onset and disease course in Spanish patients. *J. Steroid Biochem. Mol. Biol.* **2021**, *212*, 105928. [CrossRef] [PubMed]

23. Charoenngam, N.; Shirvani, A.; Reddy, N.; Vodopivec, D.M.; Apovian, C.M.; Holick, M.F. Association of Vitamin D Status With Hospital Morbidity and Mortality in Adult Hospitalized Patients With COVID-19. *Endocr. Pract.* **2021**, *27*, 271–278. [CrossRef] [PubMed]

24. Oristrell, J.; Oliva, J.C.; Casado, E.; Subirana, I.; Dominguez, D.; Toloba, A.; Balado, A.; Grau, M. Vitamin D supplementation and COVID-19 risk: A population-based, cohort study. *J. Endocrinol. Investig.* **2022**, *45*, 167–179. [CrossRef]

25. Szeto, B.; Zucker, J.E.; LaSota, E.D.; Rubin, M.R.; Walker, M.D.; Yin, M.T.; Cohen, A. Vitamin D Status and COVID-19 Clinical Outcomes in Hospitalized Patients. *Endocr. Res.* **2021**, *46*, 66–73. [CrossRef] [PubMed]

26. Pecina, J.L.; Merry, S.P.; Park, J.G.; Thacher, T.D. Vitamin D Status and Severe COVID-19 Disease Outcomes in Hospitalized Patients. *J. Prim. Care Community Health* **2021**, *12*, 21501327211041206. [CrossRef] [PubMed]

27. Bianconi, V.; Mannarino, M.R.; Figorilli, F.; Cosentini, E.; Batori, G.; Marini, E.; Lombardini, R.; Gargaro, M.; Fallarino, F.; Scarponi, A.M.; et al. Prevalence of vitamin D deficiency and its prognostic impact on patients hospitalized with COVID-19. *Nutrition* **2021**, *91–92*, 111408. [CrossRef] [PubMed]

28. Hastie, C.E.; Pell, J.P.; Sattar, N. Vitamin D and COVID-19 infection and mortality in UK Biobank. *Eur. J. Nutr.* **2021**, *60*, 545–548. [CrossRef] [PubMed]

29. Jevalikar, G.; Mithal, A.; Singh, A.; Sharma, R.; Farooqui, K.J.; Mahendru, S.; Dewan, A.; Budhiraja, S. Lack of association of baseline 25-hydroxyvitamin D levels with disease severity and mortality in Indian patients hospitalized for COVID-19. *Sci. Rep.* **2021**, *11*, 6258. [CrossRef]

30. Wenban, C.; Heer, R.S.; Baktash, V.; Kandiah, P.; Katsanouli, T.; Pandey, A.; Goindoo, R.; Ajaz, A.; Abbeele, K.V.D.; Mandal, A.K.J.; et al. Dexamethasone treatment may mitigate adverse effects of vitamin D deficiency in hospitalized Covid-19 patients. *J. Med. Virol.* **2021**, *93*, 6605–6610. [CrossRef] [PubMed]

31. Dhawan, P.; Christakos, S. Novel regulation of 25-hydroxyvitamin D3 24-hydroxylase (24(OH)ase) transcription by glucocorticoids: Cooperative effects of the glucocorticoid receptor, C/EBP beta, and the Vitamin D receptor in 24(OH)ase transcription. *J. Cell. Biochem.* **2010**, *110*, 1314–1323. [CrossRef] [PubMed]

32. Hidalgo, A.A.; Deeb, K.K.; Pike, J.W.; Johnson, C.S.; Trump, D.L. Dexamethasone enhances 1alpha,25-dihydroxyvitamin D3 effects by increasing vitamin D receptor transcription. *J. Biol. Chem.* **2011**, *286*, 36228–36237. [CrossRef] [PubMed]

33. Wang, Z.; Joshi, A.; Leopold, K.; Jackson, S.; Christensen, S.; Nayfeh, T.; Mohammed, K.; Creo, A.; Tebben, P.; Kumar, S. Association of vitamin D deficiency with COVID-19 infection severity: Systematic review and meta-analysis. *Clin. Endocrinol. (Oxf.)*, 2021; *Jun 23 Epub ahead of print.* [CrossRef]

34. Kaufman, H.W.; Niles, J.K.; Kroll, M.H.; Bi, C.; Holick, M.F. SARS-CoV-2 positivity rates associated with circulating 25-hydroxyvitamin D levels. *PLoS ONE* **2020**, *15*, e0239252. [CrossRef] [PubMed]

35. Merzon, E.; Tworowski, D.; Gorohovski, A.; Vinker, S.; Golan Cohen, A.; Green, I.; Frenkel-Morgenstern, M. Low plasma 25(OH) vitamin D level is associated with increased risk of COVID-19 infection: An Israeli population-based study. *FEBS J.* **2020**, *287*, 3693–3702. [CrossRef] [PubMed]

36. Annweiler, C.; Beaudenon, M.; Gautier, J.; Simon, R.; Dubee, V.; Gonsard, J.; Parot-Schinkel, E.; COVIT-TRIAL study group. COVID-19 and high-dose VITamin D supplementation TRIAL in high-risk older patients (COVIT-TRIAL): Study protocol for a randomized controlled trial. *Trials* **2020**, *21*, 1031. [CrossRef] [PubMed]

37. National Heart, Lung, and Blood Institute PETAL Clinical Trials Network; Ginde, A.A.; Brower, R.G.; Caterino, J.M.; Finck, L.; Banner-Goodspeed, V.M.; Grissom, C.K.; Hayden, D.; Hough, C.L.; Hyzy, R.C.; et al. Early High-Dose Vitamin D3 for Critically Ill, Vitamin D-Deficient Patients. *N. Engl. J. Med.* **2019**, *381*, 2529–2540. [CrossRef] [PubMed]

38. Murai, I.H.; Fernandes, A.L.; Sales, L.P.; Pinto, A.J.; Goessler, K.F.; Duran, C.S.C.; Silva, C.B.R.; Franco, A.S.; Macedo, M.B.; Dalmolin, H.H.H.; et al. Effect of a Single High Dose of Vitamin D3 on Hospital Length of Stay in Patients With Moderate to Severe COVID-19: A Randomized Clinical Trial. *JAMA* **2021**, *325*, 1053–1060. [CrossRef] [PubMed]

39. Rawat, D.; Roy, A.; Maitra, S.; Shankar, V.; Khanna, P.; Baidya, D.K. Vitamin D supplementation and COVID-19 treatment: A systematic review and meta-analysis. *Diabetes Metab. Syndr.* **2021**, *15*, 102189. [CrossRef] [PubMed]

40. Castillo, M.E.; Entrenas Costa, L.M.; Vaquero Barrios, J.M.; Alcala Diaz, J.F.; Lopez Miranda, J.; Bouillon, R.; Quesada Gomez, J.M. Effect of calcifediol treatment and best available therapy versus best available therapy on intensive care unit admission and mortality among patients hospitalized for COVID-19: A pilot randomized clinical study. *J. Steroid Biochem. Mol. Biol.* **2020**, *203*, 105751. [CrossRef] [PubMed]

41. Rastogi, A.; Bhansali, A.; Khare, N.; Suri, V.; Yaddanapudi, N.; Sachdeva, N.; Puri, G.D.; Malhotra, P. Short term, high-dose vitamin D supplementation for COVID-19 disease: A randomised, placebo-controlled, study (SHADE study). *Postgrad. Med. J.* 2020; *Nov 12 Epub ahead of print.* [CrossRef]

42. Institute of Medicine. *Dietary Reference Intakes for Calcium and Vitamin D*; The National Academies Press: Washington, DC, USA, 2011.

Review

A Narrative Review of the Evidence for Variations in Serum 25-Hydroxyvitamin D Concentration Thresholds for Optimal Health

William B. Grant [1,*], Fatme Al Anouti [2], Barbara J. Boucher [3], Erdinç Dursun [4], Duygu Gezen-Ak [4], Edward B. Jude [5,6,7], Tatiana Karonova [8] and Pawel Pludowski [9]

[1] Sunlight, Nutrition, and Health Research Center, P.O. Box 641603, San Francisco, CA 94164-1603, USA
[2] Department of Health Sciences, College of Natural and Health Sciences, Zayed University, Abu Dhabi 144534, United Arab Emirates; fatme.alanouti@zu.ac.ae
[3] The Blizard Institute, Barts & The London School of Medicine & Dentistry, Queen Mary University of London, London E12AT, UK; bboucher@doctors.org.uk
[4] Department of Neuroscience, Institute of Neurological Sciences, Istanbul University-Cerrahpasa, Istanbul 34098, Turkey; erdincdu@gmail.com (E.D.); duygugezenak@gmail.com (D.G.-A.)
[5] Tameside and Glossop Integrated Care NHS Foundation Trust, Fountain Street, Ashton-under-Lyne OL6 9RW, UK; Edward.jude@tgh.nhs.uk
[6] The University of Manchester, Oxford Road, Manchester M13 9PL, UK
[7] Manchester Metropolitan University, All Saints Building, Manchester M15 6BH, UK
[8] Clinical Endocrinology Laboratory, Department of Endocrinology, Almazov National Medical Research Centre, 194021 Saint-Petersburg, Russia; karonova@mail.ru
[9] Department of Biochemistry, Radioimmunology and Experimental Medicine, The Children's Memorial Health Institute, 04730 Warsaw, Poland; p.pludowski@ipczd.pl
[*] Correspondence: wbgrant@infionline.net; Tel.: +1-415-409-1980

Citation: Grant, W.B.; Al Anouti, F.; Boucher, B.J.; Dursun, E.; Gezen-Ak, D.; Jude, E.B.; Karonova, T.; Pludowski, P. A Narrative Review of the Evidence for Variations in Serum 25-Hydroxyvitamin D Concentration Thresholds for Optimal Health. *Nutrients* 2022, 14, 639. https://doi.org/10.3390/nu14030639

Academic Editors: Andrea Fabbri and Igor Sergeev

Received: 7 January 2022
Accepted: 28 January 2022
Published: 2 February 2022

Publisher's Note: MDPI stays neutral with regard to jurisdictional claims in published maps and institutional affiliations.

Abstract: Vitamin D_3 has many important health benefits. Unfortunately, these benefits are not widely known among health care personnel and the general public. As a result, most of the world's population has serum 25-hydroxyvitamin D (25(OH)D) concentrations far below optimal values. This narrative review examines the evidence for the major causes of death including cardiovascular disease, hypertension, cancer, type 2 diabetes mellitus, and COVID-19 with regard to sub-optimal 25(OH)D concentrations. Evidence for the beneficial effects comes from a variety of approaches including ecological and observational studies, studies of mechanisms, and Mendelian randomization studies. Although randomized controlled trials (RCTs) are generally considered the strongest form of evidence for pharmaceutical drugs, the study designs and the conduct of RCTs performed for vitamin D have mostly been flawed for the following reasons: they have been based on vitamin D dose rather than on baseline and achieved 25(OH)D concentrations; they have involved participants with 25(OH)D concentrations above the population mean; they have given low vitamin D doses; and they have permitted other sources of vitamin D. Thus, the strongest evidence generally comes from the other types of studies. The general finding is that optimal 25(OH)D concentrations to support health and wellbeing are above 30 ng/mL (75 nmol/L) for cardiovascular disease and all-cause mortality rate, whereas the thresholds for several other outcomes appear to range up to 40 or 50 ng/mL. The most efficient way to achieve these concentrations is through vitamin D supplementation. Although additional studies are warranted, raising serum 25(OH)D concentrations to optimal concentrations will result in a significant reduction in preventable illness and death.

Keywords: Alzheimer's disease; cancer; cardiovascular disease; COVID-19; diabetes; hypertension; Mendelian randomization; vitamin D; 25-hydroxyvitamin D

1. Introduction

Vitamin D deficiency is the most common nutritional deficiency in the world, although vitamin D is one of the most well understood compounds. It reduces risks of many adverse health outcomes through both genetic and non-genetic mechanisms; it is readily available from supplements, is safe, and inexpensive. There are over 94,000 publications on vitamin D listed on pubmed.gov as of 20 December 2021. Despite all this, vitamin D deficiency (25-hydroxyvitamin D (25(OH)D) concentration <20 ng/mL) is very common, affecting about half of the world's population [1], whereas the higher concentrations necessary for optimal non-skeletal health (25(OH)D >30 ng/mL) are not so common. The reason for this is that the beneficial effects of vitamin D for non-skeletal disorders have only received widespread research attention since 2000. Furthermore, methods for studying the many health effects of vitamin D have been evolving and researchers have had to identify their various strengths and limitations. Another factor is that conventional medicine in most countries generally focuses on treatment rather than on the prevention of disease.

This is a narrative review of what is known about the role of vitamin D supplementation and how it raises serum 25(OH)D concentration and influences health outcomes as well as achieves maximum reductions of mortality rates in developed countries from the commonest fatal diseases (e.g., cancer, cardiovascular disease including hypertension, COVID-19 and diabetes mellitus type 2). In addition, an overview is included of the types of studies used to evaluate the effects of vitamin D supplementation, and of the different serum 25(OH)D concentrations required to achieve various health outcomes, with outlines of the strengths and limitations of the various types of studies. Circulating 25(OH)D is important as the precursor of the most active vitamin D metabolite, calcitriol (1,25-dihydroxyvitamin D) formed by the 1-alpha hydroxylation of 25(OH)D in the kidneys and many other immune and metabolically active tissues as needed.

The approach taken in this report is to base recommendations on the strongest evidence available, generally from observational studies that are supported by other types of studies. The strengths and weaknesses of the various types of studies, ranging from ecological studies to meta-analyses of randomized controlled trials (RCTs) and Mendelian randomization (MR) studies are discussed at the end of the discussion. Although meta-analyses are often considered to be stronger evidence than individual studies, the fact that many studies have flaws in their design, conduct, or analysis, means that some individual studies are better suited for making recommendations than many meta-analyses.

The background for the present report is provided by data on mortality rates for major causes of death in 2016 by the World Health Organization (see Table 1 for Germany, Japan, Saudi Arabia, and the U.S.). This study will concentrate on the diseases with the highest prevalence or mortality rates.

It is apparent from the data in Table 1 that there are large differences in mortality rates by country and sex. National diet plays an important role in health outcomes. Spending on health care can affect outcomes. Obesity increases adverse health outcomes. Smoking has many adverse health effects. Males tend to smoke more than females, and this may explain why mortality rates for males are higher than for females. Although we do not discuss the effect of factors other than vitamin D and solar UVB in this study, we acknowledge that raising serum 25(OH)D concentrations may not have the same effects in all participants.

Table 1. Mortality rates (deaths/100,000/year) as well as obesity rates in 2016.

Outcome	Germany M	Germany F	Japan M	Japan F	Saudi Arabia M	Saudi Arabia F	USA M	USA F
All causes	504	328	401	217	777	608	592	404
CVD	160	106	93	54	329	261	167	104
IHD	95	54	42	22	219	154	106	56
Stroke	24	20	33	20	86	76	24	21
Cancer	148	97	139	76	64	57	132	99
Breast	0.2	19	0.1	10	0	9	0.2	18
Lung	36	17	33	10	7	3	33	23
COPD	26	15	19	7	15	12	35	28
Lower respiratory tract disease	12	7	36	17	46	42	13	10
Diabetes mellitus	12	8	5	2	30	25	19	12
Alcohol abuse	7	2	0.5	0.1	0.5	0.1	4	1
Alzheimer's disease	15	16	6	5	46	44	28	35
Obesity (%) – 2016	22		4		35		36	

Table 1. Global Health Estimates 2016: Deaths by Cause, Age, Sex, by Country and by Region, 2000–2016. Geneva, World Health Organization; 2018. Obesity data from https://obesity.procon.org/global-obesity-levels/ (accessed on 15 December 2021). COPD, chronic obstructive pulmonary disease; CVD, cardiovascular disease; F, female; IHD, ischemic heart disease; M, male.

2. Results

2.1. Cardiovascular Disease

Robert Scragg was the first to propose that the adequate provision of vitamin D might reduce the risk of cardiovascular disease (CVD) since incidence and mortality rates were highest in winter when solar UVB doses and serum 25(OH)D concentrations were lowest [2]. Markedly higher rates of CVD risk factors were found in winter than in summer in 24 population-based studies in 14 countries [3]. The risk factors considered included BMI, waist circumference systolic and diastolic blood pressure, total, high- and low-density lipoprotein cholesterol, triglycerides, and glucose levels.

Thomas Wang and colleagues published the first observational study of CVD risk with respect to serum 25(OH)D concentration in a prospective study of participants in the Framingham Offspring Study in 2008 [4]. Several vitamin D-related mechanisms appear to have protective roles for CVD including the known inhibition of vascular smooth muscle proliferation, the suppression of vascular calcification, the reduction of inflammation through the regulation of cytokines, and the regulation of blood volume and systemic vascular resistance through renin gene suppression [5].

Observational studies have shown that serum 25(OH)D concentrations correlate inversely with CVD incidence and mortality rates and also with data for coronary or ischemic heart disease, congestive heart failure and stroke. Acute cardiovascular events are commonly precipitated by plaque disruption following local inflammation and the release of destructive MMPs, especially MMP9, from invading foam macrophages. Inflammation is suppressed by vitamin D and non-skeletal MMP2/9 production is suppressed by vitamin D; circulating MMP2/9 concentrations relate inversely to serum 25(OH)D and can be suppressed by modest supplementation [6].

In clinical trials, vitamin D supplementation has been found to reduce the serum levels of total cholesterol (TC), low-density lipoprotein cholesterol (LDL-C), and triglycerides (TG) but not high-density lipoprotein cholesterol (HDL-C) [7].

A study of 20,025 patients in the U.S. Veterans Health Administration system followed for 20 years found that those with no previous myocardial infarction (MI) who had basal 25(OH)D concentrations ≤20 ng/mL and raised their concentrations to >20 ng/mL significantly reduced their risk of MI [8]. For those achieving 25(OH)D concentrations ≥30 ng/mL vs. 21 to 29 ng/mL, the hazard ratio (HR) for MI was 0.65 (95% CI, 0.49 to 0.85, p = 0.002), whereas for those who achieved 25(OH)D concentrations ≥30 ng/mL vs. ≤20 ng/mL, the HR for MI was 0.73 (95% CI, 0.55 to 0.96, p = 0.02).

Similarly, a study based on UK Biobank participants who had CVD followed for a median of 11.7 years showed a linear inverse association of CVD mortality (*P*-non-linearity = 0.07) with 25(OH)D, the adjusted HR decreasing from 1.31 (95% CI, 1.14 to 1.43) at 4 ng/mL to 0.83 (95% CI, 0.77 to 0.88) at 15 ng/mL and then (linearly) to 0.37 (95% CI, 0.22 to 0.62) at 60 ng/mL [9].

Support for the role of vitamin D in reducing the risk of CVD also comes from observational associations and MR analyses of four population-based cohort studies (UK Biobank, EPIC-CVD and two Copenhagen population-based studies) comprising 386,406 middle-aged individuals of European ancestries [10] followed after enrollment, with blood drawn from between 9 to 21 years. In the observational part of this study, coronary heart disease events were significantly increased below 12 ng/mL, stroke events below 20 ng/mL, and the CVD mortality rate below 25 ng/mL. In that MR analysis, a 4 ng/mL increase in genetically predicted 25(OH)D was associated with an OR for stroke of 0.85 (95% CI, 0.70 to 1.02, p = 0.09) and for coronary heart disease of 0.89 (95% CI, 0.76 to 1.04, p = 0.14). However, these values probably underestimate the true value due to changes in 25(OH)D over time. An analysis of HR vs. the follow-up period for observational studies of all-cause mortality rate for an 8 ng/mL difference in 25(OH)D concentration over six to fourteen years found a linear increase in HR from 0.81 (95% CI, 0.67 to 1.02) at six years to 0.96 (95% CI, 0.9 to 1.01) at fourteen years [11].

Subsequently, an MR study based on UK Biobank data for 20,805 incident CVD events amongst 267,980 subjects with both serum 25(OH)D concentration data and phenotypic 25(OH)D analyses [12] reported that each 4 ng/mL increase in phenotypic 25(OH)D was associated with a 1.6% lower risk of a CVD event (OR = 0.98 (95% CI, 0.98 to 0.99, p = 0.0001)) for 25(OH)D concentrations of up to 50 ng/mL. However, the genetic (MR) analysis using data for 35 vitamin D-related single nucleotide polymorphisms (SNPs) found in the 295,788 participants including 44,519 CVD cases, revealed that there was an L-shaped relationship, with 11% (95% CI, 1.05 to 1.18) higher odds for CVD events at 10 ng/mL than at 20 ng/mL; moreover, for 25(OH)D concentrations around 4 ng/mL, the odds were +2.3 (95% CI, 1.6 to 3.7) using the non-linear analytical approach of Staley and Burgess [13] on the 25(OH)D data after the data had been divided into 100 equal strata, with a plateau above ~20 ng/mL.

2.2. Hypertension

Good evidence now shows that vitamin D status affects the risk and prevalence of hypertension. For example, a meta-analysis of seven out of eight prospective cohort studies with 283,537 participants found the risk of developing hypertension for those in the upper third vs. lower third of 25(OH)D concentrations was reduced by 30% (RR = 0.70 (95% CI, 0.58 to 0.86)) [14] with an overall RR for incident hypertension for every 10 ng/mL increment in baseline 25(OH)D concentration of 0.88 (0.81, 0.97) using dose-response analysis.

An MR study on 142,225 participants of European descent from several countries with measurements of systolic and diastolic blood pressure and genetic analyses [15] found four variants of genes affecting 25(OH)D synthesis from vitamin D (*CYP2R1*) through involvement in the synthesis of 7-dehydrocholesterol, which is converted to vitamin D_3 in the skin via UVB irradiance followed by a thermal process involving *DHCR7*, the gene providing instructions for making an enzyme called 7-dehydrcholesterol reductase. In their meta-analysis, a 'synthesis score' was associated with a reduced risk of hypertension (OR per allele, 0.98, 0.96–0.99; $p = 0.001$). Each 10% increase in genetically determined 25(OH)D concentration was associated with a reduction of −0.29 mm Hg in diastolic blood pressure (95% CI, −0.52 to −0.07; $p = 0.01$), of −0.37 mmHg in systolic blood pressure (95% CI, −0.73 to 0.003; $p = 0.052$), and an 8% decreased risk of hypertension (OR 0.92, 95% CI, 0.87–0.97; $p = 0.002$). Although these differences in blood pressure were small, that was likely due to the limitations of the MR analyses which had few alleles of genes affecting serum 25(OH)D concentrations to consider.

An observational study of community-based participants taking ~4000 IU/day of vitamin D$_3$ to achieve 25(OH)D concentrations >40 ng/mL reported significant reductions in blood pressure and hypertension prevalence after a year [16]. At baseline, 592 participants (7.3%) were hypertensive. At follow-up (12 ± 3 months), 71% of them were no longer hypertensive. The mean 25(OH)D concentration for those hypertensive participants increased from 33 ± 16 ng/mL to 45 ± 14 ng/mL with increased vitamin D supplementation (from ~2000 IU/day to ~6000 IU/day). For those not taking hypotensive medication, systolic BP decreased by 18 ± 19 mmHg and diastolic BP fell by 12 ± 12 mmHg, whereas in those taking hypotensive medication after joining the program, systolic BP decreased by 14 ± 21 mmHg and diastolic BP decreased by 12 ± 12 mmHg, whereas there were no changes in blood pressure in the normo-tensive control participants.

2.3. Cancer

It was first proposed in 1980 that sunlight reduced the risk of colon cancer with vitamin D production being the likely reason [17]. Since then, numerous ecological studies have reported that solar UVB dose indices correlate inversely with incidence and/or mortality rates for nearly 20 types of cancer [18–20]. The best ecological studies are those from single mid-latitude countries where variations in solar UVB doses tend to be large [21,22], whereas variations in other risk-modifying factors (diet, skin pigmentation, dress, obesity, smoking, alcohol consumption, etc.) are often small or can be accounted for [19]. Thus, ecological studies provide a case for examining whether better vitamin D provision reduces cancer incidence or cancer mortality.

Observational studies have shown associations between cancer risk and solar UV radiation. A meta-analysis of 14 studies observed reduced breast cancer rates for those spending ≥1 h/day in sunlight during summer months over their lifetime or adulthood compared to <1 h/day (RR = 0.84; 95% CI: 0.77, 0.91) [23]. Another meta-analysis of six studies between 2005 and 2020 found an inverse correlation between exposure to solar UV radiation and breast cancer risk (RR: 0.70, 95% CI: 0.65, 0.75), [24]. In total, 17 case-control studies and 9 cohort studies, including 216,285 non-Hodgkin's lymphoma (NHL) and 23,017 Hodgkin's lymphoma (HL) patients, were included in the final analysis. Personal sunlight exposure was significantly associated with reduced risks of HL (OR = 0.77; 95% CI 0.68–0.87) and of all types of NHL (OR = 0.81; 95% CI 0.71–0.92) other than T-cell lymphoma [25]. Furthermore, no mechanism other than the production of vitamin D has been suggested to explain the protective effects of solar UV against cancer.

There are many observational studies of cancer incidence with serum 25(OH)D concentrations. The meta-analyses of observational studies are shown in Table 2. The reason case-control studies report greater risk reductions is probably the long follow-up times of cohort studies providing baseline 25(OH)D concentrations which become less well correlated with cancer outcomes over time [26]. Although serum 25(OH)D concentration can be reduced by acute inflammatory illness [27], cancer does not appear to have this effect.

Table 2. Meta-analyses of observational studies of individual cancer site risks in relation to serum 25(OH)D concentrations #.

Cancer Site	N Studies	Type of Study	RR (95% CI) (High vs. Low)	Reference
Bladder	5	Cc	0.70 (0.56 to 0.88)	[28]
Bladder	2	Cohort	0.80 (0.67 to 0.94)	[28]
Breast	44	Cc	0.57 (0.48 to 0.66)	[29]
Breast	6	Cohort	1.17 (0.92 to 1.48)	[29]
Colorectal	11	Cc	0.60 (0.53 to 0.68) *	[30]
Colorectal	6	Cohort	0.80 (0.66 to 0.97) *	[30]

(*) fixed effects model; # https://pubmed.ncbi.nlm.nih.gov/ (accessed on 15 December 2021).

Although the data in Table 2 demonstrate that serum 25(OH)D concentration is commonly inversely correlated with cancer incidence, they do not provide data on the relationship of 25(OH)D concentration to cancer incidence, although other studies do so. For breast cancer, two studies can be used. One is a meta-analysis of 36 case-control and four cohort studies [29] where a spline fit to the data reveals a 60% (95% CI, 45% to 70%) reduction in risk with 25(OH)D concentrations from 4 ng/mL up to 40 ng/mL and an 80% reduction with 25(OH)D concentrations of up to 80 ng/mL. However, the primary data source for 25(OH)D concentrations of >40 ng/mL is from an observational study based on data from women enrolled in vitamin D RCTs who took vitamin D supplements (1000 or 2000 IU/day) or a placebo [31,32] or were enrolled in a volunteer cohort who took doses of their choice and had serum 25(OH)D measured half-yearly [33]. This pooled cohort included 5028 women out of whom 77 developed incident breast cancer. There was an 82% lower incidence rate for 25(OH)D concentrations >60 ng/mL vs. <20 ng/mL ($p = 0.006$) and, importantly, the slope of that relationship was similar for subjects with values both below and above 40 ng/mL.

For colorectal cancer, a 2019 meta-analysis [34] using data from case-control studies of women with 25(OH)D concentrations between <15 ng/mL and >29 ng/mL found that the relative risk (RR) per 10 ng/mL increase in 25(OH)D was 0.81 (95% CI, 0.75 to 0.87), whereas for men with 25(OH)D concentrations between <16 and >30 ng/mL, the RR was 0.93 (95% CI, 0.86 to 1.00). Thus, for rises in 25(OH)D ranging from ~10 ng/mL to ~35 ng/mL, the RR was 0.59 for women and 0.83 for men. The increase in RR seen with 25(OH)D concentrations of >100 ng/mL vs. <40 ng/mL is likely due to participants starting to supplement shortly prior to enrolling in cohort studies [35] and should not necessarily be taken to indicate that higher 25(OH)D concentrations reverse the beneficial effects found at lower 25(OH)D concentrations, especially since the ecological study of cancer mortality rates in the U.S. between 1950 and 1994 showed that UVB dose–cancer mortality rate curves plateaued at the highest UVB exposures [36].

The effect of vitamin D in reducing cancer mortality rates appears to be more significant than for reducing cancer incidence. For example, the VITamin D and OmegA-3 TriaL (VITAL) study conducted by Harvard University [37] did not find a significant reduction in all-cancer incidence with vitamin D supplementation vs. placebo for the entire set of participants but did find a significantly reduced risk of mortality when the first two years of data were omitted, HR = 0.75 (0.59–0.96). Furthermore, another analysis of vitamin D RCTs also found a greater reduction for mortality rates than for incidence rates (Table 3). However, meta-analyses of observational studies found significant reduction for both all-cancer incidence and mortality rates (Table 4).

Table 3. Meta-analyses of breast cancer risk from vitamin D RCTs.

Cancer Site	*N* Studies	Outcome	RR (95% CI)	Reference
Breast	9	Incidence	0.96 (0.86 to 1.07)	[38]
Breast	5	Mortality	0.87 (0.79 to 0.96)	[38]

Table 4. Meta-analyses of cancer incidence and mortality rates from observational studies.

Cancer Site	*N* Studies	Outcome	Low, High 25(OH)D (ng/mL)	RR Low 25(OH)D	RR High 25(OH)D	Ratio High to Low	Reference
Total	8	Incidence	1, 21	1.31 (95% CI, 0.87 to 2.05)	0.71 (95% CI, 0.55 to 0.92)	0.54	[39]
	17	Mortality	10, 40	1.47 (95% CI, 1.11 to 1.88)	0.87 (95% CI, 0.75 to 1.02)	0.59	[39]

Additional support for the effects of vitamin D provision on cancer risks is that many mechanisms exist through which vitamin D reduces cancer risks, including effects on cells that are anti-proliferative, pro-differentiating, anti-inflammatory, immunomodulatory, anti-angiogenic around tumors, and anti-metastatic [40–43]. The active vitamin D metabolite (1α,25-dihydroxyvitamin D_3 or calcitriol) inhibits proliferation and promotes the epithelial differentiation of human colon carcinoma cell lines that express the vitamin D receptor (VDR) via the regulation of a large number of genes [44]. Other RCTs reported that the higher the vitamin D supplement given (up to 10,000 IU/day), the more genes whose expression is changed [45,46]. No adverse effects of supplementation with 10,000 IU/day were found in these studies. Thus, the known mechanisms support the finding that a higher 25(OH)D threshold can reduce cancer risks and cancer mortality.

Most vitamin D–cancer RCTs seem to have failed because they were designed using guidelines developed for assessing pharmaceutical drugs, not nutrients [47,48]. As a result, baseline 25(OH)D concentrations were often too high and vitamin D doses were generally too low to correct deficiency. Furthermore, changes in serum 25(OH)D concentrations differ individually in response to supplementation [49], and other sources of vitamin D, including unknown intakes from self-supplementation in both treatment and placebo arms, cannot be allowed for.

A meta-analysis of vitamin D RCTs regarding breast cancer incidence included eight trials comprising 72,275 participants with median follow-up periods ranging from 1 to 11.9 years. The doses were 400 to 1100 IU/day in four trials, 2000 IU/day in two trials, and 100,000 IU/month in two trials. This study found RRs of 1.04 (95% CI 0.85–1.29, $p = 0.68$) for vitamin D supplementation (6 trials, 33,472 participants, 246 events), and 0.99 (95% CI 0.91–1.07, $p = 0.73$) for vitamin D plus calcium (4 trials, 41,957 participants, 2195 events) [50]. The doses were too low and/or too infrequent in the trials of monthly dosing since the half-life of 25(OH)D is about 20 days; thus, it is not surprising that the RR findings were not significant.

The inspection of VITAL study results, where the mean baseline and achieved 25(OH)D concentrations for those with measured values were 27.8 and 39.7 ng/mL for males and 31.7 and 43.6 ng/mL for females, and using a vitamin D_3 dose of 2000 IU/d, the HR for overall cancer incidence was 0.96 (95% CI, 0.88 to 1.06). However, for black participants, where the mean reported baseline and achieved 25(OH)D concentrations were 25.0 and 39.7 ng/mL, respectively, the HR was 0.77 (95% CI, 0.59 to 1.01) and for participants with a BMI <25 kg/m^2 with a mean reported baseline and achieved 25(OH)D concentrations of 33.3 and 45.9 ng/mL, respectively, the HR for cancer incidence was 0.76 (95% CI, 0.63 to 0.90) [37].

In conclusion, the evidence that higher vitamin D status reduces the risk of cancer incidence includes:

1—Single-country ecological studies finding that about 20 types of cancer have incidence and mortality rates inversely correlated with various indices of solar UVB doses.

2—Observational studies reporting that several types of cancer have incidence rates inversely correlated with serum 25(OH)D in case-control or cohort observational studies.

3—An observational study using individual participant data for women taking vitamin D or placebo in two RCTs or taking vitamin D in a volunteer cohort, and with some subjects achieving 25(OH)Ds > 60 ng/mL, achieving a significant reduction in breast cancer incidence.

4—Mechanisms that explain how vitamin D reduces the risk of cancer incidence, progression, and metastasis.

5—No mechanisms are yet known that might explain how non-vitamin D mechanisms associated with UVB exposure could reduce the risk of cancer.

2.4. Type 2 Diabetes Mellitus

Type 2 diabetes mellitus (T2DM) is a condition in which there is too much glucose circulating in the blood because of long-standing increases in insulin resistance leading to the eventual deficiency in insulin responsiveness to hyperglycaemia [51,52]. Although T2DM is generally associated with obesity and 'cafeteria' diets, there is now reasonably strong evidence that risk is inversely associated with serum 25(OH)D concentration and that correcting vitamin D deficiency over time can reduce T2DM risks.

The mechanisms by which vitamin D reduces the risk of T2DM include β-cell insulin release through a rise in intracellular calcium concentration [53] and by stimulating insulin synthesis [54], which was reported in 1980 [55] (see review in [56]). Vitamin D reduces insulin resistance by reducing oxidative stress and inflammation and reducing both hepatic lipogenesis and hepatic glucose release through metformin-like effects [57,58] and by promoting increased metabolic efficiency in skeletal muscle [59,60]. Thus, it is not surprising that obesity, which is marked by increased systemic inflammation as well as lower 25(OH)D concentrations [61], and a lack of physical activity are major risk factors for T2DM [62].

Prospective observational studies support a role for vitamin D in reducing T2DM risks. A meta-analysis of 16 prospective studies published in 2013 found an OR = 1.50 (95% CI, 1.33 to 1.67) for the incidence of T2DM for low vs. high 25(OH)D status in each study [63]. Large prospective studies have shown that lower 25(OH)D status is a risk factor for metabolic syndrome [64] and for T2DM [65].

Higher 25(OH)D concentrations are associated with lower mortality rates for adults with diabetes prospectively. A study was reported on 6329 adults with diabetes from the Third National Health and Nutrition Examination Survey (NHANES III) and NHANES 2001–2014 followed up through 31 December 2015 [66]. In 55,126 person-years of follow-up, 2056 deaths were documented (605 from CVD and 309 from cancer). Compared with participants with 25(OH)D concentrations <25 nmol/L, the multivariate-adjusted HRs and 95% CI for participants with 25(OH)D concentrations >75 nmol/L were 0.59 (95% CI, 0.43, 0.83) for all-cause mortality (p_{trend} = 0.003), 0.50 (95% CI, 0.29, 0.86) for CVD mortality (p_{trend} = 0.02), and 0.49 (95% CI, 0.23, 1.04) for cancer mortality (p_{trend} = 0.12).

A meta-analysis of 24 RCTs (n = 1528 individuals with T2DM) found significant reductions in glycosylated hemoglobin (HbA1c) (mean difference: −0.30%; 95% CI: −0.45 to −0.15, $p < 0.001$), serum fasting plasma glucose (FPG) (mean difference: −4.9 mg/dL (−0.27 mmol/L); 95% CI: −8.1 to −1.6 (−0.45 to −0.09 mmol/L), p = 0.003), and the homeostatic model assessment of insulin resistance (HOMA-IR) (mean difference: −0.66; 95% CI: −1.06 to −0.26, p = 0.001) in diabetic patients [67] where a mean increase in 25(OH)D concentration in those RCTs was 17 ± 2 ng/mL, suggesting that a minimum dose of 4000 IU/day of vitamin D is advisable for improving insulin sensitivity and glycemic control in T2DM patients.

It seems that vitamin D supplementation in doses of 4000 IU/day or higher improved the manifestation of diabetic complications. An example would be the study conducted in St. Petersburg, Russia that included 62 T2DM patients with diabetic polyneuropathy [68]. The intake of 40,000 IU/week of vitamin D for 24 weeks was associated with an increased 25(OH)D concentration from 16 to 72 ng/mL and a reduction in neurological deficit and pain severity, as well as improved interleukin profiles and markers for microcirculation. No overall changes were detected in T2DM patients who received 5000 IU/week of vitamin D.

The Vitamin D and Type 2 Diabetes (D2d) Study conducted by Tufts University is the largest RCT to examine the effect of vitamin D supplementation on the risk of T2DM [69]. A total of 2423 prediabetic participants were enrolled and half were given 4000 IU/day of vitamin D_3, whereas the other half were given a placebo during a mean 2.5-year follow-up period. The HR for progression to T2DM for the treatment arm compared to the control arm was 0.88 (95% CI, 0.75 to 1.04, $p = 0.12$). Two subgroups had a significantly reduced risk when comparing treatment to control groups, those with BMI <30 kg/m^2 (HR = 0.71 (95% CI, 0.53 to 0.95)) and those not taking calcium (HR = 0.81 (95% CI, 0.66 to 0.98)). However, it was a secondary analysis related to 25(OH)D concentration maintained through the RCT that provided strong support for vitamin D supplementation reducing the progression from pre-diabetes to T2DM [70]. Over the range from 20–30 ng/mL to >50 ng/mL, for each increase in 25(OH)D by 10 ng/mL, those in the treatment arm had an HR of 0.75 (95% CI, 0.68 to 0.82) for conversion to T2DM. Therefore, secondary analysis based on serum 25(OH)D concentration is clearly an appropriate way to analyze RCT results since vitamin D, being a nutrient provided from several sources, has non-linear effects in contrast to a 'medication' [47,48].

2.5. COVID-19

The world is in the midst of the COVID-19 pandemic with nearly 5.6 million deaths reported to date (21 January 2022) (https://www.worldometers.info/coronavirus/ (accessed on 5 January 2022)). On 2 April 2020, it was suggested that vitamin D could reduce the risk of COVID-19 through several mechanisms including inducing the secretion of cathelicidins and defensins known to reduce viral replication rates and reduce the production of pro-inflammatory cytokines aggravating the inflammation that injures the lungs, leading to pneumonia. Vitamin D also increases the secretion of anti-inflammatory cytokines, overall reducing the risk of highly dangerous cytokine storms [71]. The important modulatory effect of vitamin D on immune-related genes through binding to VDR implicates a potential role in clearing SARS-CoV-2 infections [72].

The rationale for that suggestion included the observation that case fatality rates in the U.S. during the 1918–1919 influenza pandemic were mainly due to the development of pneumonia and were lowest in communities with the highest solar UVB doses [73]. In addition, a meta-analysis of RCTs had shown a reduction in acute respiratory tract infections with supplementation in vitamin D deficiency cases using analyses of individual participant data [74].

Observational studies suggest that vitamin D reduces SARS-CoV-2 infection risk. A study of >190,000 patients who had SARS-CoV-2-positive tests in the U.S. between 9 March and 19 June 2020 and had their serum 25(OH)D concentration measured during the previous twelve months (by Quest Diagnostics, Secaucus, NJ) [75] showed the following results by race/ethnicity, after adjusting for seasonal differences: black non-Hispanics, Hispanics, and white non-Hispanics had ~19%, 16% and 9% rates for a 25(OH)D concentration <20 ng/mL, respectively, and rates of 11%, 10%, and 5%, respectively for 25(OH)D values of ~55 ng/mL. These values represent reductions by about 40% for all three races/ethnicities. The higher SARS-CoV-2 positivity rates for black American non-Hispanics and Hispanics was mostly likely due to their being in lower socioeconomic strata and more likely to be unable to socially isolate or to work from home than white non-Hispanics as well as being more likely to have been aggravated by their lower vitamin D status, itself known to worsen with lower SE status as well as being increased in those with darker skin [76].

There have been many observational studies. Some were retrospective studies where 25(OH)D concentrations for those becoming ill and being diagnosed with COVID-19 were obtained from measurements made before diagnosis (seasonally adjusted). Examples include those from Israel [77] and Chicago, IL, USA [78,79]. However, the majority of such observational investigations were based on 25(OH)D concentration at the time of diagnosis upon hospital admission [80].

The most recent meta-analysis of COVID-19 risk in relation to serum 25(OH)D concentrations was published on 11 December 2021 [81]. It included results from 76 observational studies. Vitamin D deficiency/insufficiency increased the odds of developing COVID-19 (OR 1.46, 95% CI 1.28–1.65, $p < 0.0001$), developing severe disease (OR 1.90, 95% CI 1.52–2.38, $p < 0.0001$) and death (OR 2.07, 95% CI 1.28–3.35, $p = 0.003$). A major concern is that having COVID-19 must itself lower 25(OH)D concentration as is usual in severe infection, as shown experimentally [27]. This concern will eventually be clarified by examining data for COVID-19 outcomes by pre-illness and/or pre-pandemic 25(OH)D concentration in comparison with those found at the time of diagnosis. In one meta-analysis of vitamin D deficiency/insufficiency and risk of COVID-19 involving 19 studies, the OR was 1.46 (95% CI, 1.28 to 1.65). Of the 19 studies, ten had 25(OH)D concentrations measured prior to COVID-19 diagnosis, of which three were 10–15 years before the pandemic. The ORs for those three were near 1.00 and non-significant, as expected for such a long lag time. Examining the five studies with values measured in the previous year (omitting one that is a preprint), the ORs for four of them were above the mean value and one was very near the meta-analysis value, 1.46. Thus, on the basis of that analysis, there does not seem to be a significant difference between risks relating to low 25(OH)D values whether measured before or at the time of COVID-19 diagnosis.

The most recent article on vitamin D and the risk of COVID-19 hospitalization and mortality was published on 1 January 2022 [82]. It presented an analysis of 4599 veteran patients receiving care in the US Department of Veterans Affairs health care facilities who tested positive for SARS-CoV-2 during the period from 20 February to 8 November 2020 and who had serum 25(OH)D concentration data from the previous 15 to 90 days on file. Twenty one percent of the patients were hospitalized and 7.4% died within 60 days of their index SARS-CoV-2 test. Hospitalization rates decreased from 25% at 15 ng/mL to 18% at 60 ng/mL (adjusted relative risk = 1.29 (95% CI, 1.06 to 1.57), whereas morality rates decreased from 11% for vitamin D levels of 15 ng/mL to 6% at 60 ng/mL (adjusted relative risk = 1.82 (1.27 to 2.63)).

Results of a trial conducted in Turkey involving 132 COVID-19 patients with a baseline 25(OH)D concentration <30 ng/mL, of whom 80 were treated with high-dose vitamin D3 to achieve a 25(OH)D concentration >30 ng/mL, was reported recently [83]. Vitamin D_3 doses ranged from 224,000 to 500,000 IU over periods from three to 14 days. The mean 25(OH)D for the treated patients reached only 31 ± 12 ng/mL on day 7 and 35 ± 11 ng/mL on day 14. The mortality rate was 11.2% (97 out of 867) in the whole cohort. The mortality rate for patients who had comorbidities but received vitamin D treatment was 5.5% (9 out of 162). Having vitamin D treatment decreased the 14-day mortality rate significantly (OR for survival: 2.14, 95%CI: 1.06 to 4.33, $p = 0.03$).

Calcifediol (25(OH)D_3) is being used in Spain to treat COVID-19 patients, its advantage being that it increases serum 25(OH)D concentrations within hours and much faster than intact vitamin D_3. The first study of the calcifediol treatment of COVID-19 patients was conducted in Cordoba, Spain [84] in 76 consecutive patients admitted to a university hospital with clinically diagnosed COVID-19 who all received standard care with hydroxychloroquine plus azithromycin for patients with pneumonia (a broad-spectrum antibiotic). Fifty of these patients were 'randomized' to receive oral calcifediol on the day of admission at 0.532 mg followed by 0.266 mg on days 3 and 7 and then weekly until discharge or intensive care unit (ICU) admission. Calcifediol is $\sim \times 3$ times more effective in raising serum 25(OH)D concentration (after adjusting for weight) than vitamin D_3. The only significant difference in the prognostic factors for COVID-19 at baseline was previous high blood pressure (at 24% in the treatment group and 58% in the control group; $p = 0.002$). Only one of the treated patients but 13 of the untreated patients required admission to the ICU ($p < 0.001$). There was no death among treated patients, but two untreated patients requiring care in the ICU died.

Another Spanish report summarized the results of treating 537 COVID-19 patients admitted to any of the five hospitals in southern Spain between 5 February and 5 May 2020 with calcifediol (25(OH)D$_3$) [85], excluding the 76 patients already mentioned [84]. In that study, 79 patients were treated and 458 were not treated with calcifediol. The untreated patients had significantly higher rates of CURB-65 $\geq$3 (21 vs. 8%) and ARDS (25 vs. 10%), higher CRP (130 $\pm$ 100 vs. 100 $\pm$ 80 units) and blood urea nitrogen (22 $\pm$ 19 vs. 16 $\pm$ 15 units) values together with lower oxygen saturation at admission (93 $\pm$ 6% vs. 95 $\pm$ 4%). However, those differences were not mortality rate determinants. The crude OR for death in patients treated with calcifediol vs. those not so treated was 0.22 (95% CI, 0.08 to 0.61, $p < 0.01$) and was 0.16 (95% CI, 0.03 to 0.80) after adjustment for all other risk factors. On the other hand, higher age, ARDS, CURB-65 $\geq$3, cerebrovascular disease, COPD, cancer, and ratio of neutrophils to lymphocytes were significantly associated with increased mortality. Other studies have shown that low serum 25(OH)D concentration is a stronger marker of adverse outcomes than those other factors. For example, in an observational study in Iran involving 442 patients in general wards and 66 patients in the ICU, of whom 55 died, only age ($p < 0.001$, albumin ($p < 0.001$), calcium, ($p = 0.002$) and serum 25(OH)D ($p = 0.047$)) were significantly associated with mortality in multivariate analysis, whereas BMI, diabetes mellitus, hypertension, IHD, creatinine, and phosphorus were not [86].

Another review used meta-analyses on COVID-19 hospitalized patient outcomes [87]. Using two studies that gave vitamin D$_3$ and two giving calcifediol, the OR for intensive care unit admission was 0.27 (95% CI, 0.09 to 0.76); on the other hand, based on four studies giving vitamin D$_3$ and one giving calcifediol, the OR for needing mechanical ventilation was 0.34 (95% CI, 0.16 to 0.72) with these treatments. Overall, from those eight studies plus one more, the OR for mortality for any form of vitamin D/calcifediol treatment was 0.37 (95% CI, 0.21 to 0.66).

An observational study in Barcelona compared the incidence of COVID-19 for patients with respect to serum 25(OH)D concentration and whether they were being treated with vitamin D$_3$ or calcifediol [88]. For those being treated with vitamin D$_3$ and achieving >30 ng/mL, the multivariate HR compared to untreated controls with <20 ng/mL for SARS-CoV-2 infection was 0.66 (95% CI, 0.57 to 0.77), 0.72 (0.52 to 1.00) for severe COVID-19, and 0.66 (0.46 to 0.93) for COVID-19 mortality. Similarly, for those being treated with calcifediol, the multivariate HR for SARS-CoV-2 infection was 0.69 (95% CI, 0.61 to 0.79), 0.61 (0.46 to 0.81) for severe COVID-19, and 0.56 (0.42 to 0.76) for COVID-19 mortality. The small differences between vitamin D and calcifediol treatment may be due to different effective vitamin D doses and achieved 25(OH)D concentrations. Thus, there is accumulating evidence that higher serum 25(OH)D concentrations protect against COVID-19, but more research is warranted.

2.6. Alzheimer's Disease

Alzheimer's disease (AD) is a neurodegenerative disease of the brain, generally with increased beta-amyloid plaque and tau protein deposition. The mechanisms whereby vitamin D reduces the risk of AD include preventing amyloid development and clearing it from the brain [89–91]. Most of the studies on the effects of vitamin D on the risk of AD are observational. One from France followed 916 participants over 65 years old for 12 years [92]. A total of 117 dementia cases developed, of which 124 were AD. The adjusted HR for AD compared to 25(OH)D concentrations >20 ng/mL was 2.17 (95% CI, 1.37 to 5.68) for those with a 25(OH)D concentration between 12 and 25 ng/mL and 2.85 (95% CI, 1.36 to 5.97) for those with a 25(OH)D concentration <12 ng/mL.

A meta-analysis of six observational studies found the HR for AD for a 10 ng/mL increase in 25(OH)D concentration of 0.83 (95% CI, 0.68 to 0.96) [93]. Another meta-analysis based on nine observational studies found that for 25(OH)D concentrations <20 ng/mL, HR = 1.34 (95% CI, 1.13 to 1.60) [94]. A recent MR analysis based on data from the IGAP and the UK Biobank found that genetically increased 25(OH)D concentrations were significantly associated with reduced risks of AD [95].

2.7. All-Cause Mortality

Based on the effect of vitamin D on the major causes of death in developed countries, it would be expected that there would be an inverse relationship between serum 25(OH)D concentrations and subsequent all-cause mortality rate and this has been found using meta-analyses; one conducted in 2012 used 14 prospective studies with 62,548 individuals found an RR of 0.69 (95% CI, 0.60 to 0.78) for 25(OH)D = 31 ng/mL vs. 11 ng/mL, with no further decreases in mortality rates above 35 ng/mL [96]. Another meta-analysis of 32 studies found that 25(OH)D values >30 ng/mL vs. <10 ng/mL had an RR for survival of 1.9 (95% CI, 1.6 to 2.2) [97]. Another study from a European consortium of 26,916 individuals found that mortality rates increased with a lower individual participant standardized 25(OH)D concentration in a cubic spline model adjusted for age, sex, and BMI at baseline visit, compared to 30–40 ng/mL, HR = 1.06 (0.96 to 1.15) for 25(OH)D concentrations reduced to 20–30 ng/mL, 1.14 (1.03–1.24) for 25(OH)D concentrations of 16–20 ng/mL, and reaching 1.29 (1.17–1.41) for 25(OH)D concentrations of 12–16 ng/mL, and 1.72 (1.53–1.90) for 25(OH)D values <12 ng/mL [98]. In the 20-year Veterans Health Administration study regarding myocardial infarction discussed above [8], patients achieving 25(OH)D concentrations of 20–30 ng/mL vs. <20 ng/mL had an all-cause mortality HR of 0.59 (95% CI, 0.54 to 0.63), whereas in those achieving values >30 ng/mL, the mortality HR was not further reduced at 0.61 (95% CI, 0.56 to 0.67).

A meta-analysis of 52 trials with 75,454 participants in vitamin D RCTs found that supplementation was not associated with any changes in all-cause mortality rate (RR = 0.98 (95% CI, 0.95 to 1.02)) [99]. However, vitamin D supplementation did significantly reduce cancer mortality rate (RR = 0.84 (95% CI, 0.74 to 0.95)) and subgroup analyses showed all-cause mortality was significantly lower in trials giving vitamin D_3 supplementation vs. placebo (RR = 0.95 (95% CI, 0.91 to 1.00)) than in trials with vitamin D_2 supplementation vs. placebo (RR = 1.03 (95% CI, 0.98 to 1.09)). One problem with those meta-analyses was that they did not have individual participant data for analysis, as was used for the acute respiratory tract infection meta-analysis already referred to [74]. Another concern is that trials with large cohorts used in those meta-analyses were made variously between 1996 to 2018, when 25(OH)D assays used were changing over [100,101], which inevitably adds a degree of error to the reported meta-analyses.

3. Discussion

A summary of the findings reported in this review is given in Table 5. The optimal 25(OH)D concentration thresholds for these various outcomes range from 25 ng/mL to 60 ng/mL. All of these concentrations are higher than the 20 ng/mL recommended by the Institute of Medicine based on its interpretation of requirements for bone health [102]. They are in general agreement with the Endocrine Society's recommendation of >30 ng/mL [103], based on a more careful interpretation of a study of 25(OH)D concentrations and bone mineralization [104]. They are also consistent with a recommendation of 30–50 ng/mL in 2018 for the pleiotropic (non-skeletal) effects of vitamin D [105].

The 25(OH)D concentration range of 30–40 ng/mL could generally be met by the supplementation of 2000 to 4000 IU/day, which was reported as safe for all by the Institute of Medicine [102]. Achieving concentrations above 40 ng/mL could take higher doses. The Institute of Medicine noted that they did not have evidence that taking up to 10,000 IU/day of vitamin D had any adverse effects, but set the upper tolerable level at 4000 IU/day out of a concern for safety. The UK NIH also agrees that 4000 IU/day is safe (https://www.nhs.uk/conditions/vitamins-and-minerals/vitamin-d/ accessed on 4 January 2021).

Table 5. Optimal 25(OH)D concentrations for various health outcomes.

Outcome	Type of Evidence	Optimal 25OHD	Reference
All-cause mortality rate	Observational study of 25(OH)D concentration due to vitamin D supplementation	>30 ng/mL	[8]
Alzheimer's disease and dementia	Meta-analysis of observational studies	>25 ng/ml	[93]
Breast cancer	Observational study of 25(OH)D concentration due to vitamin D supplementation	>60 ng/mL	[33]
Colorectal cancer	Meta-analysis of observational studies	30–40 ng/mL	[34]
Cardiovascular disease	Observational study of the CVD mortality rate for CVD patients	>30 ng/mL	[9]
Myocardial infarction	Observational study of 25(OH)D concentration due to vitamin D supplementation'1n	>30 ng/mL	[8]
SARS-CoV-2 infection	Retrospective observational study	>50 ng/mL	[75]
COVID-19 mortality	Retrospective cohort study	>60 ng/mL	[82]
Diabetes mellitus type 2	RCT with an analysis of intratrial 25(OH)D for prediabetes patients	>50 ng/mL	[70]
Gene expression	Clinical trial	>40 ng/mL	[45]
Hypertension	Observational study of 25(OH)D concentration due to vitamin D supplementation	>40 ng/mL	[16]
Preterm delivery	Observational study of 25(OH)D concentration due to vitamin D supplementation	>40 ng/mL	[106]

It has been shown experimentally that humans can produce between 10,000 and 25,000 IU of vitamin D through whole-body exposure to one minimal erythemal dose of simulated sunlight, i.e., one instance of mid-day sun exposure without burning [107]. Thus, doses to those levels should be considered inherently safe. Recent articles have reported the safety results for high-dose vitamin D supplementation. One was a community-based, open-access vitamin D supplementation program involving 3882 participants conducted in Canada between 2013 and 2015 [108]. Participants took up to 15,000 IU/day of vitamin D_3 for between 6 and 18 months. The goal of the study was to determine vitamin D doses required to achieve a 25(OH)D concentration >40 ng/mL. It was found that participants with a normal BMI had to take at least 6000 IU/day of vitamin D, whereas overweight and obese participants had to take 7000 IU/day and 8000 IU/day, respectively. Serum 25(OH)D concentrations of up to 120 ng/mL were achieved without the perturbation of calcium homeostasis or toxicity.

Another study involved 777 long-term hospitalized patients taking 5000 to 50,000 IU/day of vitamin D_3 [109]. Subsets of those taking 5000 IU/d achieved mean 25(OH)D concentrations of 65 ± 20 ng/mL after 12 months, whereas those taking 10,000 IU/day achieved 100 ± 20 ng/mL after 12 months. No patients who achieved 25(OH)D concentrations of 40–155 ng/mL developed hypercalcemia, nephrolithiais (kidney stones), or any other symptoms of vitamin D toxicity as the result of vitamin D supplementation.

Hypersensitivity to vitamin D can develop in people with sarcoidosis and some other lymphatic disorders, causing hypercalcaemia and its complications from exposure to sunshine alone or following supplementation. See the discussion regarding vitamin D and sarcoidosis in this recent review [110].

Thus, given the multiple indications of significant health benefits from raising serum 25(OH)D concentrations above 30 or 40 ng/mL as well as the near absence of adverse effects, significant improvements in health at the individual and population levels could be achieved. Methods to achieve optimal health benefits could usefully begin with establishing effect thresholds for different disorders with reasonable certainty while allowing for variations reported with obesity, diabetes, ethnicity, age or gender and by instituting programs to encourage and facilitate raising serum 25(OH)D concentrations through a variety of approaches including sensible solar UVB exposure, vitamin D supplementation and food fortification. A vitamin D fortification program of dairy products initiated in Finland in 2003 eventually resulted in 91% of non-vitamin D supplement users reaching 25(OH)D concentrations >20 ng/mL [111], The rationale and plan for food fortification with vitamin D, which was doubled in 2010, was outlined in 2018 [112].

As for future research, the most efficient way to determine the effects of vitamin D supplementation seems to be to conduct observational studies of individual participants who supplement with vitamin D_3. A concern regarding such observational studies is that the controls might not be well matched to those supplementing with vitamin D. A way to improve such studies is to use propensity score matching of both groups, as reported in two recent vitamin D studies. One was an examination of the de novo use of vitamin D after the diagnosis of breast cancer [113]. The other was in the study from Spain regarding vitamin D_3 or calcifediol supplementation and the risk of COVID-19 [88]. Using propensity score matching in observational studies can elevate them to the level of RCTs in terms of examining causality.

Types of Studies, Strengths and Weaknesses

Many types of studies are used to help determine whether a factor modifies disease risks (incidence, survival, and/or mortality rates). A typical evidence pyramid published in 2018 showed a hierarchy with in vitro studies at the bottom, progressing upward with animal, ecological, cross-sectional, case-control studies, and randomized controlled trials (RCTs) with meta-analyses of RCTs at its apex [114]. Although this pyramid is appropriate for pharmaceutical drugs, it has various limitations when applied to nutrients. Not generally included in such pyramids is an understanding of the mechanisms by which a nutrient or agent of interest works, though that is very important when considering causality for vitamin D in each disorder of interest. For mechanisms regarding vitamin D, the reader is referred to *Vitamin D, 4th Edition* [115] as well as pubmed.gov and scholar.google.com. The main types of studies used for vitamin D are discussed here in the ascending order of classic pyramidal evidence hierarchies.

Ecological studies consider populations defined geographically and use risk-modifying factors and health outcome population averages; they can be either geographical or temporal. For example, the first indication that better provision of vitamin D reduced cancer risks came from an ecological study of colon cancer mortality rates in the U.S. in relation to annual solar radiation doses [17] and for reduction in cardiovascular disease (CVD) risks from a study showing seasonality in CVD mortality rates [2] and from a similar finding for epidemic influenza in 2006 [116]. The strengths of ecological studies include the inclusion of large numbers of participants, that solar UVB doses have large latitudinal gradients in middle-latitude countries and show large seasonal variations, and that many risk-modifying factors can be used in the analysis as were used for cancer in 2006 in an American ecological study that included alcohol consumption, Hispanic heritage, poverty level, smoking, and urban/rural residence as well as July 1992 solar UVB doses [19]. On the other hand, solar UVB is an important source of vitamin D, but is strongly associated with solar UVA (320–400 nm) radiation which has other health effects such as liberating nitric oxide from subcutaneous nitrogen compounds, which reduces arterial stiffness [117] and thereby reduces blood pressure [118] and the risk of COVID-19 [119]. However, ecological studies performed on post-2000 data generally fail to find inverse correlations between solar UVB and cancer incidence or mortality rates, most likely because people are spending less time in the sun, use more sunblock, are also more likely to be obese and no doubt also because cancer survival rates are now much improved by more effective therapies [120]. Although ecological studies cannot establish causality in isolation, they can provide strong support in combination with other types of studies.

Cross-sectional studies consider the relationship between many variables and health status at one point in time, revealing associations but unable to establish causality since variables studied may be affected by the disease of interest. Nonetheless, they can provide associational information for comparison with findings from other types of studies; for example, a cross-sectional study of women recently diagnosed with breast cancer in Brazil showed an inverse correlation between serum 25(OH)D concentration and factors used to estimate prognosis and showed an association of low 25(OH)D with increased rates of estrogen receptor-negative tumors [121].

Case-control studies involve measuring variables for those with a particular outcome (cases) with similar individuals without that outcome [122]. The study can be either retrospective, e.g., the history of solar UVB exposure, or contemporaneous with disease incidence, (e.g., serum 25(OH)D concentration). An important strength of case-control studies for cancer and all-cause mortality rate is that they generally provides concomitant serum 25(OH)D concentrations, whereas prospective studies use blood samples from the time of enrollment even though 25(OH)D concentrations change over time, thereby reducing correlations of 25(OH)D with outcomes over time [26]. This effect is especially important for breast cancer which can progress from undetectable to obvious very rapidly and is one of only a few cancers with a pronounced seasonality of incidence [123].

There are several concerns about case-control studies. One is that controls may not be well matched to cases. The way to overcome this concern is to use propensity score matching as conducted in two recent vitamin D studies [88,113]. Another is that the disease itself may affect serum 25(OH)D concentration as has been demonstrated with acute inflammatory diseases (e.g., acute respiratory tract infections [27]) but not for undiagnosed cancers, where inflammation is not generalized.

Prospective cohort studies generally enroll participants over short periods of time, measure many variables of relevance and draw blood for later analysis, before following participants, commonly over many years before outcome assessment and data analysis using a nested case-control approach. The advantages include the inclusion of many subjects that act as controls against which case risk can be assessed. An important limitation is that variable values can change over time, including serum 25(OH)D concentrations [26]. However, the results of individual cohort studies can be combined for meta-analyses, which often provide the strongest evidence for various health outcomes [124].

An important limitation of observational studies is that most participants have serum 25(OH)D concentrations between 10 and 40 ng/mL [125,126]. Most vitamin D is obtained from solar UVB exposure plus some from animal-based food including meat, fish, eggs [127], and vitamin D-fortified food [112], and supplements. However, the recommended vitamin D supplement value for adults in the U.S. is 600 IU/day up to 70 years of age and 800 IU/day for those over 70 years old [102]. According to changes in serum 25(OH)D with supplementation [12], 600 IU/day can increase 25(OH)D by about 5.6 ng/mL, and 800 IU/day by 7.5 ng/mL.

Thus, most observational studies to date include few participants with 25(OH)D concentrations >40 ng/mL unless they are supplementing with 1000 to 5000 IU/day or more as is the case for the observational studies conducted by GrassrootsHealth.net (accessed on 15 December 2021), as discussed [33,106].

Randomized controlled trials (RCTs) are considered the strongest type of evidence in medical decision making, their strength being that they examine the effect of particular substances and can, therefore, rule out many confounding factors. Unfortunately, vitamin D differs from pharmaceutical drugs in that there are several sources including solar UVB exposure, diet, and supplements, and that serum 25(OH)D concentration–health outcome relationships are non-linear. As a result, vitamin D RCTs generally fail to confirm findings from observational studies [128,129]. Robert Heaney outlined the guidelines for nutrient trials, where the important factor is that nutrient concentration, e.g., 25(OH)D, should drive both trial design and analysis and not the supplemental dosage [47,48]. As a result of using RCT design guidelines evolved for testing pharmaceutical drug efficacy most vitamin D RCTs have failed to find the beneficial effects of supplementation. The primary reasons for such predictable failures include enrolling participants with 25(OH)D concentrations that are relatively high, using relatively low vitamin D doses that cannot raise 25(OH)D values into the normal range, permitting participants (including controls) to take additional supplements, not recognizing that participants may have different vitamin D responses [49] and that there are different 25(OH)D thresholds for different health benefits.

Although not on the standard evidence pyramid, Mendelian randomization (MR) studies are also suggested to be valuable for establishing causality for vitamin D for various health outcomes. MR studies compare the estimated effect of SNPs associated with variation in 25(OH)D concentrations on the health outcomes seen in large numbers of participants, often up to 100,000. Although some MR studies report inverse correlations between the SNPs increasing serum 25(OH)D and several health outcomes [130] such as the incidence of multiple sclerosis [131] and ovarian cancer [132], no such effects were seen for eight other types of cancer [133]. The primary reasons for MRA failure are likely to include the fact that total SNP-induced variation in 25(OH)D has often been less than 25(OH)D assay variance [134] and that genome-wide association studies' (GWAS) analyses of the total percentage of SNP effects are made on the 25(OH)D data as a whole, although such data is non-linear with much of it lying in the low and high plateaus of the 25(OH)D–health outcome relationships, a problem that the GWAS analysis of 25(OH)D data stratified for different ranges of 25(OH)D efficacy might overcome [135]. That this is the case for mortality rates was shown in two recent articles, one [10] where GWAS serum 25(OH)D concentration was stratified at <10 ng/mL, 10–20 ng/mL, 20–30 ng/mL, and >30 ng/mL and significantly increased risk was only present at 25(OH)D <10 ng/mL for all-cause mortality, cardiovascular mortality, and non-CVD and non-cancer mortality with trends for increases in risk for stroke and cancer mortality. The other [12], using genetic increases in serum 25(OH)D calculated for 100 equal strata of measured serum 25(OH)D showed similar results for CVD with risk reduction for increases in 25(OH)D values up to ~20 ng/mL.

Author Contributions: Conceptualization, W.B.G.; methodology, W.B.G.; writing—original draft preparation, W.B.G., F.A.A., B.J.B., E.D., D.G.-A. and E.B.J., T.K.; writing—review and editing, W.B.G., F.A.A., B.J.B., E.D., E.B.J., T.K. and P.P.; supervision, W.B.G. All authors have read and agreed to the published version of the manuscript.

Funding: No funding was received for this study.

Conflicts of Interest: WBG's nonprofit organization, Sunlight, Nutrition and Health Research Center, receives funding from Bio-Tech Pharmacal, Inc. (Fayetteville, AR, USA). The other authors have no conflicts of interest to declare.

References

1. Lips, P.; de Jongh, R.T.; van Schoor, N.M. Trends in Vitamin D Status Around the World. *JBMR Plus* **2021**, *5*, e10585. [CrossRef] [PubMed]
2. Acharya, P.; Dalia, T.; Ranka, S.; Sethi, P.; Oni, O.A.; Safarova, M.S.; Parashara, D.; Gupta, K.; Barua, R.S. The Effects of Vitamin D Supplementation and 25-Hydroxyvitamin D Levels on the Risk of Myocardial Infarction and Mortality. *J. Endocr. Soc.* **2021**, *5*, bvab124. [CrossRef]
3. Scragg, R. Seasonality of cardiovascular disease mortality and the possible protective effect of ultra-violet radiation. *Int. J. Epidemiol.* **1981**, *10*, 337–341. [CrossRef] [PubMed]
4. Marti-Soler, H.; Gubelmann, C.; Aeschbacher, S.; Alves, L.; Bobak, M.; Bongard, V.; Clays, E.; de Gaetano, G.; Di Castelnuovo, A.; Elosua, R.; et al. Seasonality of cardiovascular risk factors: An analysis including over 230,000 participants in 15 countries. *Heart* **2014**, *100*, 1517–1523. [CrossRef]
5. Wang, T.J.; Pencina, M.J.; Booth, S.L.; Jacques, P.F.; Ingelsson, E.; Lanier, K.; Benjamin, E.J.; D'Agostino, R.B.; Wolf, M.; Vasan, R.S. Vitamin D deficiency and risk of cardiovascular disease. *Circulation* **2008**, *117*, 503–511. [CrossRef]
6. Zittermann, A.; Schleithoff, S.S.; Koerfer, R. Putting cardiovascular disease and vitamin D insufficiency into perspective. *Br. J. Nutr.* **2005**, *94*, 483–492. [CrossRef]
7. Timms, P.M.; Mannan, N.; Hitman, G.A.; Noonan, K.; Mills, P.G.; Syndercombe-Court, D.; Aganna, E.; Price, C.P.; Boucher, B.J. Circulating MMP9, vitamin D and variation in the TIMP-1 response with VDR genotype: Mechanisms for inflammatory damage in chronic disorders? *QJM* **2002**, *95*, 787–796. [CrossRef] [PubMed]
8. Li, Y.; Tong, C.H.; Rowland, C.M.; Radcliff, J.; Bare, L.A.; McPhaul, M.J.; Devlin, J.J. Association of changes in lipid levels with changes in vitamin D levels in a real-world setting. *Sci. Rep.* **2021**, *11*, 21536. [CrossRef]

9. Dai, L.; Liu, M.; Chen, L. Association of Serum 25-Hydroxyvitamin D Concentrations with All-Cause and Cause-Specific Mortality Among Adult Patients with Existing Cardiovascular Disease. *Front. Nutr* **2021**, *8*, 740855. [CrossRef]

10. Emerging Risk Factors Collaboration/EPIC-CVD/Vitamin D Studies Collaboration. Estimating dose-response relationships for vitamin D with coronary heart disease, stroke, and all-cause mortality: Observational and Mendelian randomisation analyses. *Lancet Diabetes Endocrinol.* **2021**, *9*, 837–846. [CrossRef]

11. Grant, W.B. Effect of follow-up time on the relation between prediagnostic serum 25-hydroxyvitamin D and all-cause mortality rate. *Dermatoendocrinology* **2012**, *4*, 198–202. [CrossRef]

12. Zhou, A.; Selvanayagam, J.B.; Hypponen, E. Non-linear Mendelian randomization analyses support a role for vitamin D deficiency in cardiovascular disease risk. *Eur. Heart J.* **2021**, ehab809. [CrossRef] [PubMed]

13. Staley, J.R.; Burgess, S. Semiparametric methods for estimation of a nonlinear exposure-outcome relationship using instrumental variables with application to Mendelian randomization. *Genet. Epidemiol.* **2017**, *41*, 341–352. [CrossRef]

14. Kunutsor, S.K.; Apekey, T.A.; Steur, M. Vitamin D and risk of future hypertension: Meta-analysis of 283,537 participants. *Eur. J. Epidemiol.* **2013**, *28*, 205–221. [CrossRef] [PubMed]

15. Vimaleswaran, K.S.; Cavadino, A.; Berry, D.J.; LifeLines Cohort Study, I.; Jorde, R.; Dieffenbach, A.K.; Lu, C.; Alves, A.C.; Heerspink, H.J.; Tikkanen, E.; et al. Association of vitamin D status with arterial blood pressure and hypertension risk: A mendelian randomisation study. *Lancet Diabetes Endocrinol.* **2014**, *2*, 719–729. [CrossRef]

16. Mirhosseini, N.; Vatanparast, H.; Kimball, S.M. The Association between Serum 25(OH)D Status and Blood Pressure in Participants of a Community-Based Program Taking Vitamin D Supplements. *Nutrients* **2017**, *9*, 1244. [CrossRef]

17. Garland, C.F.; Garland, F.C. Do sunlight and vitamin D reduce the likelihood of colon cancer? *Int. J. Epidemiol.* **1980**, *9*, 227–231. [CrossRef] [PubMed]

18. Boscoe, F.P.; Schymura, M.J. Solar ultraviolet-B exposure and cancer incidence and mortality in the United States, 1993–2002. *BMC Cancer* **2006**, *6*, 264. [CrossRef]

19. Grant, W.B.; Garland, C.F. The association of solar ultraviolet B (UVB) with reducing risk of cancer: Multifactorial ecologic analysis of geographic variation in age-adjusted cancer mortality rates. *Anticancer Res.* **2006**, *26*, 2687–2699.

20. Moukayed, M.; Grant, W.B. Molecular link between vitamin D and cancer prevention. *Nutrients* **2013**, *5*, 3993–4021. [CrossRef]

21. Chen, W.; Clements, M.; Rahman, B.; Zhang, S.; Qiao, Y.; Armstrong, B.K. Relationship between cancer mortality/incidence and ambient ultraviolet B irradiance in China. *Cancer Causes Control.* **2010**, *21*, 1701–1709. [CrossRef] [PubMed]

22. Fioletov, V.E.; McArthur, L.J.; Mathews, T.W.; Marrett, L. Estimated ultraviolet exposure levels for a sufficient vitamin D status in North America. *J. Photochem. Photobiol. B* **2010**, *100*, 57–66. [CrossRef] [PubMed]

23. Hiller, T.W.R.; O'Sullivan, D.E.; Brenner, D.R.; Peters, C.E.; King, W.D. Solar Ultraviolet Radiation and Breast Cancer Risk: A Systematic Review and Meta-Analysis. *Environ. Health Perspect.* **2020**, *128*, 16002. [CrossRef]

24. Li, Z.; Wu, L.; Zhang, J.; Huang, X.; Thabane, L.; Li, G. Effect of Vitamin D Supplementation on Risk of Breast Cancer: A Systematic Review and Meta-Analysis of Randomized Controlled Trials. *Front. Nutr.* **2021**, *8*, 655727. [CrossRef] [PubMed]

25. Butler-Laporte, G.; Nakanishi, T.; Mooser, V.; Morrison, D.R.; Abdullah, T.; Adeleye, O.; Mamlouk, N.; Kimchi, N.; Afrasiabi, Z.; Rezk, N.; et al. Vitamin D and COVID-19 susceptibility and severity in the COVID-19 Host Genetics Initiative: A Mendelian randomization study. *PLoS Med.* **2021**, *18*, e1003605. [CrossRef] [PubMed]

26. Grant, W.B. 25-hydroxyvitamin D and breast cancer, colorectal cancer, and colorectal adenomas: Case-control versus nested case-control studies. *Anticancer Res.* **2015**, *35*, 1153–1160.

27. Smolders, J.; van den Ouweland, J.; Geven, C.; Pickkers, P.; Kox, M. Letter to the Editor: Vitamin D deficiency in COVID-19: Mixing up cause and consequence. *Metabolism* **2021**, *115*, 154434. [CrossRef]

28. Zhao, Y.; Chen, C.; Pan, W.; Gao, M.; He, W.; Mao, R.; Lin, T.; Huang, J. Comparative efficacy of vitamin D status in reducing the risk of bladder cancer: A systematic review and network meta-analysis. *Nutrition* **2016**, *32*, 515–523. [CrossRef]

29. Song, D.; Deng, Y.; Liu, K.; Zhou, L.; Li, N.; Zheng, Y.; Hao, Q.; Yang, S.; Wu, Y.; Zhai, Z.; et al. Vitamin D intake, blood vitamin D levels, and the risk of breast cancer: A dose-response meta-analysis of observational studies. *Aging* **2019**, *11*, 12708–12732. [CrossRef]

30. Hernandez-Alonso, P.; Boughanem, H.; Canudas, S.; Becerra-Tomas, N.; Fernandez de la Puente, M.; Babio, N.; Macias-Gonzalez, M.; Salas-Salvado, J. Circulating vitamin D levels and colorectal cancer risk: A meta-analysis and systematic review of case-control and prospective cohort studies. *Crit. Rev. Food Sci. Nutr.* **2021**, 1–17. [CrossRef]

31. Lappe, J.M.; Travers-Gustafson, D.; Davies, K.M.; Recker, R.R.; Heaney, R.P. Vitamin D and calcium supplementation reduces cancer risk: Results of a randomized trial. *Am. J. Clin. Nutr.* **2007**, *85*, 1586–1591. [CrossRef] [PubMed]

32. Lappe, J.; Watson, P.; Travers-Gustafson, D.; Recker, R.; Garland, C.; Gorham, E.; Baggerly, K.; McDonnell, S.L. Effect of Vitamin D and Calcium Supplementation on Cancer Incidence in Older Women: A Randomized Clinical Trial. *JAMA* **2017**, *317*, 1234–1243. [CrossRef] [PubMed]

33. McDonnell, S.L.; Baggerly, C.A.; French, C.B.; Baggerly, L.L.; Garland, C.F.; Gorham, E.D.; Hollis, B.W.; Trump, D.L.; Lappe, J.M. Breast cancer risk markedly lower with serum 25-hydroxyvitamin D concentrations >/=60 vs. <20 ng/mL (150 vs. 50 nmol/L): Pooled analysis of two randomized trials and a prospective cohort. *PLoS ONE* **2018**, *13*, e0199265. [CrossRef] [PubMed]

34. McCullough, M.L.; Zoltick, E.S.; Weinstein, S.J.; Fedirko, V.; Wang, M.; Cook, N.R.; Eliassen, A.H.; Zeleniuch-Jacquotte, A.; Agnoli, C.; Albanes, D.; et al. Circulating Vitamin D and Colorectal Cancer Risk: An International Pooling Project of 17 Cohorts. *J. Natl. Cancer Inst.* **2019**, *111*, 158–169. [CrossRef] [PubMed]

35. Grant, W.B.; Karras, S.N.; Bischoff-Ferrari, H.A.; Annweiler, C.; Boucher, B.J.; Juzeniene, A.; Garland, C.F.; Holick, M.F. Do studies reporting 'U'-shaped serum 25-hydroxyvitamin D-health outcome relationships reflect adverse effects? *Dermatoendocrinology* **2016**, *8*, e1187349. [CrossRef] [PubMed]

36. Grant, W.B. An estimate of premature cancer mortality in the U.S. due to inadequate doses of solar ultraviolet-B radiation. *Cancer* **2002**, *94*, 1867–1875. [CrossRef]

37. Manson, J.E.; Cook, N.R.; Lee, I.M.; Christen, W.; Bassuk, S.S.; Mora, S.; Gibson, H.; Gordon, D.; Copeland, T.; D'Agostino, D.; et al. Vitamin D Supplements and Prevention of Cancer and Cardiovascular Disease. *N. Engl. J. Med.* **2019**, *380*, 33–44. [CrossRef]

38. Haykal, T.; Samji, V.; Zayed, Y.; Gakhal, I.; Dhillon, H.; Kheiri, B.; Kerbage, J.; Veerapaneni, V.; Obeid, M.; Danish, R.; et al. The role of vitamin D supplementation for primary prevention of cancer: Meta-analysis of randomized controlled trials. *J. Community Hosp. Intern. Med. Perspect* **2019**, *9*, 480–488. [CrossRef]

39. Han, J.; Guo, X.; Yu, X.; Liu, S.; Cui, X.; Zhang, B.; Liang, H. 25-Hydroxyvitamin D and Total Cancer Incidence and Mortality: A Meta-Analysis of Prospective Cohort Studies. *Nutrients* **2019**, *11*, 2295. [CrossRef]

40. Krishnan, A.V.; Feldman, D. Mechanisms of the anti-cancer and anti-inflammatory actions of vitamin D. *Annu. Rev. Pharmacol. Toxicol.* **2011**, *51*, 311–336. [CrossRef]

41. Fleet, J.C.; DeSmet, M.; Johnson, R.; Li, Y. Vitamin D and cancer: A review of molecular mechanisms. *Biochem. J.* **2012**, *441*, 61–76. [CrossRef] [PubMed]

42. Moukayed, M.; Grant, W.B. The roles of UVB and vitamin D in reducing risk of cancer incidence and mortality: A review of the epidemiology, clinical trials, and mechanisms. *Rev. Endocr. Metab. Disord.* **2017**, *18*, 167–182. [CrossRef] [PubMed]

43. Liu, W.; Zhang, L.; Xu, H.J.; Li, Y.; Hu, C.M.; Yang, J.Y.; Sun, M.Y. The Anti-Inflammatory Effects of Vitamin D in Tumorigenesis. *Int. J. Mol. Sci.* **2018**, *19*, 2736. [CrossRef] [PubMed]

44. Ferrer-Mayorga, G.; Larriba, M.J.; Crespo, P.; Munoz, A. Mechanisms of action of vitamin D in colon cancer. *J. Steroid. Biochem. Mol. Biol.* **2019**, *185*, 1–6. [CrossRef]

45. Shirvani, A.; Kalajian, T.A.; Song, A.; Holick, M.F. Disassociation of Vitamin D's Calcemic Activity and Non-calcemic Genomic Activity and Individual Responsiveness: A Randomized Controlled Double-Blind Clinical Trial. *Sci. Rep.* **2019**, *9*, 17685. [CrossRef]

46. Shirvani, A.; Kalajian, T.A.; Song, A.; Allen, R.; Charoenngam, N.; Lewanczuk, R.; Holick, M.F. Variable Genomic and Metabolomic Responses to Varying Doses of Vitamin D Supplementation. *Anticancer. Res.* **2020**, *40*, 535–543. [CrossRef]

47. Heaney, R.P. Guidelines for optimizing design and analysis of clinical studies of nutrient effects. *Nutr. Rev.* **2014**, *72*, 48–54. [CrossRef]

48. Grant, W.B.; Boucher, B.J.; Bhattoa, H.P.; Lahore, H. Why vitamin D clinical trials should be based on 25-hydroxyvitamin D concentrations. *J. Steroid Biochem. Mol. Biol.* **2018**, *177*, 266–269. [CrossRef]

49. Carlberg, C.; Haq, A. The concept of the personal vitamin D response index. *J. Steroid Biochem. Mol. Biol.* **2018**, *175*, 12–17. [CrossRef]

50. Zhou, L.; Chen, B.; Sheng, L.; Turner, A. The effect of vitamin D supplementation on the risk of breast cancer: A trial sequential meta-analysis. *Breast Cancer Res. Treat.* **2020**, *182*, 1–8. [CrossRef]

51. Kerner, W.; Bruckel, J.; German Diabetes, A. Definition, classification and diagnosis of diabetes mellitus. *Exp. Clin. Endocrinol. Diabetes* **2014**, *122*, 384–386. [CrossRef] [PubMed]

52. Boucher, B.J. Can Vitamin D Supplementation Reduce Insulin Resistance and Hence the Risks of Type 2 Diabetes? *J. Diabetes Clin. Res.* **2020**, *2*, 1–8.

53. Sergeev, I.N.; Rhoten, W.B. 1,25-Dihydroxyvitamin D3 evokes oscillations of intracellular calcium in a pancreatic beta-cell line. *Endocrinology* **1995**, *136*, 2852–2861. [CrossRef] [PubMed]

54. Borissova, A.M.; Tankova, T.; Kirilov, G.; Dakovska, L.; Kovacheva, R. The effect of vitamin D3 on insulin secretion and peripheral insulin sensitivity in type 2 diabetic patients. *Int J. Clin. Pract* **2003**, *57*, 258–261. [PubMed]

55. Norman, A.W.; Frankel, J.B.; Heldt, A.M.; Grodsky, G.M. Vitamin D deficiency inhibits pancreatic secretion of insulin. *Science* **1980**, *209*, 823–825. [CrossRef]

56. Palomer, X.; Gonzalez-Clemente, J.M.; Blanco-Vaca, F.; Mauricio, D. Role of vitamin D in the pathogenesis of type 2 diabetes mellitus. *Diabetes Obes. Metab.* **2008**, *10*, 185–197. [CrossRef]

57. Cheng, S.; So, W.Y.; Zhang, D.; Cheng, Q.; Boucher, B.J.; Leung, P.S. Calcitriol Reduces Hepatic Triglyceride Accumulation and Glucose Output Through Ca2+/CaMKKβ/AMPK Activation Under Insulin-Resistant Conditions in Type 2 Diabetes Mellitus. *Curr Mol. Med.* **2016**, *16*, 747–758. [CrossRef]

58. Leung, P.S. The Potential Protective Action of Vitamin D in Hepatic Insulin Resistance and Pancreatic Islet Dysfunction in Type 2 Diabetes Mellitus. *Nutrients* **2016**, *8*, 147. [CrossRef]

59. Montenegro, K.R.; Cruzat, V.; Carlessi, R.; Newsholme, P. Mechanisms of vitamin D action in skeletal muscle. *Nutr Res. Rev.* **2019**, *32*, 192–204. [CrossRef]

60. Szymczak-Pajor, I.; Sliwinska, A. Analysis of Association between Vitamin D Deficiency and Insulin Resistance. *Nutrients* **2019**, *11*, 794. [CrossRef]

61. De Pergola, G.; Martino, T.; Zupo, R.; Caccavo, D.; Pecorella, C.; Paradiso, S.; Silvestris, F.; Triggiani, V. 25 Hydroxyvitamin D Levels are Negatively and Independently Associated with Fat Mass in a Cohort of Healthy Overweight and Obese Subjects. *Endocr. Metab. Immune Disord. Drug Targets* **2019**, *19*, 838–844. [CrossRef] [PubMed]

62. Shoelson, S.E.; Herrero, L.; Naaz, A. Obesity, inflammation, and insulin resistance. *Gastroenterology* **2007**, *132*, 2169–2180. [CrossRef] [PubMed]

63. Afzal, S.; Bojesen, S.E.; Nordestgaard, B.G. Low 25-hydroxyvitamin D and risk of type 2 diabetes: A prospective cohort study and metaanalysis. *Clin. Chem.* **2013**, *59*, 381–391. [CrossRef] [PubMed]

64. Forouhi, N.G.; Luan, J.; Cooper, A.; Boucher, B.J.; Wareham, N.J. Baseline serum 25-hydroxy vitamin d is predictive of future glycemic status and insulin resistance: The Medical Research Council Ely Prospective Study 1990–2000. *Diabetes* **2008**, *57*, 2619–2625. [CrossRef] [PubMed]

65. Forouhi, N.G.; Ye, Z.; Rickard, A.P.; Khaw, K.T.; Luben, R.; Langenberg, C.; Wareham, N.J. Circulating 25-hydroxyvitamin D concentration and the risk of type 2 diabetes: Results from the European Prospective Investigation into Cancer (EPIC)-Norfolk cohort and updated meta-analysis of prospective studies. *Diabetologia* **2012**, *55*, 2173–2182. [CrossRef]

66. Wang, C.; Wang, S.; Li, D.; Chen, P.; Han, S.; Zhao, G.; Chen, Y.; Zhao, J.; Xiong, J.; Qiu, J.; et al. Human Cathelicidin Inhibits SARS-CoV-2 Infection: Killing Two Birds with One Stone. *ACS Infect. Dis.* **2021**, *7*, 1545–1554. [CrossRef]

67. Mirhosseini, N.; Rainsbury, J.; Kimball, S.M. Vitamin D Supplementation, Serum 25(OH)D Concentrations and Cardiovascular Disease Risk Factors: A Systematic Review and Meta-Analysis. *Front. Cardiovasc. Med.* **2018**, *5*, 87. [CrossRef]

68. Karonova, T.; Stepanova, A.; Bystrova, A.; Jude, E.B. High-Dose Vitamin D Supplementation Improves Microcirculation and Reduces Inflammation in Diabetic Neuropathy Patients. *Nutrients* **2020**, *12*, 2518. [CrossRef]

69. Pittas, A.G.; Dawson-Hughes, B.; Sheehan, P.; Ware, J.H.; Knowler, W.C.; Aroda, V.R.; Brodsky, I.; Ceglia, L.; Chadha, C.; Chatterjee, R.; et al. Vitamin D Supplementation and Prevention of Type 2 Diabetes. *N. Engl. J. Med.* **2019**, *381*, 520–530. [CrossRef]

70. Dawson-Hughes, B.; Staten, M.A.; Knowler, W.C.; Nelson, J.; Vickery, E.M.; LeBlanc, E.S.; Neff, L.M.; Park, J.; Pittas, A.G.; Group, D.d.R. Intratrial Exposure to Vitamin D and New-Onset Diabetes Among Adults with Prediabetes: A Secondary Analysis From the Vitamin D and Type 2 Diabetes (D2d) Study. *Diabetes Care* **2020**, *43*, 2916–2922. [CrossRef]

71. Grant, W.B.; Lahore, H.; McDonnell, S.L.; Baggerly, C.A.; Franch, C.B.; Aliano, J.L.; Bhattoa, H.P. Evidence That Vitamin D Supplementation Could Reduce Risk of Influenza and COVID-19 Infections and Deaths. *Nutrients* **2020**, *12*, 988. [CrossRef] [PubMed]

72. Evans, R.M.; Lippman, S.M. Shining Light on the COVID-19 Pandemic: A Vitamin D Receptor Checkpoint in Defense of Unregulated Wound Healing. *Cell Metab.* **2020**, *32*, 704–709. [CrossRef] [PubMed]

73. Grant, W.B.; Giovannucci, E. The possible roles of solar ultraviolet-B radiation and vitamin D in reducing case-fatality rates from the 1918–1919 influenza pandemic in the United States. *Dermatoendocrinology* **2009**, *1*, 215–219. [CrossRef] [PubMed]

74. Martineau, A.R.; Jolliffe, D.A.; Hooper, R.L.; Greenberg, L.; Aloia, J.F.; Bergman, P.; Dubnov-Raz, G.; Esposito, S.; Ganmaa, D.; Ginde, A.A.; et al. Vitamin D supplementation to prevent acute respiratory tract infections: Systematic review and meta-analysis of individual participant data. *BMJ* **2017**, *356*, i6583. [CrossRef]

75. Kaufman, H.W.; Niles, J.K.; Kroll, M.H.; Bi, C.; Holick, M.F. SARS-CoV-2 positivity rates associated with circulating 25-hydroxyvitamin D levels. *PLoS ONE* **2020**, *15*, e0239252. [CrossRef] [PubMed]

76. Lo, C.H.; Nguyen, L.H.; Drew, D.A.; Warner, E.T.; Joshi, A.D.; Graham, M.S.; Anyane-Yeboa, A.; Shebl, F.M.; Astley, C.M.; Figueiredo, J.C.; et al. Race, ethnicity, community-level socioeconomic factors, and risk of COVID-19 in the United States and the United Kingdom. *EClinicalMedicine* **2021**, *38*, 101029. [CrossRef]

77. Merzon, E.; Tworowski, D.; Gorohovski, A.; Vinker, S.; Golan Cohen, A.; Green, I.; Frenkel-Morgenstern, M. Low plasma 25(OH) vitamin D level is associated with increased risk of COVID-19 infection: An Israeli population-based study. *FEBS J.* **2020**, *287*, 3693–3702. [CrossRef]

78. Meltzer, D.O.; Best, T.J.; Zhang, H.; Vokes, T.; Arora, V.; Solway, J. Association of Vitamin D Status and Other Clinical Characteristics With COVID-19 Test Results. *JAMA Netw. Open* **2020**, *3*, e2019722. [CrossRef]

79. Meltzer, D.O.; Best, T.J.; Zhang, H.; Vokes, T.; Arora, V.M.; Solway, J. Association of Vitamin D Levels, Race/Ethnicity, and Clinical Characteristics With COVID-19 Test Results. *JAMA Netw. Open* **2021**, *4*, e214117. [CrossRef]

80. Radujkovic, A.; Hippchen, T.; Tiwari-Heckler, S.; Dreher, S.; Boxberger, M.; Merle, U. Vitamin D Deficiency and Outcome of COVID-19 Patients. *Nutrients* **2020**, *12*, 2757. [CrossRef]

81. Dissanayake, H.A. Prognostic and therapeutic role of vitamin D in COVID-19: Systematic review and meta-analysis. *J. Clin. Endocrinol. Metab.* **2021**, dgab892. [CrossRef] [PubMed]

82. Seal, K.H.; Bertenthal, D.; Carey, E.; Grunfeld, C.; Bikle, D.D.; Lu, C.M. Association of Vitamin D Status and COVID-19-Related Hospitalization and Mortality. *J. Gen. Intern. Med.* **2022**. [CrossRef] [PubMed]

83. Gonen, M.S.; Alaylioglu, M.; Durcan, E.; Ozdemir, Y.; Sahin, S.; Konukoglu, D.; Nohut, O.K.; Urkmez, S.; Kucukece, B.; Balkan, I.I.; et al. Rapid and Effective Vitamin D Supplementation May Present Better Clinical Outcomes in COVID-19 (SARS-CoV-2) Patients by Altering Serum INOS1, IL1B, IFNg, Cathelicidin-LL37, and ICAM1. *Nutrients* **2021**, *13*, 4047. [CrossRef] [PubMed]

84. Entrenas-Castillo, M.; Entrenas Costa, L.M.; Vaquero Barrios, J.M.; Alcala Diaz, J.F.; Lopez Miranda, J.; Bouillon, R.; Quesada Gomez, J.M. Effect of calcifediol treatment and best available therapy versus best available therapy on intensive care unit admission and mortality among patients hospitalized for COVID-19: A pilot randomized clinical study. *J. Steroid Biochem. Mol. Biol.* **2020**, *203*, 105751. [CrossRef] [PubMed]

85. Alcala-Diaz, J.F.; Limia-Perez, L.; Gomez-Huelgas, R.; Martin-Escalante, M.D.; Cortez-Rodriguez, A.L.; Lopez-Carmona, M.D. Calcifediol Treatment and Hospital Mortality Due to COVID-19: A Cohort Study. *Nutrients* **2021**, *13*, 1760. [CrossRef]

86. Vasheghani, M.; Jannati, N.; Baghaei, P.; Rezaei, M.; Aliyari, R.; Marjani, M. The relationship between serum 25-hydroxyvitamin D levels and the severity of COVID-19 disease and its mortality. *Sci. Rep.* **2021**, *11*, 17594. [CrossRef]

87. Hariyanto, T.I.; Intan, D.; Hananto, J.E.; Harapan, H.; Kaurniawan, A. Vitamin D supplementation and Covid-19 outcomes: A systematic review, meta-analysis and meta-regression. *Rev. Med. Virol.* **2020**, *2021*, e2269. [CrossRef]

88. Oristrell, J.; Oliva, J.C.; Casado, E.; Subirana, I.; Dominguez, D.; Toloba, A.; Balado, A.; Grau, M. Vitamin D supplementation and COVID-19 risk: A population-based, cohort study. *J. Endocrinol. Invest.* **2022**, *45*, 167–179. [CrossRef]

89. Dursun, E.; Gezen-Ak, D. Vitamin D basis of Alzheimer's disease: From genetics to biomarkers. *Hormones* **2019**, *18*, 7–15. [CrossRef]

90. Gezen-Ak, D.; Atasoy, I.L.; Candas, E.; Alaylioglu, M.; Yilmazer, S.; Dursun, E. Vitamin D Receptor Regulates Amyloid Beta 1–42 Production with Protein Disulfide Isomerase A3. *ACS Chem. Neurosci.* **2017**, *8*, 2335–2346. [CrossRef]

91. Gezen-Ak, D.; Dursun, E. Molecular basis of vitamin D action in neurodegeneration: The story of a team perspective. *Hormones* **2019**, *18*, 17–21. [CrossRef] [PubMed]

92. Feart, C.; Helmer, C.; Merle, B.; Herrmann, F.R.; Annweiler, C.; Dartigues, J.F.; Delcourt, C.; Samieri, C. Associations of lower vitamin D concentrations with cognitive decline and long-term risk of dementia and Alzheimer's disease in older adults. *Alzheimers Dement.* **2017**, *13*, 1207–1216. [CrossRef] [PubMed]

93. Jayedi, A.; Rashidy-Pour, A.; Shab-Bidar, S. Vitamin D status and risk of dementia and Alzheimer's disease: A meta-analysis of dose-response (dagger). *Nutr. Neurosci.* **2019**, *22*, 750–759. [CrossRef] [PubMed]

94. Chai, B.; Gao, F.; Wu, R.; Dong, T.; Gu, C.; Lin, Q.; Zhang, Y. Vitamin D deficiency as a risk factor for dementia and Alzheimer's disease: An updated meta-analysis. *BMC Neurol.* **2019**, *19*, 284. [CrossRef]

95. Wang, L.; Qiao, Y.; Zhang, H.; Zhang, Y.; Hua, J.; Jin, S.; Liu, G. Circulating Vitamin D Levels and Alzheimer's Disease: A Mendelian Randomization Study in the IGAP and UK Biobank. *J. Alzheimers Dis.* **2020**, *73*, 609–618. [CrossRef]

96. Zittermann, A.; Iodice, S.; Pilz, S.; Grant, W.B.; Bagnardi, V.; Gandini, S. Vitamin D deficiency and mortality risk in the general population: A meta-analysis of prospective cohort studies. *Am. J. Clin. Nutr.* **2012**, *95*, 91–100. [CrossRef] [PubMed]

97. Garland, C.F.; Kim, J.J.; Mohr, S.B.; Gorham, E.D.; Grant, W.B.; Giovannucci, E.L.; Baggerly, L.; Hofflich, H.; Ramsdell, J.W.; Zeng, K.; et al. Meta-analysis of all-cause mortality according to serum 25-hydroxyvitamin D. *Am. J. Public Health* **2014**, *104*, e43–e50. [CrossRef] [PubMed]

98. Gaksch, M.; Jorde, R.; Grimnes, G.; Joakimsen, R.; Schirmer, H.; Wilsgaard, T.; Mathiesen, E.B.; Njolstad, I.; Lochen, M.L.; Marz, W.; et al. Vitamin D and mortality: Individual participant data meta-analysis of standardized 25-hydroxyvitamin D in 26916 individuals from a European consortium. *PLoS ONE* **2017**, *12*, e0170791. [CrossRef] [PubMed]

99. Zhang, Y.; Fang, F.; Tang, J.; Jia, L.; Feng, Y.; Xu, P.; Faramand, A. Association between vitamin D supplementation and mortality: Systematic review and meta-analysis. *BMJ* **2019**, *366*, l4673. [CrossRef]

100. Brouwer-Brolsma, E.M.; Bischoff-Ferrari, H.A.; Bouillon, R.; Feskens, E.J.; Gallagher, C.J.; Hypponen, E.; Llewellyn, D.J.; Stoecklin, E.; Dierkes, J.; Kies, A.K.; et al. Vitamin D: Do we get enough? A discussion between vitamin D experts in order to make a step towards the harmonisation of dietary reference intakes for vitamin D across Europe. *Osteoporos Int.* **2013**, *24*, 1567–1577. [CrossRef]

101. Wise, S.A.; Camara, J.E.; Sempos, C.T.; Lukas, P.; Le Goff, C.; Peeters, S.; Burdette, C.Q.; Nalin, F.; Hahm, G.; Durazo-Arvizu, R.A.; et al. Vitamin D Standardization Program (VDSP) intralaboratory study for the assessment of 25-hydroxyvitamin D assay variability and bias. *J. Steroid Biochem. Mol. Biol.* **2021**, *212*, 105917. [CrossRef] [PubMed]

102. Ross, A.C.; Manson, J.E.; Abrams, S.A.; Aloia, J.F.; Brannon, P.M.; Clinton, S.K.; Durazo-Arvizu, R.A.; Gallagher, J.C.; Gallo, R.L.; Jones, G.; et al. The 2011 report on dietary reference intakes for calcium and vitamin D from the Institute of Medicine: What clinicians need to know. *J. Clin. Endocrinol. Metab.* **2011**, *96*, 53–58. [CrossRef] [PubMed]

103. Holick, M.F.; Binkley, N.C.; Bischoff-Ferrari, H.A.; Gordon, C.M.; Hanley, D.A.; Heaney, R.P.; Murad, M.H.; Weaver, C.M.; Endocrine, S. Evaluation, treatment, and prevention of vitamin D deficiency: An Endocrine Society clinical practice guideline. *J. Clin. Endocrinol. Metab.* **2011**, *96*, 1911–1930. [CrossRef]

104. Priemel, M.; von Domarus, C.; Klatte, T.O.; Kessler, S.; Schlie, J.; Meier, S.; Proksch, N.; Pastor, F.; Netter, C.; Streichert, T.; et al. Bone mineralization defects and vitamin D deficiency: Histomorphometric analysis of iliac crest bone biopsies and circulating 25-hydroxyvitamin D in 675 patients. *J. Bone Miner. Res.* **2010**, *25*, 305–312. [CrossRef]

105. Pludowski, P.; Holick, M.F.; Grant, W.B.; Konstantynowicz, J.; Mascarenhas, M.R.; Haq, A.; Povoroznyuk, V.; Balatska, N.; Barbosa, A.P.; Karonova, T.; et al. Vitamin D supplementation guidelines. *Steroid Biochem. Mol. Biol.* **2018**, *175*, 125–135. [CrossRef] [PubMed]

106. McDonnell, S.L.; Baggerly, K.A.; Baggerly, C.A.; Aliano, J.L.; French, C.B.; Baggerly, L.L.; Ebeling, M.D.; Rittenberg, C.S.; Goodier, C.G.; Mateus Nino, J.F.; et al. Maternal 25(OH)D concentrations >/=40 ng/mL associated with 60% lower preterm birth risk among general obstetrical patients at an urban medical center. *PLoS ONE* **2017**, *12*, e0180483. [CrossRef] [PubMed]

107. Holick, M.F. Environmental factors that influence the cutaneous production of vitamin D. *Am. J. Clin. Nutr.* **1995**, *61*, 638S–645S. [CrossRef]

108. Kimball, S.M.; Mirhosseini, N.; Holick, M.F. Evaluation of vitamin D3 intakes up to 15,000 international units/day and serum 25-hydroxyvitamin D concentrations up to 300 nmol/L on calcium metabolism in a community setting. *Dermatoendocrinology* **2017**, *9*, e1300213. [CrossRef]

109. McCullough, P.J.; Lehrer, D.S.; Amend, J. Daily oral dosing of vitamin D3 using 5000 TO 50,000 international units a day in long-term hospitalized patients: Insights from a seven year experience. *J. Steroid Biochem. Mol. Biol.* **2019**, *189*, 228–239. [CrossRef]

110. Gianella, F.; Hsia, C.C.; Sakhaee, K. The role of vitamin D in sarcoidosis. *Fac. Rev.* **2020**, *9*, 14. [CrossRef]

111. Jaaskelainen, T.; Itkonen, S.T.; Lundqvist, A.; Erkkola, M.; Koskela, T.; Lakkala, K.; Dowling, K.G.; Hull, G.L.; Kroger, H.; Karppinen, J.; et al. The positive impact of general vitamin D food fortification policy on vitamin D status in a representative adult Finnish population: Evidence from an 11-y follow-up based on standardized 25-hydroxyvitamin D data. *Am. J. Clin. Nutr.* **2017**, *105*, 1512–1520. [CrossRef] [PubMed]

112. Pilz, S.; Marz, W.; Cashman, K.D.; Kiely, M.E.; Whiting, S.J.; Holick, M.F.; Grant, W.B.; Pludowski, P.; Hiligsmann, M.; Trummer, C.; et al. Rationale and Plan for Vitamin D Food Fortification: A Review and Guidance Paper. *Front. Endocrinol.* **2018**, *9*, 373. [CrossRef] [PubMed]

113. Madden, J.M.; Murphy, L.; Zgaga, L.; Bennett, K. De novo vitamin D supplement use post-diagnosis is associated with breast cancer survival. *Breast Cancer Res. Treat.* **2018**, *172*, 179–190. [CrossRef] [PubMed]

114. Lucas, R.M.; Rodney Harris, R.M. On the Nature of Evidence and 'Proving' Causality: Smoking and Lung Cancer vs. Sun Exposure, Vitamin D and Multiple Sclerosis. *Int. J. Environ. Res. Public Health* **2018**, *15*, 1726. [CrossRef]

115. Numerous. *Vitamin D. Biochemistry, Physiology and Diagnostics*, 4th ed.; Academic Press: Cambridge, MA, USA, 2018; p. 1141.

116. Cannell, J.J.; Vieth, R.; Umhau, J.C.; Holick, M.F.; Grant, W.B.; Madronich, S.; Garland, C.F.; Giovannucci, E. Epidemic influenza and vitamin D. *Epidemiol. Infect.* **2006**, *134*, 1129–1140. [CrossRef]

117. Andrukhova, O.; Slavic, S.; Zeitz, U.; Riesen, S.C.; Heppelmann, M.S.; Ambrisko, T.D.; Markovic, M.; Kuebler, W.M.; Erben, R.G. Vitamin D is a regulator of endothelial nitric oxide synthase and arterial stiffness in mice. *Mol. Endocrinol.* **2014**, *28*, 53–64. [CrossRef]

118. Weller, R.B.; Wang, Y.; He, J.; Maddux, F.W.; Usvyat, L.; Zhang, H.; Feelisch, M.; Kotanko, P. Does Incident Solar Ultraviolet Radiation Lower Blood Pressure? *J. Am. Heart Assoc.* **2020**, *9*, e013837. [CrossRef]

119. Cherrie, M.; Clemens, T.; Colandrea, C.; Feng, Z.; Webb, D.J.; Weller, R.B.; Dibben, C. Ultraviolet A radiation and COVID-19 deaths in the USA with replication studies in England and Italy. *Br. J. Dermatol.* **2021**, *185*, 363–370. [CrossRef]

120. Grant, W.B. Vitamin D Levels Affect Breast Cancer Survival Rates. *Ann. Surg. Oncol.* **2017**, *24*, 570–571. [CrossRef]

121. de Sousa Almeida-Filho, B.; De Luca Vespoli, H.; Pessoa, E.C.; Machado, M.; Nahas-Neto, J.; Nahas, E.A.P. Vitamin D deficiency is associated with poor breast cancer prognostic features in postmenopausal women. *J. Steroid Biochem. Mol. Biol.* **2017**, *174*, 284–289. [CrossRef]

122. Lewallen, S.; Courtright, P. Epidemiology in practice: Case-control studies. *Community Eye Health* **1998**, *11*, 57–58. [PubMed]

123. Oh, E.Y.; Ansell, C.; Nawaz, H.; Yang, C.H.; Wood, P.A.; Hrushesky, W.J. Global breast cancer seasonality. *Breast Cancer Res. Treat.* **2010**, *123*, 233–243. [CrossRef] [PubMed]

124. Scragg, R. Limitations of vitamin D supplementation trials: Why observational studies will continue to help determine the role of vitamin D in health. *J. Steroid Biochem. Mol. Biol.* **2018**, *177*, 6–9. [CrossRef] [PubMed]

125. Yetley, E.A.; Pfeiffer, C.M.; Schleicher, R.L.; Phinney, K.W.; Lacher, D.A.; Christakos, S.; Eckfeldt, J.H.; Fleet, J.C.; Howard, G.; Hoofnagle, A.N.; et al. NHANES monitoring of serum 25-hydroxyvitamin D: A roundtable summary. *J. Nutr.* **2010**, *140*, 2030S–2045S. [PubMed]

126. Haq, A.; Svobodova, J.; Imran, S.; Stanford, C.; Razzaque, M.S. Vitamin D deficiency: A single centre analysis of patients from 136 countries. *J. Steroid Biochem. Mol. Biol.* **2016**, *164*, 209–213. [CrossRef] [PubMed]

127. Crowe, F.L.; Steur, M.; Allen, N.E.; Appleby, P.N.; Travis, R.C.; Key, T.J. Plasma concentrations of 25-hydroxyvitamin D in meat eaters, fish eaters, vegetarians and vegans: Results from the EPIC-Oxford study. *Public Health Nutr.* **2011**, *14*, 340–346. [CrossRef]

128. Autier, P.; Mullie, P.; Macacu, A.; Dragomir, M.; Boniol, M.; Coppens, K.; Pizot, C.; Boniol, M. Effect of vitamin D supplementation on non-skeletal disorders: A systematic review of meta-analyses and randomised trials. *Lancet Diabetes Endocrinol.* **2017**, *5*, 986–1004. [CrossRef]

129. Rejnmark, L.; Bislev, L.S.; Cashman, K.D.; Eiriksdottir, G.; Gaksch, M.; Grubler, M.; Grimnes, G.; Gudnason, V.; Lips, P.; Pilz, S.; et al. Non-skeletal health effects of vitamin D supplementation: A systematic review on findings from meta-analyses summarizing trial data. *PLoS ONE* **2017**, *12*, e0180512. [CrossRef]

130. Xia, Z.; Man, Q.; Li, L.; Song, P.; Jia, S.; Song, S.; Meng, L.; Zhang, J. Vitamin D receptor gene polymorphisms modify the association of serum 25-hydroxyvitamin D levels with handgrip strength in the elderly in Northern China. *Nutrition* **2019**, *57*, 202–207. [CrossRef]

131. Mokry, L.E.; Ross, S.; Ahmad, O.S.; Forgetta, V.; Smith, G.D.; Goltzman, D.; Leong, A.; Greenwood, C.M.; Thanassoulis, G.; Richards, J.B. Vitamin D and Risk of Multiple Sclerosis: A Mendelian Randomization Study. *PLoS Med.* **2015**, *12*, e1001866. [CrossRef]

132. Ong, J.S.; Cuellar-Partida, G.; Lu, Y.; Australian Ovarian Cancer Study; Fasching, P.A.; Hein, A.; Burghaus, S.; Beckmann, M.W.; Lambrechts, D.; Van Nieuwenhuysen, E.; et al. Association of vitamin D levels and risk of ovarian cancer: A Mendelian randomization study. *Int. J. Epidemiol.* **2016**, *45*, 1619–1630. [CrossRef] [PubMed]

133. Ong, J.S.; Dixon-Suen, S.C.; Han, X.; An, J.; Esophageal Cancer, C.; Me Research, T.; Liyanage, U.; Dusingize, J.C.; Schumacher, J.; Gockel, I.; et al. A comprehensive re-assessment of the association between vitamin D and cancer susceptibility using Mendelian randomization. *Nat. Commun.* **2021**, *12*, 246. [CrossRef] [PubMed]

134. Boucher, B.J. No evidence that vitamin D is able to prevent or affect the severity of COVID-19 in individuals with European ancestry: A Mendelian randomisation study of open data, by Amin et al. *BMJ Nutr. Prev. Health* **2021**, *4*, 352–353. [CrossRef] [PubMed]
135. Zhou, A.; Selvanayagam, J.B.; Hyppönen, E. Non-linear Mendelian randomization analyses support a role for vitamin D deficiency in cardiovascular disease risk. *Eur. Heart J.* **2021**. Epub ahead of print. [CrossRef]

nutrients

MDPI

Article

Vitamin D Deficiency and Its Associated Factors among Female Migrants in the United Arab Emirates

Fatme Al Anouti [1], Luai A. Ahmed [2], Azmat Riaz [3], William B. Grant [4], Nadir Shah [5], Raghib Ali [6], Juma Alkaabi [7] and Syed M. Shah [2,8,*]

[1] Department of Health Sciences, College of Natural and Health Sciences, Zayed University, Abu Dhabi 144534, United Arab Emirates; fatme.alanouti@zu.ac.ae
[2] Institute of Public Health, College of Medicine and Health Sciences, United Arab Emirates University, Al Ain 17666, United Arab Emirates; luai.ahmed@uaeu.ac.ae
[3] Department of Obstetrics and Gynecology, Ajman University, Ajman 20550, United Arab Emirates; azmatriazkhan@gmail.com
[4] Sunlight, Nutrition and Health Research Center, P.O. Box 641603, San Francisco, CA 94164-1603, USA; wbgrant@infionline.net
[5] Planning and Development Division, Government of Gilgit Baltistan, Gilgit 15100, Pakistan; nadir.shah.sun@gmail.com
[6] Public Health Research Center, New York University Abu Dhabi, Abu Dhabi 129188, United Arab Emirates; ra107@nyu.edu
[7] Department of Internal Medicine, College of Medicine and Health Sciences, United Arab Emirates University, Al Ain 17666, United Arab Emirates; j.kaabi@uaeu.ac.ae
[8] Department of Family Medicine, Aga Khan University, Karachi 3500, Pakistan
[*] Correspondence: syeds@uaeu.ac.ae; Tel.: +92-971-3-713-7458

Citation: Anouti, F.A.; Ahmed, L.A.; Riaz, A.; Grant, W.B.; Shah, N.; Ali, R.; Alkaabi, J.; Shah, S.M. Vitamin D Deficiency and Its Associated Factors among Female Migrants in the United Arab Emirates. *Nutrients* **2022**, *14*, 1074. https://doi.org/10.3390/nu14051074

Academic Editor: Connie Weaver

Received: 26 January 2022
Accepted: 22 February 2022
Published: 3 March 2022

Publisher's Note: MDPI stays neutral with regard to jurisdictional claims in published maps and institutional affiliations.

Abstract: Vitamin D is important for bone health, and vitamin D deficiency could be linked to noncommunicable diseases, including cardiovascular disease. The purpose of this study was to determine the prevalence of vitamin D deficiency and its associated risk factors among female migrants from Philippines, Arab, and South Asian countries residing in the United Arab Emirates (UAE). We used a cross-sectional study to recruit a random sample ($N = 550$) of female migrants aged 18 years and over in the city of Al Ain, UAE. Vitamin D deficiency was defined as serum 25-hydroxyvitamin D concentrations ≤ 20 ng/mL (50 nmol/L). We used multivariable logistic regression analysis to identify risk factors associated with vitamin D deficiency. The mean age of participants was 35 years (SD $\pm$ 10). The overall prevalence rate of vitamin D deficiency was 67% (95% CI 60–73%), with the highest rate seen in Arabs (87%), followed by South Asians (83%) and the lowest in Filipinas (15%). Multivariate analyses showed that low physical activity (adjusted odds ratio (aOR) = 4.59; 95% CI 1.98, 10.63), having more than 5 years duration of residence in the UAE (aOR = 4.65; 95% CI: 1.31, 16.53) and being obese (aOR = 3.56; 95% CI 1.04, 12.20) were independently associated with vitamin D deficiency, after controlling for age and nationality. In summary, vitamin D deficiency was highly prevalent among female migrants, especially Arabs and South Asians. It is crucial that health professionals in the UAE become aware of this situation among this vulnerable subpopulation and provide intervention strategies aiming to rectify vitamin D deficiency by focusing more on sun exposure, physical activity, and supplementation.

Keywords: vitamin D status; prevalence; female migrants; United Arab Emirates

1. Introduction

Vitamin D deficiency remains a worldwide public health problem, affecting large proportions of the population in developed and the developing countries [1]. Vitamin D deficiency remains common in children and adults and the musculoskeletal consequences of inadequate vitamin D are well established. In utero and during childhood, vitamin D deficiency can lead to growth retardation and skeletal deformities including childhood rickets [2]. Vitamin D supplementation during pregnancy is associated with improved infant growth and reduction of fetal or neonatal mortality [3]. In adults, vitamin D deficiency can cause osteomalacia, muscle weakness and increased risk of fracture [4]. Although the strongest evidence for the effect of vitamin D deficiency is related to skeletal disorders, low concentrations of vitamin D are associated with several non-skeletal disorders including cardiovascular diseases, several types of cancer, neurodegenerative diseases, disorders of glucose metabolism, and a possible role in the recently emerging pandemic of COVID-19 [5–8].

The extreme consequences of vitamin D deficiency, such as rickets in children and osteomalacia in adults, have been almost eliminated in some developed countries through adequate diet, food fortification, and the encouragement of moderate sunlight exposure [9]. Paradoxically, populations in the Middle East and North Africa (MENA) region have some of the lowest serum 25-hydroxy-vitamin D [25(OH)D] concentrations worldwide, despite the abundance of sunshine throughout the year [10,11].

Vitamin D deficiency among women in Arab countries has been attributed to inadequate exposure of skin to sunlight due to a very conservative style of dress that covers most of the body when they are outdoors [11,12]. In a study of Arab-American women, there was a significantly higher prevalence of vitamin D deficiency in women practicing a conservative style of dress, compared to their counterparts practicing a less conservative dressing [13,14]. We previously examined the prevalence of vitamin D deficiency among adolescents aged 15 to 18 years in the city of Al Ain, UAE. A higher proportion (32.0%) of female adolescents had vitamin D deficiency as compared to their male counterparts (8.0%) [15].

Given the remarkably high percentage of working migrants, known as 'guest workers' or 'expatriates', the UAE is one of the most culturally diverse countries. Expatriate foreign workers account for almost 80% of the UAE's total population [16]. A significant proportion of female migrant workers, including women from various Arab countries, South Asia, and Philippines are typically employed for indoor work activities, such as domestic or office work, sales, and beauty salons.

Migrant workers are among the most vulnerable populations that could be afflicted with vitamin D deficiency [17]. This study aimed to estimate the prevalence of vitamin D deficiency and examine the correlates of low levels of serum 25(OH)D concentrations among Arab, South Asian, and Filipina migrants residing in the UAE. To our knowledge this is the first research investigation that targets this under-studied subpopulation to assess the burden of vitamin D deficiency and pave for future studies that could aim to address such an important public health issue.

2. Methods

2.1. Study Design and Ethics

The study employed a cross-sectional design. We obtained a College of Medicine and Health Sciences, UAE individual faculty grant for the project, entitled "Chronic Diseases Prevention in Immigrants: putting CVD risk factors on surveillance screen". Ethical approval was obtained from the Al Ain Medical District Human Research (AAMDHREC 10/21) and study participants provided written informed consent.

2.2. Selection of Study Participants

The target study population consisted of female Arab, South Asian, and Filipina migrant workers aged 18 years and older. We used the formula for binomial distribution $(n= {}^{z}z\alpha^{2}\ p\ (1-p)/d^{2})$ to estimate the sample size, where (n) is the sample size, (zα) is the normal deviate (1.96) at 5% level of significance, (p) is the prevalence, and (d) is the precision. Assuming a precision or tolerable variation of $\pm$0.06 around an estimated prevalence of 70% for vitamin D deficiency in women [9], a sample of 200 participants would be needed.

All expatriate workers seeking employment or renewing their visa in the UAE are required by law to undergo health and communicable disease screening. The sampling frame in this study was a list of all expatriate workers from the Arab region, South Asia, and Philippines who were enrolled for medical examination at the only visa screening center in the city of Al Ain, Abu Dhabi emirate, during the process of either obtaining a new visa or renewing it over a period of six months [18].

2.3. Inclusion Criteria

Female migrant workers who were of an Arab, South Asian, or Philippines' nationality, aged ≥18 years, were able to read and speak Arabic, Urdu, Hindi, Bengali, or Filipino, and to provide a written informed consent were eligible to participate in the study. Due to the low literacy rate among the South Asian expatriate population in the UAE [16], the study questionnaire was interviewer-administered, and all interviews were conducted in Urdu, Hindi, Bengali, or Filipino, and led by native Urdu, Hindi, Bengali or Filipino speaking research assistants who had received appropriate training. Eligible participants registered for the visa screening were invited to participate in the study which spanned over six months to complete recruitment of the required sample.

2.4. Measures

We used an adapted version of the World Health Organization (WHO) global standard "STEPS" survey questionnaire entitled "Chronic Diseases Prevention in Immigrants: putting CVD risk factors on surveillance screen" for population-based assessment of the prevalence of noncommunicable diseases (NCDs) risk factors including vitamin D deficiency [19].

A 5 mL venous fasting blood sample was obtained from the study participants by qualified nurses using standardized tubes. Blood samples were immediately transferred to Tawam Hospital (Al Ain) laboratory where they underwent standardized (quality controlled) analyses. Serum 25(OH)D concentrations were measured by radioimmunoassay (DiaSorin, Stillwater, Minnesota, MN, USA). The intra-assay and inter-assay for coefficients of variation were 8.3% and 3.2% respectively. We used the reference value for serum 25(OH)D concentrations <20 ng/mL (50 nmol/L) to define vitamin D deficiency [4].

We collected information on demographics, lifestyle factors, family and personal disease history, home country residence setting (rural, urban), occupation, and monthly salary in UAE dirham or AED (USD1.00 ~ AED3.67).

Studies in migrants from Western developed countries indicated a decline in health with the increased duration of stay, which could be attributed to the adoption of local behaviors and norms and diet, also known as acculturation [20]. We used the duration of residence as a marker to evaluate the effect of acculturation on vitamin D status among the participants [21].

Participants' weight and height measurements were performed using standard weight and height scales (SECA, Hamburg, Germany). Body mass index (BMI) was calculated as weight in kilograms divided by the square of the height in meters and BMI categorization was based on WHO recommendations: being overweight (BMI 25.0 to 29.9 kg/m^2), and obesity (≥30.0 kg/m^2) [22]. Resting brachial blood pressure (BP) was measured using a calibrated automated BP measurement device (Omron HEM-705cp) in sitting position using the right upper arm and an appropriately sized cuff after a period of five minutes' rest. The average of two measures taken was used for analysis. Hypertension (HTN) was defined as being on anti-hypertensive medications or having a systolic blood pressure ≥140 mmHg or a diastolic blood pressure ≥90 mmHg [23]. Study participants were classified as current smokers if they answered yes to the question, "have you ever smoked cigarettes, cigars or shisha?". Information on physical activity was obtained using the International Physical Activity Questionnaire (IPAQ-short version) [24]. We measured the frequency (days per week) and duration (minutes per day) of moderate- and vigorous-intensity physical activity in a period of seven-days prior to the survey. Physical activity was based on recall of daily activity patterns in the previous 7 days. Using the US guideline for physical activity, recommended by the Centers for Disease Control and Prevention (CDC) and the American College of Sports Medicine (ACSM), we identified the proportion of participants reporting moderate-intensity physical activity for a minimum of 30 min on five days each week or vigorous-intensity physical activity for a minimum of 20 min on three days each week [25].

2.5. Statistical Analysis

Data were analyzed using SPSS version 27.0 (IBM, Armonk, NY, USA). Categorical variables were presented as frequencies and percentages. Continuous variables were presented as mean ± standard deviation. Participants were grouped into three categories based on serum 25(OH)D levels and all variables were compared using Chi-Square test for categorical variables and One-Way Analysis of Variance (ANOVA) for continuous variables. Moreover, differences in results among the participants (Arabs, Asians, Filipinas) were explored using Chi-Square test for categorical variables and Independent-Samples *t*-Test for continuous variables. Simple and multivariate ordered logistic regression models were constructed to determine the predictors of vitamin D deficiency. Simple and multivariate binary logistic regression models were considered to identify the variables associated with vitamin D deficiency. A *p*-value of <0.05 was considered statistically significant in our analyses.

3. Results

The mean age of study participants was 35 years (SD ± 10), 33 years (SD ± 8) for Filipinas, 37 years (SD ± 11) for Arabs, and 34 years (SD ± 10) for South Asian. Table 1 shows the characteristics of the study population.

Table 1. Characteristics of the female migrant participants living in Al Ain, UAE.

Variable		Filipinas		Arab		South Asian	
	N	n	%	n	%	n	%
All	553	290	52.4	136	24.6	127	23.0
Age, (years)							
18–30	186	123	(43.0)	43	(31.6)	55	(43.7)
31–40	171	119	(41.6)	45	(33.3)	39	(39.9)
≥41	191	44	(15.4)	47	(34.8)	32	(25.4)
Education of the participant							
No formal schooling	39	15	(5.2)	15	(11.2)	9	(7.3)
Up to secondary	203	124	(43.2)	40	(29.8)	59	(31.4)
College or higher	303	148	(51.6)	79	(59.0)	76	(61.3)
Marital status							
Unmarried	194	137	(49.6)	34	(25.6)	23	(18.6)
Married	306	118	(42.8)	89	(66.9)	99	(79.8)
Divorced, or widowed	33	21	(7.6)	10	(7.5)	2	(1.6)
Occupation							
Housemaid	192	176	63.6	8	6.2	8	6.8
Housewife	122	4	1.4	64	49.6	54	45.8
Driver	10	10	3.6	0	0	0	0.0
Cook	10	10	3.6	0	0	0	0.0
Administrator, supervisor	35	21	7.6	7	5.4	7	5.9
Teacher	36	5	1.8	16	12.4	15	12.7
Health care worker	55	23	8.3	13	10.1	19	16.1
Other	64	28	10.1	21	16.3	15	12.7
Monthly income, AED (1 USD = 3.6 AED)							
Lowest (812.6)	133	125	(52.5)	5	(13.2)	3	(6.5)
Middle (1365.8)	85	72	(30.3)	5	(13.2)	8	(17.4)
Highest (7422.8)	104	41	(17.2)	28	(73.6)	35	(76.1)
Residence in home country, *n* (%)							
Urban	263	129	(50.8)	54	(43.2)	80	(67.2)
Rural	235	125	(49.2)	71	(56.8)	39	(32.8)
Duration of residence in UAE							
<1 year	185	133	(53.0)	27	(22.9)	25	(22.9)
1 to 5 years	208	106	(42.2)	43	(36.4)	59	(54.1)
≥5 years	85	12	(4.8)	48	(40.7)	25	(23.0)

Table 1. *Cont.*

Variable		Filipinas		Arab		South Asian	
	N	*n*	%	*n*	%	*n*	%
Moderate or vigorous physical activity							
Yes	161	133	(45.9)	15	(11.0)	13	(10.2)
No	391	157	(54.1)	121	89.0)	114	(89.8)
Body mass index categories							
<25.0	294	187	(64.5)	44	(32.4)	63	(49.6)
25–30	166	83	(28.6)	41	(30.1	42	(33.1)
≥30.0	93	20	(6.9)	51	(37.5)	22	(17.3)
Cigarette smoking, currently							
No	514	268	(92.4)	122	(89.7)	124	(97.6)
Yes	39	22	(7.6)	14	(10.3)	3	(2.4)
Alcohol consumption							
No	448	193	(66.6)	135	(99.3)	120	(94.5)
Yes	105	97	(33.4)	1	(0.7)	7	(5.5)
Hemoglobin A1c level							
<5.7%	151	52	(91.2)	64	(71.1)	35	(53.9)
5.7–6.4%	38	4	(7.0)	15	(16.7)	19	(29.2)
≥6.5%	23	1	(1.8)	11	(12.2)	11	(16.9)
Blood pressure, mm Hg	363	188	(64.8)	93	(68.4)	82	(64.6)
<140/90	190	102	(35.2)	43	(31.6)	45	(35.4)
≥140/90 or on hypertension medication							
Levels of the serum 25(OH)D concentrations (ng/mL)							
Mean (ng/mL, (± SD)	20 (±12)	30	(±11)	14	(±10)	15	(±9)
Serum 25(OH)D concentrations by category							
>20 ng/mL (50 nmol/L), *n* (%)	71 (33.3)	48	(84.2)	12	(13.3)	11	(16.7)
≤20 ng/mL (50 nmol/L)	142 (66.7)	9	(15.8)	78	(86.7)	55	(83.3)

Data are presented as *N* and *n* (%); data presented as mean ± SD.

A significant proportion (7.2%) had no formal schooling, 37.2% had secondary level education and 55.6% had college or higher levels of education. A high proportion (57.4%) were married, 36.4% were unmarried and 6.2% were divorced or widowed. A high proportion of study participants worked as housemaids (36.6%) and housewives (23.3%). The rest were health care workers (10.5%), teachers (6.9%), administrators or supervisors (6.7%), drivers or cooks (4%), and multitude of other activities (10.5%). With regards to background in home country, 52.8% were from urban settings while 47.2% had rural backgrounds. The prevalence of obesity in Filipinas was notably low (6.9%) as compared to their Arabs (37.5%) and South Asians (17.3%) counterparts. Moderate and vigorous physical activity was reported by a higher proportion of Filipinas (45.9%) as compared to Arab (11.0%) and South Asian (10.2%) female immigrants.

Table 2 shows the mean ± SD of 25(OH)D concentrations and vitamin D deficiency in the study population.

Table 2. Distribution of inadequate 25(OH)D levels (ng/mL) among the female migrant participants according to sociodemographic, lifestyle, and clinical characteristics.

Variable	25(OH)D Levels (ng/mL)						
					<20 ng/mL)		
	N	Mean	(±SD)	*p*	%	(95% CI)	*p*
All	213	19	(11)		66.7	(60.0–72.7)	
Age, (years)							
18–34	113	18	(12)	0.37	68.1	(58.9–76.1)	0.38
35–44	57	20	(14)		59.6	(47.1–72.2)	
≥45	43	18	(11)		72.1	(55.8–83.2)	

Table 2. *Cont.*

Variable	25(OH)D Levels (ng/mL)						
					<20 ng/mL		
	N	**Mean**	**(±SD)**	*p*	**%**	**(95% CI)**	*p*
Nationality							
Filipinas	57	30	(12)	<0.001	15.8	(8.3–27.8)	<0.001
Arab	90	14	(10)		86.7	(77.8–92.3)	
South Asian	66	15	(9)		83.3	(72.2–90.6)	
Education of the participant							
No formal schooling	18	17	(15)	0.02	77.8	(52.6–91.7)	<0.001
Up to secondary	91	22	(13)		49.4	(39.2–59.7)	
College or higher	101	17	(10)		79.2	(70.1–86.1)	
Marital status							
Unmarried	69	21	(13)	0.26	56.5	(44.5–67.58)	0.06
Married	129	18	(11)		72.9	(64.5–79.8)	
Divorced, or widowed	10	19	(12)		60.0	(28.2–85.1)	
Occupation							
Housemaid	64	29	(12)	<0.001	21.9	(13.3–33.8)	<0.001
Housewife	71	14	(9)		88.7	(78.9–94.3)	
Driver	8	12	(7)		87.5	(31.9–98.1)	
Administrator, supervisor	9	14	(5)		77.8	(39.6–94.9)	
Teacher	21	15	(10)		80.9	(58.0–92.9)	
Health care worker	18	18	(8)		77.8	(52.6–91.7)	
Other	22	14	(13)		90.9	(69.1–97.8)	
Monthly income, Dirham (AED) (US dollar = 3.7 AED)							
Lowest (801.2)	42	29	12	<0.001	23.8	(13.1–39.3)	<0.001
Middle (1386.3)	19	24	13		36.8	(18.1–60.6)	
Highest (8397.5)	40	17	13		80.0	(64.4–89.8)	
Residence, n (%)							
Urban	84	19	13	0.74	65.5	(54.6–74.9)	0.57
Rural	114	19	12		69.3	(60.1–77.1)	
Duration of residence in UAE							
<1 year	62	22	14	0.005	56.4	(43.8–68.2)	<0.001
1 to 5 years	72	18	11		65.3	(53.5–75.4)	
≥5 years	49	14	14		91.8	(79.69–96.9)	
Physical activity (mod/vigorous.)							
Yes	52	25	13	<0.001	42.3	(29.5–56.2)	<0.001
No	161	17	11		75.5	(67.2–80.7)	
Body mass index categories				0.002			
<25.0	94	21	13		57.4	(47.62–67.1)	0.007
25–30	74	19	12		67.6	(55.9–77.3)	
≥30.0	45	14	8		84.4	(70.5–92.4)	
Cigarette smoking, currently							
No	199	19	12	0.99	65.3	(17.1–20.5)	0.12
Yes	14	19	16		85.7	(9.5–28.0)	
Alcohol use							
No	197	18	12	<0.001	71.1	(64.3–77.0)	<0.001
Yes	16	32	13		12.5	(2.9–39.9)	
Hemoglobin A1c level							
<5.7%	151	20	13	0.29	62.2	(54.2–69.7)	0.11
5.7–6.4%	38	16	8		79.9	(62.8–89.2)	
≥6.5%	23	19	13		73.9	(52.1.3–88.1)	
Blood pressure, mm Hg							
<140/90	139	18	13	0.45	70.5	(62.3–77.5)	0.10
≥140/90 or on hypertension medication	74	20	11		59.5	(47.8–70.1)	

Data are presented as frequencies (%) and mean ± SD and odds ratio (OR) (95% CI); *p* < 0.05 considered significant (shown in boldface).

Overall, the 25(OH)D concentrations were 20 ± 11 ng/mL in the study population, specifically 30 ± 11 in the Filipinas, 14 ± 10 in Arabs, and 15 ± 9 in South Asians. The overall prevalence rate of vitamin D deficiency (25(OH)D ≤ 20 ng/mL) was 66.7% (95% CI, 60.0%–72.7%). Examination of the prevalence rate of vitamin D deficiency by age revealed insignificant variations. There were significant differences in 25(OH)D concentrations as well as the prevalence rate of vitamin D deficiency by nationality, education level, occupation, income, duration of residence in the UAE, physical activity, body mass index categories, and self-reported alcohol use. There was an inverse correlation between earnings and vitamin D status and similarly an inverse correlation between a high education level and lower vitamin D status. This was due to the fact that vitamin D deficiency is much lower (15.8%) when compared to Arab (86.7%) and South Asian (83.3), as shown in Table 1. A high proportion of Filipinas (52.5%) were in the lowest tertile of monthly earning when compared to Arab (13.2%) and South Asian (6.5%) in the lowest tertile of monthly earning. Similarly, a high proportion of Arab and South Asian had college or higher level of education. A high proportion of Filipinas reported alcohol consumption (33.4%) compared to their (Arab (0.7%) and South Asian (5.55) counterparts.

The boxplot of participants' serum 25(OH)D concentrations (ng/mL) according to BMI categories in Figure 1 demonstrates the low levels of 25(OH)D concentrations (ng/dL) among overweight and obese participants. Figure 2 shows participants' 25(OH)D concentrations levels overall, and by the nationality.

Table 3 summarizes the results from multivariate analyses. After adjusting for all related factors, including age and nationality, low physical activity (AOR = 4.59; 95% CI 1.1.98–10.63), having more than 5 years duration of residence in UAE (AOR = 4.65; 95% CI 1.31–16.53), and being obese (AOR = 3.56; 95% CI 1.04–12.20) were independently associated with vitamin D deficiency.

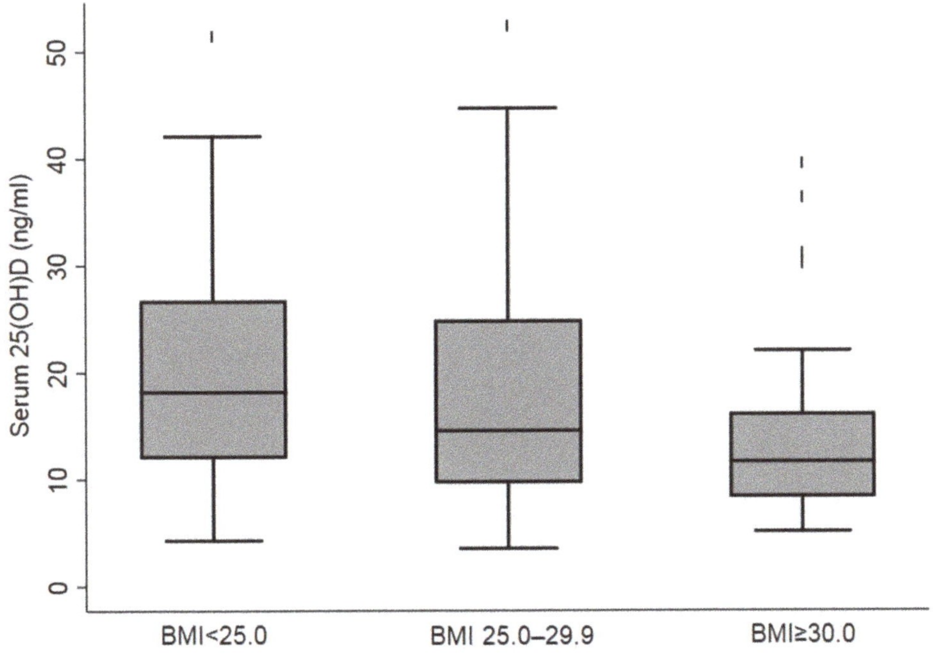

Figure 1. Boxplot of Serum 25(OH)D Concentrations according to Body Mass Index in the Female Migrant Participants Living in Al Ain, UAE.

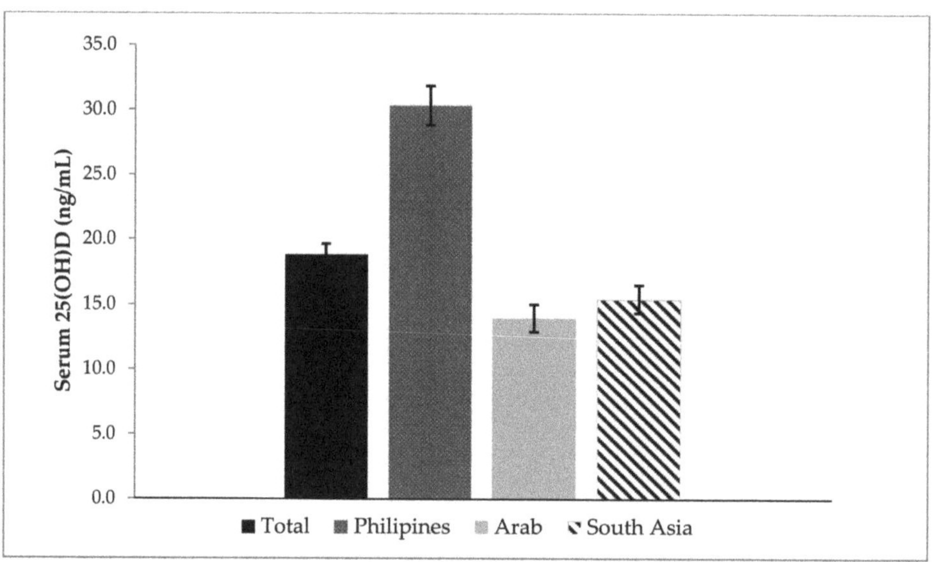

Figure 2. Mean Serum 25(OH)D Concentrations ± SEM in the Females Migrant Participants Living in Al Ain, UAE.

Table 3. Multivariate logistic regression analysis for vitamin D deficiency correlates among the female migrant participants.

Determinants	Adjusted		
	OR	**95% CI**	*p*
Age in years	0.97	(0.92–1.01)	0.18
Nationality			
Filipinas	0.06	(0.01–0.42)	0.004
Arab	Reference		
South Asian	0.79	(0.32–1.98)	0.62
Education			
No formal education	Reference		
Up to secondary	0.24	(0.05–1.04)	0.06
College or higher	0.66	0.15–2.89)	0.59
Duration of residence in UAE			
≤1 year	Reference		
>1 to 5 years	0.78	(0.33–1.83)	0.67
>5 years	4.65	(1.31–16.53)	0.02
Body mass index categories BMI			
≤24.99	Reference		
25–29.99	2.21	(0.92–5.30)	0.08
≥30.0	3.56	(1.04–12.20)	0.04
Low level of physical activity *			
No	Reference		
Yes	4.59	(1.98–10.63)	<0.001

* Self-reported (IPAQ questionnaire) moderate activity of 30 min on five days or vigorous intensity. Physical activity for a minimum of 20 min on three days each week. Data are presented as frequencies (%) and odds ratio (OR) (95% CI); $p < 0.05$ considered significant (shown in boldface).

4. Discussion

This study showed the extent of vitamin D deficiency among female migrants. Vitamin D deficiency was especially common among Arab and South Asian migrants. Using serum 25(OH)D concentrations ≤20 ng/mL as a cutoff to define vitamin D deficiency, our data demonstrated that over 80% of female migrants from Arab and South Asian countries were vitamin D deficient. A retrospective study of the UAE population on 60,979 subjects originating from 136 different countries revealed that 78.9% of expatriates were suffering from hypovitaminosis, with more females being afflicted with severe deficiency [26]. Previous studies have shown lower prevalence among adults in developed countries, 24% in US [27], 37% in Canada [28], 32% in Australia [29], and 40% in the European countries [30]. Paradoxically, vitamin D deficiency of up to 80% was observed in Middle Eastern Arab countries despite the sunny weather [10,31]. Moreover, a systematic review and meta-analysis of the prevalence of vitamin D deficiency among the population of Africa on 21,474 individuals from 23 countries within the continent revealed the prevalence was higher than speculated at a rate of 34%; with women having remarkably lower 25(OH)D concentrations compared to men. It was concluded that being a female and living in urban areas in the northern and southern parts of Africa were associated with a higher risk of developing vitamin D deficiency [32]. The findings in the current study were in concordance with previous studies, and similarly documented a high prevalence of vitamin D deficiency among Arab female migrants originating from African and Middle Eastern countries. The reasons for vitamin D deficiency in Arab female migrants might include wearing conservative skin-concealing clothes, relatively high obesity, and low dietary vitamin D intake [9,10]. An additional factor might be related to dark skin complexion since the skin pigment melanin absorbs sunlight and dark skin color reduces the capacity of skin to synthesize vitamin D_3 [33]. National examination surveys in the US indicated that over 80% of non-Hispanic black American adults, including men and women, had $\leq$20 ng/mL serum 25(OH)D concentrations [34,35].

Among South Asian females, 83.3% were vitamin D deficient in the present investigation. A recent systematic review and meta-analysis of vitamin D deficiency in Asia involving 472 studies on 746,564 individuals demonstrated that region and altitude were important correlates of vitamin D deficiency [36]. Data from different parts in India revealed a 70% prevalence of vitamin D deficiency among adult with a higher prevalence of up to 79% among females [37]. Studies conducted among South Asians in Australia, Canada, European countries, UK, and USA have found a prevalence of vitamin D deficiency in epidemic proportions compared to their native counterparts [34]. The proposed causes of vitamin D deficiency among South Asians in Western countries included low vitamin D intake, relatively high obesity prevalence, less exposure to sunlight, and wearing conservative clothes for cultural and religious reasons [38].

The low prevalence of vitamin D deficiency among Filipina migrants as compared to the South Asian and Arabs might be due to the lower prevalence of both obesity and physical inactivity [39]. However, additional confounders could also be involved including differences in dietary intake of vitamin D (as Filipinas highly consume oily fish) and dissimilarities in sartorial and lifestyle habits, which might entail more sun exposure and hence concomitant photosynthesis of vitamin D [40].

In our study, low physical activity, acculturation as measured by more than 5 years of residency in the UAE, and obesity were all independently associated with vitamin D deficiency. Physical activity was a significant independent predictor of vitamin D deficiency in several previous studies. Additionally, some researchers showed that serum 25(OH)D levels are affected by both physical activity and BMI in the context of obesity [41,42]. Pragmatically speaking, the positive impact of physical activity on vitamin D status could be attributed to metabolism upon energy disbursement with muscle contractions; however, when performed alfresco, it stimulates the conversion of dermal 7-dehydrocholesterol into vitamin D by solar UV-B rays [43]. The effect of physical activity could be complex, since physical inactivity is a well-established risk factor for overweight and obesity, which, in turn, could predispose to vitamin D deficiency through adipose sequestration and volumetric dilution of 25(OH)D [44]. Nevertheless, a study on older adults revealed a favorable correlation between physical activity and serum 25(OH)D levels over time regardless of exposure to sun [45]. In agreement with our results, numerous population-based investigations had demonstrated that obesity and reduced physical activity were independent determinants of vitamin D deficiency in South Asia [46], Philippines [40], Middle East [47], Europe, Australia, [29] and the US [35].

Obese females (BMI ≥ 30) in our study had a high prevalence of vitamin D deficiency. It is worth mentioning that obesity has been well established as a risk factor for vitamin D deficiency. Several studies have demonstrated a close association between vitamin D deficiency and obesity among the Asian and Arab adult population [48]. Data from the World Health Organization (WHO) reveal that the UAE currently ranks fifth in the world in obesity, at a prevalence rate of 36% (33% for males and 39% for females) [49].

The rapid socioeconomic transition in MENA countries has resulted in increased urbanization and drastic lifestyle changes and has also manifested in sedentary and unhealthy dietary practices. These factors combined with the growing fast-food industry has significantly impacted the prevalence of obesity among this population. Vitamin D deficiency and the rapid increase in the prevalence of obesity are both considered important public health issues contributing to morbidity and mortality in the UAE [50].

A series of evidence support the fact that obesity might be driving low serum 25(OH)D concentrations due to decreased bioavailability of vitamin D through sequestration within body fat and volumetric dilution [36,42]. It has been proposed that dietary calcium and vitamin D status might play a role in weight and fat regulation [51]. According to this hypothesis, lower calcium or vitamin D results in increased parathyroid hormone (PTH), which in turn decreases 25(OH)D and increases intracellular calcium into adipocytes, consequently inhibiting lipolysis and stimulating lipogenesis [52]. Furthermore, there are profound indications for an association between vitamin D levels and obesity, and that serum level of 25(OH)D are reduced in obese subjects in adults [53].

The data we analyzed highlighted that length of stay of greater than 5 years in the UAE was a strong predictor of vitamin D deficiency among female immigrants. This parameter is often used as an indicator of adjustment to a new culture acculturation. The relation between acculturation and immigrants' health is quite complex given the diversity of cultures and how temporal adaptation might affect health outcomes by embracing new habits that could act as risk factors for diseases, such asreduced physical activity and obesity [54]. A cross-sectional study in Canada that examined the prevalence of vitamin D deficiency among first generation immigrants (n = 11,579) concluded that the length of time lived in Canada, lifestyle, and ethnicity were important correlates. Overall, immigrants had significantly lower levels of 25(OH)D compared to native-born Canadians with Arabs and Southeast Asians in comparison to other white ethnic groups having the highest levels of vitamin D deficiency [55].

Immigration studies exploring the impact of acculturation on vitamin D status have demonstrated a significant positive correlation with the length of stay and serum 25(OH)D among immigrants in Canada [56]. This was in contrast with our findings and could be explained by the fact that integration into a new modernized life in the UAE with a thriving fast-food industry and more sedentary life could be the major culprit for the negative aspect of acculturation [55]. Importantly, food fortification with vitamin D in the UAE, unlike Canada, is not strictly implemented and this might also explain the conflicting results we obtained [10,27].

However, our results are in concordance with reported findings from numerous other studies which examined the relationship between culture and health in the context of acculturation on the health of Asian immigrants in the US, Australia, and Europe [55]. A recent review confirmed that vitamin D deficiency was at epidemic levels among South Asian immigrants living in Western countries mainly due to low dietary intake of vitamin D, high obesity, avoidance of sun exposure, and having covered sartorial dress style either due to cultural or religious reasons [38,57]. Moreover, investigators concluded that being a female, living in an urbanized setting, and limiting physical activity in Southeast Asia were strongly correlated with a low vitamin D status. In addition, religious, behavioral practices, and nutritional dissimilarities between Muslims and non-Muslims contributed to the remarkable differences in vitamin D status among these two subgroups of the same population in Thailand, with non-Muslims having 10 nmol/L higher levels of 25(OH)D [56].

The effect of clothing preference on vitamin D status of females has been well described by other researchers [58].

Vitamin D deficiency is a global public health crisis to which immigrants are predominantly susceptible especially with sub-populations that have darker skin and who move to the Western regions that have high latitudes [59]. In general, however, the vitamin D status of northern African countries was similar to populations in the Middle East, most likely due to similar environmental and behavioral styles [9]. In comparison to other ethnicities, immigrants from the African descent who settled in temperate areas in the US and Europe had a higher prevalence of vitamin D deficiency [60,61]. Moreover, national surveys in Europe reported that approximately 40% of these sub-populations had 25(OH)D concentrations below the sufficiency cutoff [62]. This pattern had been ascribed to the reduced adaptability of their darker skin color for vitamin D photosynthesis in regions that are not as sunny as Africa. Concurrently, a remarkable prevalence of 82.1% for vitamin D deficiency was documented among African American in the US in comparison to a nationwide mean value of 41.9% [60]. Moreover, a decrease in 25(OH)D concentrations was consistently reported with more northern latitudes and the duration of time since moving from Africa [59]. In a large sample (n = 12,346) of native Emirati population, a high proportion (84%) of native Emirati females had vitamin D deficiency (<50 nmol/L) [63]. The predisposing factors for the vitamin D deficiency were high BMI, central obesity, high blood pressure, high cholesterol, and impaired blood glucose levels [64].

5. Strengths and Limitations

This study is the first to examine the prevalence of vitamin D deficiency among an under-studied and vulnerable subpopulation of female migrants in the UAE. Low vitamin D status along with other biomarkers among this subpopulation could signify health deterioration and the need for action. Our findings have valuable global health implications regarding the physical wellbeing of foreign immigrants and could inform future studies aiming to respond to chronic and infectious diseases including COVID-19 [63]. Moreover, the data obtained was for immigrants from different countries and could present a unique understanding to the cultural diversity and its intricate relation to immigrants' health within the UAE.

Despite the strength of the study, there are several limitations that should be acknowledged. The cross-sectional nature of design does not confirm causation of vitamin D deficiency, and hence findings remain correlational. Serum 25(OH)D concentrations are frequently used as clinical indicators for the assessment of vitamin D status, which is determined by sun exposure, dietary intake, and supplementation. In this research, such data was not available for analysis. However, the impact of seasonal variation in this study was minimal since the research was conducted in the summer, during which the contribution to vitamin D status by sun exposure is often restricted, since residents of the UAE avoid the sun and heat during this very hot season and spend their time indoor in cool air-conditioned settings [27,51]. We did not collect information on dietary habits, and/or supplement use, Fitzpatrick skin type, or tanning/outdoor habits. Vitamin D status is affected by diseases pertaining to impaired digestion and malabsorption like Crohn's and Celiac diseases; we did not collect this data in our study.

Examining differences in dietary habits, as well as behavioural practices towards sun exposure among the different subpopulations of female immigrants, are worth further investigation. The attitude and practices of the participants towards sun exposure were not explored but might have deterred them from securing an adequate vitamin D status. Whiter skin complexion had been associated with beauty, attractiveness, higher socio-economic advantages, and prestige among females from some populations, particularly in South Asia and Africa [65–68]. Lastly, genetic factors are also an important element to consider in such research owing to the contribution of genetic variants within vitamin D metabolism to the vitamin D status [68].

6. Conclusions

In conclusion, vitamin D deficiency is highly prevalent among female immigrant workers in the UAE. A prevention strategy including Vitamin D supplementation and educating this vulnerable subpopulation is urgently needed. Currently, the vitamin D recommendations for the UAE do not contain any specific guidelines for the immigrant subpopulations. Until food fortification with vitamin D becomes effective in the UAE, sun exposure and the use of supplements remain the most significant determinants of vitamin D status for the UAE population. Our study might have important connotations for introducing culturally acceptable approaches of increasing skin exposure to the sun and a subpopulation specific vitamin D dietary intake.

Author Contributions: S.M.S.: conceptualization, project administration, data analysis and interpretation, writing of original draft. F.A.A.: writing of original draft and interpretation of results. A.R.: formal analysis. N.S.: writing of methodology. R.A.: funding acquisition. W.B.G., L.A.A. and J.A.: reviewing and editing. All other authors contributed to reviewing final version. All authors have read and agreed to the published version of the manuscript.

Funding: This study was supported by a faculty grant number (NP09-30) from the College of Medicine and Health Sciences, UAE University, to S.K. The faculty grant for the project is entitled "Chronic Diseases Prevention in Immigrants: putting CVD risk factors on surveillance screen".

Institutional Review Board Statement: The study was conducted according to the guidelines of the Declaration of Helsinki and was approved by the Institutional Review Board (or Ethics Committee) of UAEU, Al Ain, UAE (AAMDHREC 10/21).

Informed Consent Statement: Informed consent was obtained from all subjects involved in the study.

Data Availability Statement: The data presented in this study are available on request from the corresponding author.

Acknowledgments: We thank the participants of the study for their willingness to participate and support research. We acknowledge the assistance of the health care workers; without their assistance this study would not have been possible.

Conflicts of Interest: William B. Grant receives funding from Bio-Tech Pharmacal, Inc. (Fayetteville, AR, USA). All other authors declare no conflict of interest.

References

1. Lips, P.; de Jongh, R.T.; Van Schoor, N.M. Trends in Vitamin D Status Around the World. *JBMR Plus* **2020**, *12*, e10585. [CrossRef] [PubMed]
2. Saraf, R.; Morton, S.M.B.; Camargo, C.A.; Grant, C.C. Global summary of maternal and newborn vitamin D status—A systematic review. *Matern. Child Nutr.* **2016**, *12*, 647–668. [CrossRef] [PubMed]
3. Bi, W.G.; Nuyt, A.M.; Weiler, H.; Leduc, L.; Santamaria, C.; Wei, S.Q. Association between vitamin D supplementation during pregnancy and offspring growth, morbidity, and mortality: A systematic review and meta-analysis. *JAMA Pediatr.* **2018**, *7*, 635–645. [CrossRef] [PubMed]
4. Holick, M.F.; Binkley, N.C.; Bischoff-Ferrari, H.A.; Gordon, C.M.; Hanley, D.A.; Heaney, R.P.; Murad, M.H.; Weaver, C.M. Evaluation, treatment, and prevention of vitamin D deficiency: An Endocrine Society clinical practice guideline. *J. Clin. Endocrinol. Metab.* **2011**, *96*, 1911–1930. [CrossRef] [PubMed]
5. Oristrell, J.; Oliva, J.C.; Casad, E.; Subirana, I.; Dominguez, D.; Toloba, A.; Balado, A.; Grau, M.J. Vitamin D supplementation and COVID-19 risk: A population-based, cohort study. *J. Endocrinol. Investig.* **2021**, *45*, 167–179. [CrossRef]
6. Dissanayake, H.A.; de Silva, N.L.; Sumanatilleke, M.; de Silva, S.D.N.; Gamage, K.K.K.; Dematapitiya, C.; Kuruppu, D.C.; Ranasimghe, P.; Sumanathan, S.; Katulanda, P.J. Prognostic and therapeutic role of vitamin D in COVID-19: Systematic review and meta-analysis. *J. Clin. Endocrinol. Metab.* **2021**, *20*, dgab892. [CrossRef] [PubMed]
7. Rejnmark, L.; Bislev, L.S.; Cashman, K.D.; Eirlksdottir, G.; Gaksch, M.; Grubler, M.; Grimnes, G.; Gudnason, V.; Lips, P.; Pilz, S.; et al. Non-skeletal healtheffects of vitamin D supplementation: A systematic review on finding from meta-analyses summarizing trial data. *PLoS ONE* **2017**, *12*, e0180512. [CrossRef]
8. Pludowski, P.; Holick, M.F.; Grant, W.B.; Konstantynowicz, J.; Mascarenhas, M.R.; Haq, A.; Povoroznyuk, V.; Balatska, N.; Barbosa, A.P.; Karonova, T.; et al. Vitamin D supplementation guidelines. *J. Steroid Biochem. Mol. Biol.* **2018**, *175*, 125–135. [CrossRef]
9. Chakhtoura, M.; Rahme, M.; Chamoun, N.; Fuleihan, G.E. Vitamin D in the Middle East, and North Africa. *Bone Rep.* **2018**, *8*, 135–146. [CrossRef]

10. Grant, W.B.; Fakhoury, H.M.A.; Karras, S.N.; Al Anouti, F.; Bhattoa, H.P. Variations in 25-Hydroxyvitamin D in countries from the Middle East and Europe: The Roles of UVB Exposure and Diet. *Nutrients* **2019**, *11*, 2065. [CrossRef]
11. Welch, T.R.; Bergstrom, W.H.; Tsang, R.C. Vitamin D-deficient rickets: The re-emergence of a once conquered disease. *J. Pediatr.* **2000**, *137*, 143–145. [CrossRef] [PubMed]
12. Yared, M.N.; Chemali, R.; Yaacoub, N.; Halaby, G. Hypovitaminosis D in a sunny country: Relation to lifestyle and bone markers. *J. Bone Miner. Res.* **2000**, *15*, 1856–1862. [CrossRef]
13. Sedrani, K.M.; Elidrissy, A.W.T.H.; El-Arabi, K.M. Sunlight and vitamin D status in normal Saudi subjects. *Am. J. Clin. Nutr.* **1983**, *38*, 129–132. [CrossRef]
14. Hobbs, R.D.; Habib, Z.; Alromaihi, D.; Ldi, L.; Parikh, N.; Blocki, F.; Rao, D.S. Severe vitamin D deficiency in Arab-American women living in Dearborn, Michigan. *Endocr. Pract.* **2009**, *15*, 35–40. [CrossRef] [PubMed]
15. Muhairi, S.J.; Mehairi, A.E.; Khouri, A.A.; Naqbi, M.; Maskari, F.; Al Kaabi, J.; Al Dhaheri, A.; Naglekerke, N.; Shah, S. Vitamin D deficiency among healthy adolescents in Al Ain, United Arab Emirates. *BMC Public Health* **2013**, *13*, 33. [CrossRef] [PubMed]
16. United Arab Emirates Ministry of Economy. Preliminary Results of Population, Housing, and Establishment Census 2005, United Arab Emirates 2006. Available online: http://www.cscc.unc.edu/uaee/public/UNLICOMMUAE2005CensusResults07282008.pdf (accessed on 10 June 2020).
17. Yousef, S.; Elliott, J.; Manuel, D.; Colman, I.; Papadimitropoulos, M.; Hossain, A.; Leclair, N.; Well, G.A. Study protocol: Worldwide comparison of vitamin D status of immigrants from different ethnic origins and native-born populations-a systematic review and meta-analysis. *Syst. Rev.* **2019**, *8*, 211. [CrossRef] [PubMed]
18. Shah, S.M.; Ali, R.; Loney, T.; Aziz, F.; ElBarazi, I.; Al Dhaheri, S.; Farooqi, M.H.; Blair, I. Prevalence of diabetes among migrant women and duration of residence in the United Arab Emirates: A cross sectional study. *PLoS ONE* **2017**, *12*, e0169949. [CrossRef] [PubMed]
19. World Health Organization. The WHO STEPwise Approach to Noncommunicable Disease Risk Factor Surveillance. Available online: https://www.who.int/ncds/surveillance/steps/STEPS_Manual.pdf (accessed on 16 May 2020).
20. Herrick, K.A.; Storandt, R.J.; Afful, J.; Pfeiffer, C.M.; Schleicher, R.L.; Gahche, J.J.; Potischman, N. Vitamin D status in the United States 2011–2014. *Am. J. Clin. Nutr.* **2019**, *110*, 150–157. [CrossRef]
21. Pottie, K.; Greenaway, C.; Feightner, J.; Welch, V.; Swinkels, H.; Rashid, M.; Narasiah, L.; Kirmayer, L.J.; Ueffing, E.; MacDonald, N.E.; et al. Evidence-based clinical guidelines for immigrants and refugees. *CMAJ* **2011**, *183*, E824–E925. [CrossRef]
22. World Health Organization. *Obesity: Preventing and Managing the Global Epidemic: Report of a WHO Consultation*; World Health Organization: Geneva, Switzerland, 2000.
23. 1999 World Health Organization-International Society of Hypertension Guidelines for the Management of Hypertension. Guidelines Subcommittee. *J. Hypertens.* **1999**, *17*, 151–183.
24. International Physical Activity Questionnaire (IPAQ). Guidelines for Data Processing and Analysis of the International Physical Activity Questionnaire (IPAQ)-Short and Long Forms. 2005. Available online: http://www.ipaq.ki.se/scoring.pdf (accessed on 16 May 2020).
25. Haskell, W.L.; Lee, I.M.; Pate, R.R.; Powell, K.E.; Blair, S.N.; Franklin, B.A.; Macera, C.A.; Heath, G.W.; Thompson, P.D.; Bauman, A. Physical activity and public health: Updated recommendation for adults from the American College of Sports Medicine and the American Heart Association. *Med. Sci. Sports Exerc.* **2007**, *39*, 1423–1434. [CrossRef] [PubMed]
26. Haq, A.; Svobodova, J.; Imran, S.; Stanford, C.; Razzaque, M.S. Vitamin D deficiency: A single center analysis of patients from 136 countries. *J. Steroid Biochem. Mol. Biol.* **2016**, *164*, 209–213. [CrossRef] [PubMed]
27. Schleicher, R.L.; Sternberg, M.R.; Looker, A.C.; Yetley, E.A.; Lacher, D.A.; Sempos, C.T.; Taylor, C.L.; Durazo-Arvizu, R.A.; Maw, K.L.; Chaudhary-Webb, M.; et al. National estimates of serum 25(OH)D and metabolite concentrations measured by liquid chromatography-Tandem, mass spectrometry in the US population during 2007–2010. *J. Nutr.* **2016**, *146*, 1051–1061. [CrossRef] [PubMed]
28. Sarafin, K.; Durazo-Arvizu, R.; Tian, L.; Phinney, K.W.; Tai, S.; Camara, J.E.; Merkel, J.; Green, E.; Sempos, C.T.; Brooks, S.P. Standardizing 25-hydoxyvitamin D values from the Canadian Health Measures Survey. *Am. J. Clin. Nutr.* **2015**, *102*, 1044–1050. [CrossRef] [PubMed]
29. Horton-French, K.; Dunlop, E.; Lucas, R.M.; Pereira, G.; Black, L.J. Prevalence and predictors of vitamin D deficiency in a nationally representative sample of Australian adolescents and young adults. *Eur. J. Clin. Nutr.* **2021**, *75*, 1627–1636. [CrossRef]
30. Lips, P.; Cashman, K.D.; Lamberg-Allardt, C.; Bischoff-Ferrari, H.A.; Obermayer-Pietsch, B.R.; Bianchi, M.; Stepan, J.; El-Hajj Fuleihan, G.; Bouillon, R. Current vitamin D status in European and Middle East countries and strategies to prevent vitamin D deficiency: A position statement of the European Calcified Tissue Society. *Eur. J. Endocrinol.* **2019**, *180*, 23–54. [CrossRef]
31. Mogire, R.M.; Mutua, A.; Kimita, W.; Kamau, A.; Bejon, P.; Pettifor, J.M.; Adeyemo, A.; Williams, T.N.; Atkinson, S.H. Prevalence of vitamin D deficiency in Africa: A systematic review and meta-analysis. *Lancet Glob. Health* **2020**, *8*, e134-42. [CrossRef]
32. Clemens, T.L.; Henderson, S.L.; Adams, J.S.; Holick, M.F. Increased skin pigment reduces the capacity skin to synthesise vitamin D3. *Lancet* **1982**, *319*, 4–6. [CrossRef]
33. Nesby-O'Dell, S.; Scanlon, K.S.; Cogswell, M.E.; Gillespie, C.; Hollis, B.W.; Looker, A.C.; Allen, C.; Doughertly, C.; Gunter, E.W.; Bowman, B.A. Hypovitaminosis D prevalence and determinants among White women of reproductive age: Third National and Nutrition Examination Survey 1988–1994. *Am. J. Clin. Nutr.* **2002**, *76*, 187–192. [CrossRef]
34. Yetley, E.A. Assessing the Vitamin D status of the US population. *Am. J. Clin. Nutr.* **2008**, *88*, 558S–564S. [CrossRef]

35. Jiang, Z.; Pu, R.; Li, N.; Chen, C.; Li, J.; Dai, W.; Wang, Y.; Hu, J.; Zhu, D.; Yu, Q.; et al. High prevalence of vitamin D deficiency in Asia: A systematic review and meta-analysis. *Crit. Rev. Food Sci. Nutr.* **2021**, *16*, 1–10. [CrossRef] [PubMed]

36. Kambo, P.; Dwivedi, S.; Toteja, G.S. Prevalence of hypovitaminosis D in India & wayfoward. *Indian J. Med. Res.* **2018**, *148*, 548–556. [CrossRef]

37. Darling, A.L. Vitamin D deficiency in western dwelling South Asian populations: An unrecognized epidemic. *Proc. Nutr. Soc.* **2020**, *79*, 259–271. [CrossRef] [PubMed]

38. Brock, K.; Huang, W.Y.; Fraser, D.R.; Ke, L.; Tseng, M.; Stolzenberg-Solomon, R.; Peters, U.; Ahn, J.; Purdue, M.; Mason, R.S.; et al. Low vitamin D is associated with physical inactivity, obesity and low vitamin D intake in a large US sample of healthy middle-aged men and women. *J. Steroid Biochem. Mol. Biol.* **2010**, *121*, 462–466. [CrossRef] [PubMed]

39. Zumaraga, M.P.; Concepcion, M.A.; Duante, C.; Rodriguez, M. Next Generation Sequencing of Lifestyle and Nutrition related genetic polymorphisms reveals independent loci for low serum 25-hydroxyvitamin D levels among adult respondents of the 2013 Philippine National Nutrition Survey. *J. Asean Fed. Endocr. Soc.* **2021**, *36*, 56–63. [CrossRef] [PubMed]

40. Valentini, A.; Perrone, M.A.; Cianfarani, M.A.; Tarantino, U.; Massoud, R.; Merra, G.; Bernardini, S.; Morris, H.A.; Bertoli, A. Obesity, vitamin D status and physical activity: 1,25(OH)2D as a potential marker of vitamin D deficiency in obese subjects. *Panminerva Med.* **2020**, *62*, 83–92. [CrossRef]

41. De Oliveira, L.F.; De Azevedo, L.G.; da Mota Santana, J.; de Sales, L.P.C.; Pereira Santos, M. Obesity and overweight decreases the effect of 25(OH) D supplementation in adults: Systematic review and meta-analysis of randomized controlled trials. *Rev. Endocr. Metab. Disord.* **2020**, *21*, 61. [CrossRef]

42. Holick, M.F. Biological Effects of Sunlight, Ultraviolet Radiation, Visible Light, Infrared Radiation and Vitamin D for Health. *Anticancer Res.* **2016**, *36*, 1345–1356.

43. Pietiläinen, K.H.; Kaprio, J.; Borg, P.; Plasqui, G.; Yki-Järvinen, H.; Kujala, U.M.; Rose, R.J.; Westerterp, K.R.; Rissanen, A. Physical inactivity and obesity: A vicious circle. *Obesity* **2008**, *16*, 409–414. [CrossRef]

44. Scott, D.; Blizzard, L.; Fell, J.; Ding, C.; Winzenberg, T.; Jones, G. A prospective study of the associations between 25-hydroxy-vitamin D, sarcopenia progression and physical activity in older adults. *Clin. Endocrinol.* **2010**, *73*, 581–587. [CrossRef]

45. Pang, Y.; Kim, O.; Choi, J.A.; Jung, H.; Kim, J.; Lee, H.; Lee, H. Vitamin D deficiency and associated factors in south Korean childbearing women: A cross-sectional study. *BMC Nurs.* **2021**, *20*, 218. [CrossRef] [PubMed]

46. Al-Othman, A.; Al-Musharaf, S.; Al-Daghri, N.M.; Krishnaswamy, S.; Yusuf, D.S.; Alkharfy, K.M.; Al-Saleh, Y.; Al-Attas, O.S.; Alokail, M.S.; Moharram, O.; et al. Effect of physical activity and sun exposure on vitamin D status of Saudi children and adolescents. *BMC Pediatr.* **2012**, *12*, 92.

47. Liu, X.; Baylin, A.; Levy, P.D. Vitamin D deficiency and insufficiency among US adults: Prevalence, predictors and clinical implications. *Br. J. Nutr.* **2018**, *119*, 928–936. [CrossRef] [PubMed]

48. World Health Organization. Obesity and Overweight. 2021. Available online: https://www.who.int/news-room/fact-sheets/detail/obesity-and-overweight (accessed on 10 June 2021).

49. Haq, A.; Wimalawansa, S.J.; Pludowski, P.; Anouti, F.A. Clinical practice guidelines for vitamin D in the United Arab Emirates. *J. Steroid Biochem. Mol. Biol.* **2018**, *175*, 4–11. [CrossRef]

50. Zemel, M.B.; Shi, H.; Greer, B.; Dirienzo, D.; Zemel, P.C. Regulation of adiposity by dietary calcium. *FASEB J.* **2000**, *14*, 1132–1138. [CrossRef]

51. Chang, E.; Kim, Y. Vitamin D decreases adipocyte lipid storage and increases NAD-SIRT1 pathway in 3T3-L1 adipocytes. *Nutrition* **2016**, *32*, 702–708. [CrossRef]

52. De Pergola, G.; Martino, T.; Zupo, R.; Caccavo, D.; Pecorella, C.; Paradiso, S.; Silvestris, F.; Triggiani, V. 25 Hydroxyvitamin D Levels are Negatively and Independently Associated with Fat Mass in a Cohort of Healthy Overweight and Obese Subjects. *Endocr. Metab. Immune Disord. Drug Targets* **2019**, *19*, 838–844. [CrossRef]

53. Salant, T.; Lauderdale, D.S. Measuring culture: A critical review of acculturation and health in Asian immigrant populations. *Soc. Sci. Med.* **2003**, *57*, 71–90. [CrossRef]

54. Yousef, S.; Manuel, D.; Colman, I.; Papadimitropoulos, M.; Hossain, A.; Faris, M.; Wells, G.A. Vitamin D Status among First-Generation Immigrants from Different Ethnic Groups and Origins: An Observational Study Using the Canadian Health Measures Survey. *Nutrients* **2021**, *13*, 2702. [CrossRef]

55. Sanou, D.; O'Reilly, E.; Ngnie-Teta, I.; Batal, M.; Mondain, N.; Andrew, C.; Newbold, B.K.; Bourgeault, I.L. Acculturation and nutritional health of immigrants in Canada: A scoping review. *J. Immigrant Minority Health* **2014**, *16*, 24–34. [CrossRef]

56. Buyukuslu, N.; Esin, K.; Hizli, H.; Sunal, N.; Yigit, P.; Garipagaoglu, M. Clothing preference affects vitamin D status of young women. *Nutr. Res.* **2014**, *34*, 688–693. [CrossRef] [PubMed]

57. Chailurkit, L.O.; Aekplakorn, W.; Ongphiphadhanakul, B. Regional variation and determinants of vitamin D status in sunshine-abundant Thailand. *BMC Public Health* **2011**, *11*, 853. [CrossRef] [PubMed]

58. Martin, C.A.; Gowda, U.; Renzaho, A.M. The prevalence of vitamin D deficiency among dark-skinned populations according to their stage of migration and region of birth: A meta-analysis. *Nutrition* **2016**, *32*, 21–32. [CrossRef] [PubMed]

59. Harris, S.S. Vitamin D and African Americans. *J. Nutr.* **2006**, *136*, 1126–1129. [CrossRef] [PubMed]

60. Arabi, A. El Rassi, R.; El-Hajj, F.G. Hypovitaminosis D in developing countries-prevalence, risk factors and outcomes. *Nat. Rev. Endocrinol.* **2010**, *6*, 550–561. [CrossRef] [PubMed]

61. Durazo-Arvizu, R.A.; Camacho, P.; Bovet, P.; Forrester, T.; Lambert, E.V.; Plange-Rhule, J.; Hoofnagle, A.N.; Aloia, J.; Tayo, B.; Dugas, L.R.; et al. 25-Hydroxyvitamin D in African-origin populations at varying latitudes challenges the construct of a physiologic norm. *Am. J. Clin. Nutr.* **2014**, *100*, 908–914. [CrossRef] [PubMed]
62. AlZarooni, A.A.R.; AlMarzouqi, F.I.; AlDramaki, S.H.; Prinsloo, E.A.M.; Nagelkerke, N. Prevalence of vitamin D deficiency and associated comorbidities among Abu Dhabi Emirates population. *BMC Res. Notes* **2019**, *12*, 503. [CrossRef]
63. Alrayyes, S.F.; Alrayyes, S.F.; Farooq, U.D. Skin-lightening patterns among female students: A cross-sectional study in Saudi Arabia. *Int. J. Women's Dermatol.* **2019**, *5*, 246–250. [CrossRef] [PubMed]
64. Janzi, S.; Padilla, E.C.; Najafi KRamne, S.; Ahlqvist, E.; Borne, Y.; Sonestedt, E. Single Nucleotide Polymorphisms in close proximity to the fibroblast growth factor 11 (FG21) Gene Found to be associated with sugar intake in a Swedish population. *Nutrients* **2021**, *13*, 3954. [CrossRef] [PubMed]
65. Coetzee, V.; Faerber, S.J.; Greeff, J.M.; Lefevre, C.E.; Re, D.E.; Perrett, D.I. African Perceptions of Female Attractiveness. *PLoS ONE* **2012**, *7*, e48116. [CrossRef] [PubMed]
66. Shroff, H.; Diedrichs, P.C.; Craddock, N. Skin Color, Cultural Capital, and Beauty Products: An Investigation of the Use of Skin Fairness Products in Mumbai, India. *Front. Public Health* **2018**, *5*, 365. [CrossRef] [PubMed]
67. Li, E.P.H.; Min, H.J.; Belk, R.W.; Kimura, J.; Bahl, S. Skin Lightening and Beauty in Four Asian Cultures. *Adv. Consum. Res.* **2008**, *35*, 444–449.
68. Carlberg, C. Nutrigenomics of vitamin D. *Nutrients* **2019**, *11*, 676. [CrossRef] [PubMed]

nutrients

Article

Effects of Vitamin D Supplementation on 24-Hour Blood Pressure in Patients with Low 25-Hydroxyvitamin D Levels: A Randomized Controlled Trial

Verena Theiler-Schwetz [1], Christian Trummer [1], Martin R. Grübler [1,2], Martin H. Keppel [3], Armin Zittermann [4], Andreas Tomaschitz [5], Spyridon N. Karras [6], Winfried März [7,8,9], Stefan Pilz [1,*] and Stephanie Gängler [2,10]

[1] Division of Endocrinology and Diabetology, Department of Internal Medicine, Medical University of Graz, 8036 Graz, Austria; verena.schwetz@medunigraz.at (V.T.-S.); christian.trummer@medunigraz.at (C.T.); martin.gruebler@gmx.net (M.R.G.)

[2] Center on Aging and Mobility, University Hospital Zurich, City Hospital Waid&Triemli and University of Zurich, 8006 Zurich, Switzerland; stephanie.gaengler@usz.ch

[3] Department of Laboratory Medicine, Paracelsus Medical University Salzburg, 5020 Salzburg, Austria; keppel.martin@gmail.com

[4] Clinic for Thoracic and Cardiovascular Surgery, Herz- und Diabeteszentrum Nordrhein-Westfalen (NRW), Ruhr University Bochum, 32545 Bad Oeynhausen, Germany; azittermann@hdz-nrw.de

[5] Health Center Trofaiach-Gössgrabenstrasse, 8793 Trofaiach, Austria; andreas.tomaschitz@gmx.at

[6] National Scholarship Foundation, 55535 Thessaloniki, Greece; karraspiros@yahoo.gr

[7] Clinical Institute of Medical and Chemical Laboratory Diagnostics, Medical University of Graz, 8036 Graz, Austria; winfried.maerz@synlab.com

[8] SYNLAB Academy, Synlab Holding Deutschland GmbH, 68159 Mannheim, Germany

[9] V th Department of Medicine (Nephrology, Hypertensiology, Rheumatology, Endocrinology, Diabetology, Lipidology), Medical Faculty Mannheim, University of Heidelberg, 68167 Mannheim, Germany

[10] Department of Aging Medicine and Aging Research, University Hospital Zurich and University of Zurich, 8006 Zurich, Switzerland

* Correspondence: stefan.pilz@medunigraz.at

Citation: Theiler-Schwetz, V.; Trummer, C.; Grübler, M.R.; Keppel, M.H.; Zittermann, A.; Tomaschitz, A.; Karras, S.N.; März, W.; Pilz, S.; Gängler, S. Effects of Vitamin D Supplementation on 24-Hour Blood Pressure in Patients with Low 25-Hydroxyvitamin D Levels: A Randomized Controlled Trial. *Nutrients* **2022**, *14*, 1360. https://doi.org/10.3390/nu14071360

Academic Editor: Andrea Fabbri

Received: 2 February 2022
Accepted: 21 March 2022
Published: 24 March 2022

Publisher's Note: MDPI stays neutral with regard to jurisdictional claims in published maps and institutional affiliations.

Abstract: Accumulating evidence suggests that potential cardiovascular benefits of vitamin D supplementation may be restricted to individuals with very low 25-hydroxyvitamin D (25(OH)D) concentrations; the effect of vitamin D on blood pressure (BP) remains unclear. We addressed this issue in a post hoc analysis of the double-blind, randomized, placebo-controlled Styrian Vitamin D Hypertension Trial (2011–2014) with 200 hypertensive patients with 25(OH)D levels <30 ng/mL. We evaluated whether 2800 IU of vitamin D3/day or placebo (1:1) for 8 weeks affects 24-hour systolic ambulatory BP in patients with 25(OH)D concentrations <20 ng/mL, <16 ng/mL, and <12 ng/mL and whether achieved 25(OH)D concentrations were associated with BP measures. Taking into account correction for multiple testing, p values < 0.0026 were considered significant. No significant treatment effects on 24-hour BP were observed when different baseline 25(OH)D thresholds were used (all p-values > 0.30). However, there was a marginally significant trend towards an inverse association between the achieved 25(OH)D level with 24-hour systolic BP (-0.196 per ng/mL 25(OH)D, 95% CI (-0.325 to -0.067); $p = 0.003$). In conclusion, we could not document the antihypertensive effects of vitamin D in vitamin D-deficient individuals, but the association between achieved 25(OH)D concentrations and BP warrants further investigations on cardiovascular benefits of vitamin D in severe vitamin D deficiency.

Keywords: blood pressure; cardiovascular risk; vitamin D deficiency; cholecalciferol

1. Introduction

Hypertension is a major cardiovascular risk factor and one of three leading risk factors for global disease burden [1]. Vitamin D deficiency is highly prevalent worldwide and associated with poor skeletal health [2–4]. In addition, epidemiological studies have shown inverse associations between serum 25-hydroxyvitamin D (25(OH)D) and cardio-vascular events and/or mortality [5,6]. With reference to major cardiovascular risk factors, mechanistic studies have suggested several pathways linking hypertension and vitamin D deficiency. These include the finding that vitamin D receptor-null mice had increased renin expression and plasma angiotensin II production, leading to hypertension [7]. Furthermore, possible antihypertensive effects of vitamin D receptor activation include improvements in endothelial/vascular function and nephroprotection [6].

In spite of epidemiological data linking vitamin D deficiency to pathways of blood pressure (BP) regulation, clinical studies investigating the effect of vitamin D supplementation on BP and cardiovascular risk have shown inconclusive and often negative [8,9] results. For participants in our single-center, double-blind, placebo-controlled, parallel-group Styrian Vitamin D Hypertension Trial, no significant effect on BP could be observed after supplementing 2800 IU of vitamin D3 per day over a period of 8 weeks in patients with 25(OH)D levels <30 ng/mL [10]. Most meta-analyses of randomized controlled trials (RCTs) concluded that there is no significant antihypertensive effect of vitamin D, but the data are inconsistent with some RCTs reporting on moderate yet statistically significant antihypertensive effects of vitamin D [11–16].

Four large RCTs on cardiovascular risk and mortality have been published recently: the ViDA, the VITAL, the Do-Health, and the D-Health study, failing to show an effect of vitamin D supplementation on cardiovascular disease [17,18], mortality [19], and BP [20]. Limitations of the above-mentioned and many other vitamin D RCTs include the fact that vitamin D supplementation was administered in the general population, including unscreened and possibly vitamin D replete participants, rather than focusing on deficient populations only. In line with these findings, observational studies show that vitamin D supplementation regardless of the prevailing vitamin D status in apparently healthy individuals is likely to show no significant cardiovascular benefit. These studies show that over a wide range of 25(OH)D concentrations, an association of 25(OH)D with various health outcomes is lacking [21–23]. Only at very low 25(OH)D concentrations, a steep increase in risk was apparent [21–23]. Therefore, the missing effect of vitamin D supplementation on clinical outcomes in these four RCTs was not surprising and unanticipated [24,25]. Mendelian randomization (MR) studies, using genetically determined 25(OH)D concentrations as an instrumental variable to evaluate associations with clinical outcomes [26], support the hypothesis that cardiovascular benefits of vitamin D may exist but are restricted to individuals with severe vitamin D deficiency [27,28].

The aforementioned limitations of currently published RCTs, along with the findings of epidemiological and MR studies, underline the necessity to elucidate the effect of vitamin D supplementation in patients with severe vitamin D deficiency. We hypothesize that vitamin D supplementation has an effect on 24-hour systolic (the primary study outcome parameter) and (as secondary outcome parameters) diastolic BP, aldosterone, renin, and pulse wave velocity in patients with 25(OH)D levels <20 ng/mL, <16 ng/mL, and <12 ng/mL in a post hoc analysis from the Styrian Vitamin D Hypertension Trial.

Further, in an exploratory analysis, we aim to assess possible associations between achieved vitamin D status and 24-hour systolic and diastolic BP in all patients included. Primary outcome analyses of this RCT had not reported any beneficial effect of vitamin D on cardiovascular risk factors, including BP, but data on severely vitamin D deficient patients or analyses on associations of achieved 25(OH)D with BP had not been carried out [10].

2. Materials and Methods

2.1. Study Design

This study is a post hoc analysis of the single-center, double-blind, placebo-controlled, parallel-group Styrian Vitamin D Hypertension Trial [10] performed at the Medical University of Graz. The trial was registered at EU Clinical Trials Register (http://www.clinicaltrialsregister.eu, accessed on 16 February 2011, EudraCT number 2009-018125-70) and at clinicaltrials.gov (ClinicalTrials.gov Identifier NCT02136771). The publications of this trial adhere to the Consolidated Standards of Reporting Trials (CONSORT) 2010 statement [29].

2.2. Study Participants

The study participants of the Styrian Vitamin D Hypertension Trial were adults aged 18 years or older. They had arterial hypertension and a 25(OH)D serum concentration <30 ng/mL (multiply by 2.496 to convert ng/mL to nmol/L). We defined the diagnosis of arterial hypertension according to current guidelines at the time of inclusion [30], i.e., an office systolic BP of $\geq$140 mmHg or office diastolic BP $\geq$ 90 mmHg, a mean 24-hour ABPM of systolic $\geq$125 mmHg or diastolic $\geq$80 mmHg, a home BP of systolic $\geq$130 mmHg or diastolic $\geq$85 mmHg, or if patients were receiving antihypertensive treatment. As previously published [10], exclusion criteria included elevated levels of calcium, acute coronary syndrome, or cerebrovascular events within the previous two weeks. Further, pregnant or lactating women were excluded. Additionally, exclusion criteria included drug intake due to participation in another clinical study or an estimated glomerular filtration rate <15 mL/min per 1.73 m^2. A change in antihypertensive treatment during the previous four weeks or planned change of antihypertensive treatment was also part of the exclusion criteria, as were diseases with an estimated life expectancy of fewer than 12 months. Furthermore, 24-hour systolic blood pressure >160 mmHg or <120 mmHg, 24-hour diastolic blood pressure >100 mmHg, any relevant acute diseases requiring drug therapy, chemotherapy, or radiation, or a regular daily intake of more than 880 IU of vitamin D during the last four weeks in addition to the study medication were part of the list of exclusion criteria. All subjects participating in the trial gave their written informed consent prior to study inclusion. The ethics committee of the Medical University of Graz, Austria, approved the study which was designed to comply with the Declaration of Helsinki. Participants were recruited from the outpatient clinics of the Division of Cardiology and the Division of Endocrinology and Diabetology, Department of Internal Medicine, Medical University of Graz, Graz, Austria, between June 2011 and August 2014.

2.3. Intervention

The study medication was filled into numbered bottles; this was performed according to a computer-generated randomization list. Randomization was based on a web-based software called Randomizer (http://www.randomizer accessed on 24 June 2014) provided by the Institute for Medical Informatics, Statistics, and Documentation, Medical University of Graz, Graz, Austria, with good clinical practice compliance as confirmed by the Austrian Agency for Health and Food Safety. Eligible participants were randomly allocated in a 1:1 ratio. Subjects received either 2800 IU of vitamin D3 (Oleovit D3, Fresenius Kabi Austria, Graz, Austria) or a matching placebo (coconut oil) administered both orally by seven oily drops per day for a duration of eight weeks. To ensure adequate study medication intake, patients were asked to return the empty bottles at the follow-up study visits. We carried out a permuted block randomization with a block size of 10 and stratification according to gender. Investigators/authors enrolling participants, collecting data, and assigning intervention were blinded to participant allocation.

2.4. Outcome Measure

The primary outcome measure in this post hoc analysis was the between-group difference in 24-hour systolic BP, which resembled the primary end point of the main trial [10]. Secondary outcome measures included between-group differences in 24-hour diastolic BP, plasma renin concentration, plasma aldosterone concentration, and pulse wave velocity [10]. Initially, pulse wave velocity had not been listed as an outcome during the first trial registration (EudraCT number, 2009-018125-70) but, as with all other outcomes, had been prespecified before the beginning of the trial [10].

In the present post hoc investigation, we re-analyzed the dataset to examine the effect of vitamin D supplementation on the aforementioned primary and secondary outcome measures in patients with 25(OH)D concentrations <20 ng/mL, <16 ng/mL, and <12 ng/mL, as it remains to be elucidated whether subgroups with severe vitamin D deficiency would benefit from vitamin D administration [6]. These cut-offs were chosen based on considerations by the Institute of Medicine, stating that serum 25(OH)D concentrations between 12 ng/mL and 20 ng/mL are considered insufficient for some in the population, and concentrations greater than 20 ng/mL sufficient for nearly all [31]. 25(OH)D levels <12 ng/mL are classified as severe vitamin D deficiency due to the steep increase in the risk of osteomalacia and nutritional rickets below those values [32–36]. The additional threshold of <16 ng/mL was included, as 25(OH)D levels of 16 ng/mL are needed by 50% of individuals aged > 1 year to achieve bone health [31] and due to the fact that MR studies show inverse associations with all-cause mortality and 25(OH)D concentrations up until this cut-off [27].

2.5. Measurements

Patient interviews, physical examinations, and sampling of blood were carried out at study visits between 7 a.m. and 11 a.m. after an overnight fast throughout the year. Ambulatory BP monitoring (ABPM) measurements and 24-hour urine collections started after the visit. Patients returned to the outpatient clinic the following day. At this point, eligible study participants were randomized and started taking the study medication. ABPM measurements were repeated after 8 weeks. We used a validated 24-hour ABPM device (Spacelabs 90217A; Spacelabs Healthcare, Inc, Issaquah, WA, USA) for the measurement of 24-hour systolic and diastolic BP. The circumference of the patient's upper arm was measured to choose the appropriate cuff for BP recordings. BP measurements were performed every 15 min during the day (from 6 a.m. to 10 p.m.) and every 30 min during the night (from 10 p.m. to 6 a.m.). ABPM followed the recommendations of the European Society of Hypertension [37]. The Spacelabs 90217 device is in compliance with the Association for the Advancement of Medical Instrumentation's standard and has earned the highest British Hypertension Society grade of "A" both for systolic and for diastolic blood pressures [38]. A comparison study of brachial blood pressure recordings with Spacelabs 90217A used in this study versus sphygmomanometer did not show significant differences in systolic and diastolic BP measurements [39].

Serum levels of 25(OH)D were measured by a chemiluminescence assay (IDS-iSYS 25-hydroxyvitamin assay; Immunodiagnostic Systems Ltd., Boldon, UK) with an intra-assay and inter-assay coefficient of variation (CV) of 6.2% and 11.6%, respectively. Plasma aldosterone concentration (PAC) was determined by means of an RIA (Active Aldosterone RIA DSL-8600; Diagnostic Systems Laboratories, Inc, Webster, TX) with an intra-assay and inter-assay CV of 3.3–4.5% and 5.9–9.8%, respectively. Plasma renin concentrations (PRC) were determined by a "RENIN III GENERATION" (GEN. III) RIA assay (Renin IRMA RIA-4541, DRG Instruments GmbH, Marburg, Germany) with an intra-assay and inter-assay CV of 0.6–4.5% and 2.7–14.5%, respectively. All other parameters were determined by routine laboratory procedures. All parameters were measured on a daily basis.

2.6. Statistical Methods

Sample size calculations were based on the primary outcome of the trial, as previously described [10]. Continuous data following a normal distribution (determined by Kolmogorov–Smirnov tests and data visualization by histogram) are shown as means with standard deviations, variables with a non-normal distribution are shown as medians with interquartile ranges (IQR), and categorical data are presented as percentages. Skewed variables were log(e) transformed before their use in parametric statistical analyses. Group comparisons at baseline were analyzed with either unpaired Student's *t*-tests or chi-squared tests or, in case of a non-normal distribution, by use of the Mann–Whitney U test. Analyses of outcome variables were performed according to the intention-to-treat principle without data imputation and with the inclusion of all participants with baseline and follow-up values of the respective outcome variable.

In order to test for differences in the outcome variables between the treatment and the placebo group at the follow-up visit in patients with severe vitamin D deficiency (i.e., with 25(OH)D levels <20 ng/mL, <16 ng/mL, and <12 ng/mL), analysis of covariance (ANCOVA) with adjustments for baseline values was used.

Additionally, in an exploratory analysis, the relationship between achieved serum 25(OH)D concentrations and 24-hour systolic BP and 24-hour diastolic BP was assessed using linear mixed models with a random intercept for participants. This was performed for 25(OH)D as continuous measurement and measurements above or below the cut-off level of ≤20 ng/mL. These models were adjusted for treatment, time, and their interaction.

To further investigate the relationship between achieved vitamin D concentration and the change in 24-hour systolic BP, LOESS smoothing (local regression, span = 0.75) was applied to the data as previously published [40] to visualize a possible non-linear relationship of change in 24-hour systolic BP and achieved 25(OH)D concentrations after 8 weeks.

We used Bonferroni correction for p-values to account for multiple testing ($p = 0.05/19$). Thus, p-values < 0.0026 were considered statistically significant. Statistical analyses were performed with SPSS version 27 (SPSS, Chicago, IL, USA) and R (version 4.1.1).

3. Results

Two hundred patients were randomized, and 188 completed the study (with an age range from 18.8 to 86.0 years). A significant increase in 25(OH)D and a significant decrease in parathyroid hormone were observed and previously published [10]. Of these 188 patients, 70 (37%) had 25(OH)D levels below 20 ng/mL, 42 (21%) had 25(OH)D levels below 16 ng/mL, and 14 patients (7%) had 25(OH) levels below 12 ng/mL. Previously published baseline characteristics of all 188 patients [10] are shown in Table 1. The 70 patients with 25(OH)D levels <20 ng/mL, of whom 32 (45.7%) were women, had a mean age of 59.9 ± 3.1 and median 25(OH)D levels of 14.7 ng/mL (IQR 12.3–18.2).

Table 1. Selected baseline characteristics of all randomized study participants, as previously published [10].

	Vitamin D Group (n = 100)	Placebo Group (n = 100)
Females (%)	46	48
Age (years)	60.5 ± 10.9	59.7 ± 11.4
Body mass index (kg/m^2)	30.4 ± 4.4	30.4 ± 6.2
Active smoker (%)	19	14
Previous MI (%)	8	5
Office systolic BP (mm Hg)	143.7 ± 15.2	142.3 ± 15.4
Office diastolic BP (mm Hg)	87.1 ± 10.3	86.8 ± 10.8
24-hour systolic BP (mmHg)	132.1 ± 8.4	131.7 ± 9.7
24-hour diastolic BP (mmHg)	78.6 ± 7.5	77.8 ± 8.4
NT-proBNP (pg/mL)	69 (35–142)	99 (51–169)
PRC (μU/mL)	16.3 (10.1–39.0)	16.1 (9.5–52.0)
PAC (ng/mL)	15.4 (9.7–19.4)	14.7 (10.6–19.9)
eGFR (mL/min/1.73m2) CKI-EPI	80.0 ± 17.9	77.2 ± 17.9
PWV (m/s)	8.42 ± 1.90	8.28 ± 2.26
25(OH)D (ng/mL)	22.0 ± 5.7	20.5 ± 5.7
25(OH)D < 20 ng/mL (n)	33	42
25(OH)D < 16 ng/mL (n)	18	27
25(OH)D < 12 ng/mL (n)	6	8
PTH (pg/mL)	48.9 (40.0–61.7)	51.6 (39.5–65.8)
Plasma calcium (mmol/L)	2.37 ± 0.10	2.37 ± 0.11
Antihypertensive drugs (n)	2 (1–3)	2 (1–3)
ACE-inhibitor (%)	25	38
AT II blocker (%)	33	31
Thiazide diuretic (%)	39	45
Beta-blocker (%)	44	49
Calcium channel blocker (%)	27	25

Data are presented as means with standard deviation, medians with interquartile ranges, or percentages. Comparisons between the vitamin D and placebo groups were calculated with Student's t-test or Chi-square test. MI—myocardial infarction; BP—blood pressure; NT-proBNP—N-terminal pro-B-type natriuretic peptide; PRC—plasma renin concentration; PAC—plasma aldosterone concentration; eGFR—estimated glomerular filtration rate; PWC—pulse wave velocity; 25(OH)D—25-hydroxyvitamin D; PTH—parathyroid hormone.

In the group of patients with 25(OH)D levels <20 ng/mL, median 25(OH)D was 14.5 (12.1–16.9) in the placebo group and 14.7 (13.0–19.1) in the vitamin D group at baseline. At follow-up, 25(OH)D was 16.3 (IQR 13.3–22.0) in the placebo group and 31.6 (28.6–40.1) in the vitamin D group. In the group of patients with 25(OH)D <16 ng/mL, median 25(OH)D was 12.7 (11.8–14.6) in the placebo group and 13.1 (11.7–13.6) in the vitamin D group. At follow-up, median 25(OH)D was 16.3 (12.4–20.0) in the placebo group and 31.7 (28.1–42.5) in the vitamin D group. In the group of patients with 25(OH)D <12 ng/mL, median 25(OH)D was 11.5 (9.7–11.8) in the placebo group and 11.3 (9.0–11.7) in the vitamin D group. At follow-up, 25(OH)D was 13.3 (12.1–21.9) in the placebo group and 30.3 (27.3–34.6) in the vitamin D group.

There was no significant effect of vitamin D supplementation on 24-hour systolic or diastolic BP in either of the subgroups with low 25(OH)D concentrations (see Tables 2–4). When analyzing the raw data, there was a decrease in 24-hour systolic and diastolic BP in the vitamin D group (Table 4) of patients with 25(OH)D levels <12 ng/mL, which was, however, non-significant and therefore has to be interpreted with great caution. No significant treatment effect was observed for aldosterone or pulse wave velocity in all subgroups or renin in the subgroups with 25(OH)D <16 ng/mL or <12 ng/mL (Tables 2–4). A significant effect was only seen for renin in the subgroup <20 ng/mL (Table 2). However, after correction for multiple testing (Bonferroni), these results did not remain significant.

Evaluating the achieved 25(OH)D concentrations as a continuous variable in all patients, higher achieved 25(OH)D concentrations were significantly associated with lower levels of 24-hour systolic BP (-0.196, 95% CI (from -0.325 to -0.067); $p = 0.003$). Further, achieved 25(OH)D concentrations $\leq$20 ng/mL were significantly associated with higher 24-hour systolic BP (2.275, 95%CI (from 0.281 to 4.278); $p = 0.026$). After Bonferroni correction for multiple testing, however, both associations did not remain significant (corrected p-value = 0.0026). No significant association was seen for 24-hour diastolic BP with achieved 25(OH)D assessed as continuous measurement (-0.022; 95% CI (from -0.107 to 0.063); $p = 0.613$) or for 25(OH)D using the cut-off of $\leq$20 ng/mL (0.871 95% CI (from -0.433 to 2.166); $p = 0.19$).

Table 2. Outcome variables at baseline and follow-up and treatment effects in 70 study participants with 25(OH)D < 20 ng/mL who completed the trial (of 75 participants with 25(OH)D < 20 ng/mL who had initially been randomized).

	Baseline	Follow-Up	Treatment Effect	*p*-Value
		24-hour systolic blood pressure, mm Hg *		
Vitamin D (*n* = 31)	131.5 (124.0–141.3)	130.0 (125.1–137.7)	0.0 (−4.7 to 4.7)	0.971
Placebo (*n* = 39)	131.2 (126.4–137.9)	128.3 (121.4–140.6)		
		24-hour diastolic blood pressure, mm Hg		
Vitamin D (*n* = 31)	77.7 ± 7.0	76.9 ± 8.6	0.7 (−1.8 to 3.3)	0.572
Placebo (*n* = 39)	77.2 ± 8.3	75.6 ± 8.5		
		Plasma renin concentration, µU/Ml *		
Vitamin D (*n* = 31)	18.6 (11.6–52.4)	15.8 (9.9–33.5)	−13 (−28 to 1)	0.014
Placebo (*n* = 39)	16.6 (11.3–50.9)	21.5 (13.2–44.2)		
		Plasma aldosterone concentration, ng/dL *		
Vitamin D (*n* = 31)	15.7 (9.0–21.0)	14.0 (10.4–19.6)	−1.3 (−4.5 to 2.0)	0.274
Placebo (*n* = 39)	12.8 (9.2–18.1)	16.4 (12.5–21.9)		
		Pulse wave velocity, m/s *		
Vitamin D (*n* = 30)	8.35 (6.99–9.44)	7.90 (7.10–9.35)	0.21 (−0.62 to 1.04)	0.486
Placebo (*n* = 32)	8.43 (7.33–10.30)	8.05 (7.00–9.98)		

Data at baseline and follow-up are shown as medians with SD or as medians with interquartile range. Treatment effects (with 95% confidence intervals) and *p*-values were calculated by ANCOVA for group differences at follow-up with adjustment for baseline values. * Skewed variables for which logarithmically transformed values were used in ANCOVA. Untransformed values are shown in the Table.

Table 3. Outcome variables at baseline and follow-up and treatment effects in 42 study participants with 25(OH)D < 16 ng/mL who completed the trial (of 45 participants with 25(OH)D < 16 ng/mL who had initially been randomized).

	Baseline	**Follow-Up**	**Treatment Effect**	*p*-**Value**
	24-hour systolic blood pressure, mm Hg			
Vitamin D (n = 16)	132.1 ± 8.8	131.9 ±6.5	0.3 (−6.0 to 6.5)	0.931
Placebo (n = 26)	134.9 ± 9.0	134.1 ± 13.2		
	24-hour diastolic blood pressure, mm Hg *			
Vitamin D (n = 16)	77.3 (71.6–80.3)	73.3 (70.1–77.9)	0.187 (−2.622 to 2.996)	0.862
Placebo (n = 26)	74.3 (68.8–83.4)	73.1 (67.0–80.0)		
	Plasma renin concentration, µU/Ml *			
Vitamin D (n = 16)	25.7 (13.0–112.4)	20.1 (11.9–45.9)	−14.0 (−37.0 to 9.1)	0.150
Placebo (n = 26)	16.4 (11.0–74.2)	19.1 (11.7–50.8)		
	Plasma aldosterone concentration, ng/dL			
Vitamin D (n = 16)	17.8 ± 14.2	16.9 ± 6.9	−1.6 (−5.7 to 2.4)	0.422
Placebo (n = 26)	13.6 ± 4.6	17.1 ± 6.8		
	Pulse wave velocity, m/s *			
Vitamin D (n = 16)	8.73 (7.83–9.35)	8.40 (7.70–10.60)	0.68 (−0.40 to 1.75)	0.204
Placebo (n = 21)	8.70 (7.85–10.80)	8.23 (7.43–9.75)		

Data at baseline and follow-up are shown as medians with SD or as medians with interquartile range. Treatment effects (with 95% confidence intervals) and *p*-values were calculated by ANCOVA for group differences at follow-up with adjustment for baseline values. * Skewed variables for which logarithmically transformed values were used in ANCOVA. Untransformed values are shown in the Table.

Table 4. Outcome variables at baseline and follow-up and treatment effects in 14 study participants with 25(OH)D < 12 ng/mL who completed the trial (i.e., all of 14 participants with 25(OH)D < 12 ng/mL who had initially been randomized).

	Baseline	**Follow-Up**	**Treatment Effect**	*p*-**Value**
	24-hour systolic blood pressure, mm Hg *			
Vitamin D (n = 6)	135.1 (128.0–141.7)	130.3 (125.9–136.8)	−5.9 (−20.1 to 8.3)	0.347
Placebo (n = 8)	139.8 (136.7–143.0)	140.7 (136.3–152.4)		
	24-hour diastolic blood pressure, mm Hg			
Vitamin D (n = 6)	77.8 ± 6.3	74.9 ± 6.8	−2.0 (−8.1 to 4.1)	0.477
Placebo (n = 8)	73.4 ± 7.7	73.5 ± 7.5		
	Plasma renin concentration, µU/Ml *			
Vitamin D (n = 6)	69.1 (14.1–157.1)	24.8 (12.6–96.2)	−1.5 (−48.6 to 45.6)	0.859
Placebo (n = 8)	39.5 (11.6–101.4)	34.6 (14.9–104.3)		
	Plasma aldosterone concentration, ng/dL			
Vitamin D (n = 6)	12.1 ± 8.7	12.5 ± 5.3	−0.9 (−6.8 to 5.0)	0.747
Placebo (n = 8)	13.3 ± 5.8	14.1 ± 7.1		
	Pulse wave velocity, m/s *			
Vitamin D (n = 6)	8.88 (8.05–11.65)	8.98 (7.85–12.05)	−0.60 (−2.85 to 1.65)	0.759
Placebo (n = 7)	7.90 (7.60–10.00)	7.90 (6.95–9.10)		

Data at baseline and follow-up are shown as medians with SD or as medians with interquartile range. Treatment effects (with 95% confidence intervals) and *p*-values were calculated by ANCOVA for group differences at follow-up with adjustment for baseline values. * Skewed variables for which logarithmically transformed values were used in ANCOVA. Untransformed values are shown in the Table.

An exploratory, observational analysis to assess the change in 24-hour systolic BP with regards to achieved 25(OH)D levels suggests that a target concentration of 23.0 ng/mL may be associated with maximal BP reduction (see Figure 1).

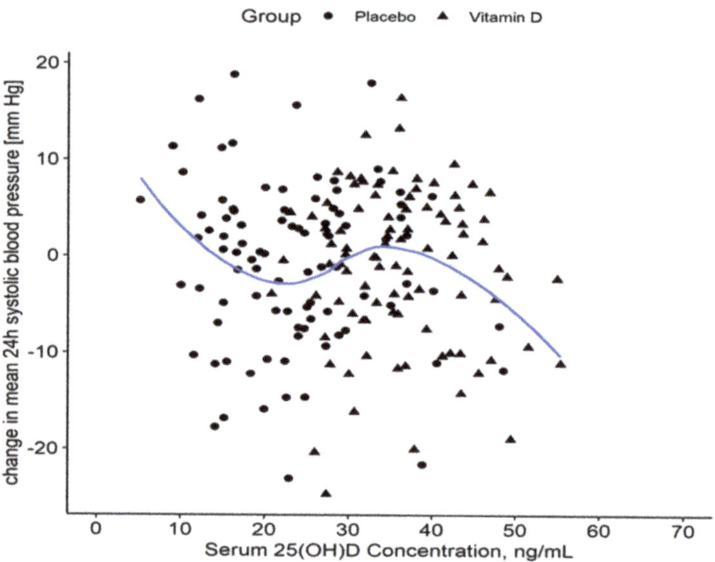

Figure 1. Change in 24-hour systolic blood pressure by serum 25(OH)D concentration. The LOESS curve visualizes a possible non-linear relationship of change in 24-hour systolic blood pressure at 8 weeks from baseline as a function of achieved 25(OH)D. The shaded area represents the 95% CI bands. The total sample size is n = 200. LOESS—locally estimated scatterplot smoothing; 25(OH)D—25-hydroxyvitamin D.

4. Discussion

No significant treatment effects of vitamin D supplementation on 24-hour systolic or diastolic BP, aldosterone, renin, or pulse wave velocity were observed in hypertensive patients with 25(OH)D levels <20 ng/mL, <16 ng/mL, and <12 ng/mL. However, an exploratory analysis of the entire study cohort showed a marginally significant trend towards an association between achieved 25(OH)D concentrations and 24-hour systolic BP.

The findings observed in our subgroup with 25(OH)D levels <20 ng/mL are partly in line with the subgroup analyses from the ViDA study investigating vitamin D effects on cardiovascular disease occurrence [17]. While the ViDA study had included community-resident adults aged from 50 to 84 years and not hypertensive patients, subgroup analysis for patients with 25(OH)D levels <20 ng/mL had been prespecified. There was no significant difference in cardiovascular disease occurrence in participants with baseline vitamin D deficiency (i.e., 25(OH)D < 20 ng/mL) or without as well as for secondary outcomes, including the number of participants developing hypertension over 3.3 years [17]. However, in a prespecified subsample of participants, monthly, high-dose, 1-year vitamin D supplementation lowered central BP parameters among adults with 25(OH)D levels <20 ng/mL but not in the total sample. However, effects on brachial BP were non-significant [41].

Subgroup analyses for even more severely vitamin D deficient people are largely unavailable in published RCTs. The post hoc subgroup analyses in the Styrian Hypertension Trial for patients with 25(OH)D levels <16 ng/mL and <12 ng/mL are, however, accompanied by the limitation of very small sample sizes. Particularly with regard to observational data [5,6], though, it thus remains to be elucidated whether patients with

25(OH)D concentrations even lower than 20 ng/mL would show reductions in BP after vitamin D supplementation in RCTs designed for this patient group.

Data from MR studies support the assumption that future RCTs should focus on patients with severe vitamin D deficiency. In vitamin D-deficient individuals with 25(OH)D concentrations <10 ng/mL, genetic analyses provided strong evidence for an inverse association with all-cause mortality (odds ratio, OR) per 4 ng/mL increase in genetically predicted 25(OH)D concentration of 0.69 (95% CI from 0.59 to 0.80; $p < 0.0001$) and a non-significant inverse association for coronary heart disease of 0.89 (95% CI from 0.76 to 1.04; $p = 0.14$) [27]. After a finer stratification, the association with all-cause mortality was observed up to 25(OH)D levels of 16 ng/mL [27]. A MR study including >140 000 individuals documented that each 10% increase in genetically determined/instrumented 25(OH)D levels was associated with a reduction in systolic BP of −0.37 mmHg (95% CI from −0.73 to 0.003; $p = 0.052$) and a reduction in diastolic BP of –0.29 mmHg (95% CI from −0.52 to −0.07; $p = 0.01$) [42]. Zhou and colleagues reported an L-shaped association between genetically predicted serum 25(OH)D and cardiovascular disease risk and a similar association for systolic and diastolic BP [28]. Individuals with 25(OH)D concentrations of 10 ng/mL were estimated to have 0.70 mmHg (95% CI from 0.15–1.26) and 0.25 mmHg (95% CI from −0.02 to 0.51) higher BP compared with 20 ng/mL [28]. These findings are in line with our data, implying that a very large sample size would have been needed to detect significant treatment effects of vitamin D on BP in deficient individuals. These MR studies thus highlight that the magnitude of a possible BP-lowering effect of vitamin D might be small on an individual level but could be relevant on a population level. If this benefit for people with vitamin D deficiency is confirmed by future studies, implications may be to support a population approach for preventing vitamin D deficiency via food fortification [43] rather than population-wide screening for deficiency followed by supplementation, especially in light of the high prevalence of vitamin D deficiency worldwide [44].

Future RCTs should not only focus on including patients with severe vitamin D deficiency but also investigate achieved 25(OH)D levels, as suggested by our data showing a possible association of achieved 25(OH)D levels with 24-hour systolic BP. One study investigating achieved 25(OH)D concentrations found a reduction of mean systolic BP over 2 years following supplementation with daily 2000 IU and 800 IU of vitamin D [40]. Due to the lack of a placebo group, a BP-lowering effect of vitamin D could not firmly be established. In this study, an achieved concentration of 28.7 ng/mL of 25(OH)D was associated with a 6.01 mm Hg (95% CI from −8.20 to −3.82) reduction in daytime systolic BP. At 25(OH)D concentrations below and above 28.7 ng/mL, an increasing loss of benefit was observed. These findings are in line with a meta-analysis showing a dose–response relationship between 25(OH)D concentrations and hypertension risk, pointing at a substantial increment in the risk of hypertension at <28.4 ng/mL [45], while our exploratory analysis suggested a threshold of 23 ng/mL.

We want to emphasize that taking into account the achieved concentrations of 25(OH)D is particularly important with regard to vitamin D RCTs. A true placebo group totally lacking vitamin D exposure is biologically impossible, so between-group comparisons of vitamin D RCTs will always be based on lower versus higher vitamin D exposure [25]. With the inverse association of achieved 25(OH)D levels with 24-hour systolic BP presented in this post hoc analysis, we highlight the importance of this consideration and the necessity to take achieved or intra-trial 25(OH)D concentrations into account when interpreting vitamin D RCT results.

Our study has strengths and several limitations worth mentioning. The latter include that we report findings from a single center with a cohort of white hypertensive patients, which may not be generalizable to other study populations [10]. Further, the very low prevalence of patients with severe vitamin D deficiency, especially when performing subgroup analysis, is a major limitation, so our findings should only be regarded as explorative and hypothesis generating. Additionally, it has to be regarded as a major limitation that in the subgroup analyses, patients with 25(OH)D levels <20 ng/mL, <16 ng/mL, and <12 ng/mL are in part double and triple reported, as patients in the very severely deficient and severely deficient groups are also reported in the less severely deficient subgroup(s). Strengths are the well-validated BP assessment with ABPM and the fact that treatment caused a significant increment in 25(OH)D levels [10].

To conclude, we could not document a significant treatment effect of vitamin D supplementation on 24-hour systolic or diastolic BP, aldosterone, renin, or pulse wave velocity in a small number of hypertensive patients with 25(OH)D levels <20 ng/mL, <16 ng/mL, and <12 ng/mL. However, marginally significant trends in our analyses suggest an inverse association of achieved 25(OH)D concentrations with 24-hour systolic BP. Taken together, further trials to investigate whether vitamin D supplementation might have beneficial effects in severely vitamin D deficient patients also taking into account achieved 25(OH)D levels after supplementation are warranted. Such investigations are of high importance as BP-lowering effects in vitamin D deficient individuals may improve public health when considering the relatively high prevalence of vitamin D deficiency, the efficacy of antihypertensive effects in reducing morbidity and mortality, and the relatively cheap and effective approaches for improving vitamin D status such as food fortification.

Author Contributions: Conceptualization, V.T.-S., A.T. and S.P.; Data curation, M.R.G., M.H.K. and S.P.; Formal analysis, V.T.-S., C.T., M.R.G., S.P. and S.G.; Funding acquisition, S.P.; Investigation, M.R.G., M.H.K. and S.P.; Methodology, S.P.; Project administration, S.P.; Writing—original draft, V.T.-S.; Writing—review and editing, V.T.-S., C.T., M.R.G., M.H.K., A.Z., A.T., S.N.K., W.M., S.P. and S.G. All authors have read and agreed to the published version of the manuscript.

Funding: The Styrian Vitamin D Hypertension Trial was supported by funding from the Austrian National Bank (Jubilaeumsfond: project no.: 13878 and 13905).

Institutional Review Board Statement: The study was conducted in accordance with the Declaration of Helsinki and approved by the Ethics Committee of the Medical University of Graz, Graz, Austria (EK-Nr.: 21-155 ex 09/10.).

Informed Consent Statement: Informed consent was obtained from all subjects involved in the study.

Data Availability Statement: Information available upon request to the corresponding author.

Acknowledgments: We thank all study participants. We also thank Fresenius Kabi for providing the study medication and BioPersMed (COMET K-project 825329, funded by the Austrian Federal Ministry of Transport, Innovation and Technology, the Austrian Federal Ministry of Economics and Labour/Federal Ministry of Economy, Family and Youth, and by the Styrian Business Promotion Agency) for analysis equipment.

Conflicts of Interest: The authors declare no conflict of interest.

References

1. Lim, S.S.; Vos, T.; Flaxman, A.D.; Danaei, G.; Shibuya, K.; Adair-Rohani, H.; Amann, M.; Anderson, H.R.; Andrews, K.G.; Aryee, M.; et al. A Comparative Risk Assessment of Burden of Disease and Injury Attributable to 67 Risk Factors and Risk Factor Clusters in 21 Regions, 1990–2010: A Systematic Analysis for the Global Burden of Disease Study 2010. *Lancet* **2012**, *380*, 2224–2260. [PubMed]
2. Pilz, S.; Zittermann, A.; Trummer, C.; Theiler-Schwetz, V.; Lerchbaum, E.; Keppel, M.H.; Grubler, M.R.; Marz, W.; Pandis, M. Vitamin D Testing and Treatment: A Narrative Review of Current Evidence. *Endocr. Connect.* **2019**, *8*, R27–R43. [PubMed]
3. Giustina, A.; Bouillon, R.; Binkley, N.; Sempos, C.; Adler, R.A.; Bollerslev, J.; Dawson-Hughes, B.; Ebeling, P.R.; Feldman, D.; Heijboer, A.; et al. Controversies in Vitamin D: A Statement from the Third International Conference. *JBMR Plus* **2020**, *4*, e10417.
4. Cashman, K.D.; Dowling, K.G.; Skrabakova, Z.; Gonzalez-Gross, M.; Valtuena, J.; De Henauw, S.; Moreno, L.; Damsgaard, C.T.; Michaelsen, K.F.; Molgaard, C.; et al. Vitamin D Deficiency in Europe: Pandemic? *Am. J. Clin. Nutr.* **2016**, *103*, 1033–1044.
5. Bouillon, R.; Marcocci, C.; Carmeliet, G.; Bikle, D.; White, J.H.; Dawson-Hughes, B.; Lips, P.; Munns, C.F.; Lazaretti-Castro, M.; Giustina, A.; et al. Skeletal and Extraskeletal Actions of Vitamin D: Current Evidence and Outstanding Questions. *Endocr. Rev.* **2019**, *40*, 1109–1151. [PubMed]
6. Zittermann, A.; Trummer, C.; Theiler-Schwetz, V.; Lerchbaum, E.; Marz, W.; Pilz, S. Vitamin D and Cardiovascular Disease: An Updated Narrative Review. *Int. J. Mol. Sci.* **2021**, *22*, 2896. [CrossRef] [PubMed]
7. Li, Y.C.; Kong, J.; Wei, M.; Chen, Z.F.; Liu, S.Q.; Cao, L.P. 1,25-Dihydroxyvitamin D(3) is a Negative Endocrine Regulator of the Renin-Angiotensin System. *J. Clin. Investig.* **2002**, *110*, 229–238.

8. Kubiak, J.; Thorsby, P.M.; Kamycheva, E.; Jorde, R. Vitamin D Supplementation Does Not Improve CVD Risk Factors in Vitamin D-Insufficient Subjects. *Endocr. Connect.* **2018**, *7*, 840–849.
9. Swart, K.M.; Lips, P.; Brouwer, I.A.; Jorde, R.; Heymans, M.W.; Grimnes, G.; Grubler, M.R.; Gaksch, M.; Tomaschitz, A.; Pilz, S.; et al. Effects of Vitamin D Supplementation on Markers for Cardiovascular Disease and Type 2 Diabetes: An Individual Participant Data Meta-Analysis of Randomized Controlled Trials. *Am. J. Clin. Nutr.* **2018**, *107*, 1043–1053.
10. Pilz, S.; Gaksch, M.; Kienreich, K.; Grubler, M.; Verheyen, N.; Fahrleitner-Pammer, A.; Treiber, G.; Drechsler, C.; O Hartaigh, B.; Obermayer-Pietsch, B.; et al. Effects of Vitamin D on Blood Pressure and Cardiovascular Risk Factors: A Randomized Controlled Trial. *Hypertension* **2015**, *65*, 1195–1201.
11. Bouillon, R.; Manousaki, D.; Rosen, C.; Trajanoska, K.; Rivadeneira, F.; Richards, J.B. The Health Effects of Vitamin D Supplementation: Evidence from Human Studies. *Nat. Rev. Endocrinol.* **2022**, *18*, 96–110. [PubMed]
12. Manson, J.E.; Bassuk, S.S.; Buring, J.E.; VITAL Research Group. Principal Results of the VITamin D and OmegA-3 TriaL (VITAL) and Updated Meta-Analyses of Relevant Vitamin D Trials. *J. Steroid Biochem. Mol. Biol.* **2020**, *198*, 105522. [PubMed]
13. Barbarawi, M.; Kheiri, B.; Zayed, Y.; Barbarawi, O.; Dhillon, H.; Swaid, B.; Yelangi, A.; Sundus, S.; Bachuwa, G.; Alkotob, M.L.; et al. Vitamin D Supplementation and Cardiovascular Disease Risks in More than 83,000 Individuals in 21 Randomized Clinical Trials: A Meta-Analysis. *JAMA Cardiol.* **2019**, *4*, 765–776. [PubMed]
14. Ganmaa, D.; Enkhmaa, D.; Nasantogtokh, E.; Sukhbaatar, S.; Tumur-Ochir, K.E.; Manson, J.E. Vitamin D, Respiratory Infections, and Chronic Disease: Review of Meta-Analyses and Randomized Clinical Trials. *J. Intern. Med.* **2022**, *291*, 141–164.
15. Scragg, R.; Sluyter, J.D. Is there Proof of Extraskeletal Benefits from Vitamin D Supplementation from Recent Mega Trials of Vitamin D? *JBMR Plus* **2021**, *5*, e10459.
16. Beveridge, L.A.; Khan, F.; Struthers, A.D.; Armitage, J.; Barchetta, I.; Bressendorff, I.; Cavallo, M.G.; Clarke, R.; Dalan, R.; Dreyer, G.; et al. Effect of Vitamin D Supplementation on Markers of Vascular Function: A Systematic Review and Individual Participant Meta-Analysis. *J. Am. Heart Assoc.* **2018**, *7*, e008273. [CrossRef]
17. Scragg, R.; Stewart, A.W.; Waayer, D.; Lawes, C.M.M.; Toop, L.; Sluyter, J.; Murphy, J.; Khaw, K.T.; Camargo, C.A., Jr. Effect of Monthly High-Dose Vitamin D Supplementation on Cardiovascular Disease in the Vitamin D Assessment Study: A Randomized Clinical Trial. *JAMA Cardiol.* **2017**, *2*, 608–616.
18. Manson, J.E.; Cook, N.R.; Lee, I.M.; Christen, W.; Bassuk, S.S.; Mora, S.; Gibson, H.; Gordon, D.; Copeland, T.; D'Agostino, D.; et al. Vitamin D Supplements and Prevention of Cancer and Cardiovascular Disease. *N. Engl. J. Med.* **2019**, *380*, 33–44.
19. Neale, R.E.; Baxter, C.; Romero, B.D.; McLeod, D.S.A.; English, D.R.; Armstrong, B.K.; Ebeling, P.R.; Hartel, G.; Kimlin, M.G.; O'Connell, R.; et al. The D-Health Trial: A Randomised Controlled Trial of the Effect of Vitamin D on Mortality. *Lancet Diabetes Endocrinol.* **2022**, *10*, 120–128.
20. Bischoff-Ferrari, H.A.; Vellas, B.; Rizzoli, R.; Kressig, R.W.; da Silva, J.A.P.; Blauth, M.; Felson, D.T.; McCloskey, E.V.; Watzl, B.; Hofbauer, L.C.; et al. Effect of Vitamin D Supplementation, Omega-3 Fatty Acid Supplementation, or a Strength-Training Exercise Program on Clinical Outcomes in Older Adults: The DO-HEALTH Randomized Clinical Trial. *JAMA* **2020**, *324*, 1855–1868.
21. Gaksch, M.; Jorde, R.; Grimnes, G.; Joakimsen, R.; Schirmer, H.; Wilsgaard, T.; Mathiesen, E.B.; Njolstad, I.; Lochen, M.L.; Marz, W.; et al. Vitamin D and Mortality: Individual Participant Data Meta-Analysis of Standardized 25-Hydroxyvitamin D in 26916 Individuals from a European Consortium. *PLoS ONE* **2017**, *12*, e0170791.
22. Durazo-Arvizu, R.A.; Dawson-Hughes, B.; Kramer, H.; Cao, G.; Merkel, J.; Coates, P.M.; Sempos, C.T. The Reverse J-Shaped Association between Serum Total 25-Hydroxyvitamin D Concentration and all-Cause Mortality: The Impact of Assay Standardization. *Am. J. Epidemiol.* **2017**, *185*, 720–726. [PubMed]
23. Schottker, B.; Jorde, R.; Peasey, A.; Thorand, B.; Jansen, E.H.; Groot, L.; Streppel, M.; Gardiner, J.; Ordonez-Mena, J.M.; Perna, L.; et al. Vitamin D and Mortality: Meta-Analysis of Individual Participant Data from a Large Consortium of Cohort Studies from Europe and the United States. *BMJ* **2014**, *348*, g3656. [PubMed]
24. Pilz, S.; Rutters, F.; Dekker, J.M. Disease Prevention: Vitamin D Trials. *Science* **2012**, *338*, 883.
25. Pilz, S.; Trummer, C.; Theiler-Schwetz, V.; Grubler, M.R.; Verheyen, N.D.; Odler, B.; Karras, S.N.; Zittermann, A.; Marz, W. Critical Appraisal of Large Vitamin D Randomized Controlled Trials. *Nutrients* **2022**, *14*, 303. [CrossRef]
26. Mukamal, K.J.; Stampfer, M.J.; Rimm, E.B. Genetic Instrumental Variable Analysis: Time to Call Mendelian Randomization what it is. the Example of Alcohol and Cardiovascular Disease. *Eur. J. Epidemiol.* **2020**, *35*, 93–97.
27. Emerging Risk Factors Collaboration/EPIC-CVD/Vitamin D Studies Collaboration. Estimating Dose-Response Relationships for Vitamin D with Coronary Heart Disease, Stroke, and all-Cause Mortality: Observational and Mendelian Randomisation Analyses. *Lancet Diabetes Endocrinol.* **2021**, *9*, 837–846.
28. Zhou, A.; Selvanayagam, J.B.; Hypponen, E. Non-Linear Mendelian Randomization Analyses Support a Role for Vitamin D Deficiency in Cardiovascular Disease Risk. *Eur. Heart J.* **2021**, ehab809. [CrossRef]
29. Moher, D.; Hopewell, S.; Schulz, K.F.; Montori, V.; Gotzsche, P.C.; Devereaux, P.J.; Elbourne, D.; Egger, M.; Altman, D.G. CONSORT 2010 Explanation and Elaboration: Updated Guidelines for Reporting Parallel Group Randomised Trials. *BMJ* **2010**, *340*, c869.

30. Mancia, G.; Fagard, R.; Narkiewicz, K.; Redon, J.; Zanchetti, A.; Bohm, M.; Christiaens, T.; Cifkova, R.; De Backer, G.; Dominiczak, A.; et al. 2013 ESH/ESC Guidelines for the Management of Arterial Hypertension: The Task Force for the Management of Arterial Hypertension of the European Society of Hypertension (ESH) and of the European Society of Cardiology (ESC). *Eur. Heart J.* **2013**, *34*, 2159–2219.

31. Ross, A.C.; Taylor, C.L.; Yaktine, A.L.; Del Valle, H.B. (Eds.) *Dietary Reference Intakes for Calcium and Vitamin D. Institute of Medicine (US) Committee to Review Dietary Reference Intakes for Vitamin D and Calcium;* The National Academies Collection, Reports Funded by National Institutes of Health; The National Academies Press: Washington, DC, USA, 2011.

32. EFSA. EFSA Panel on Dietetic Products, Nutrition and Allergies (NDA). Dietary Reference Values for Vitamin D. *EFSA J.* **2016**, *14*, 4547.

33. Braegger, C.; Campoy, C.; Colomb, V.; Decsi, T.; Domellof, M.; Fewtrell, M.; Hojsak, I.; Mihatsch, W.; Molgaard, C.; Shamir, R.; et al. Vitamin D in the Healthy European Paediatric Population. *J. Pediatr. Gastroenterol. Nutr.* **2013**, *56*, 692–701. [PubMed]

34. Munns, C.F.; Shaw, N.; Kiely, M.; Specker, B.L.; Thacher, T.D.; Ozono, K.; Michigami, T.; Tiosano, D.; Mughal, M.Z.; Makitie, O.; et al. Global Consensus Recommendations on Prevention and Management of Nutritional Rickets. *J. Clin. Endocrinol. Metab.* **2016**, *101*, 394–415. [PubMed]

35. Martineau, A.R.; Jolliffe, D.A.; Hooper, R.L.; Greenberg, L.; Aloia, J.F.; Bergman, P.; Dubnov-Raz, G.; Esposito, S.; Ganmaa, D.; Ginde, A.A.; et al. Vitamin D Supplementation to Prevent Acute Respiratory Tract Infections: Systematic Review and Meta-Analysis of Individual Participant Data. *BMJ* **2017**, *356*, i6583.

36. Cashman, K.D. Vitamin D Deficiency: Defining, Prevalence, Causes, and Strategies of Addressing. *Calcif. Tissue Int.* **2020**, *106*, 14–29.

37. O'Brien, E.; Parati, G.; Stergiou, G. Ambulatory Blood Pressure Measurement: What is the International Consensus? *Hypertension* **2013**, *62*, 988–994.

38. Baumgart, P.; Kamp, J. Accuracy of the SpaceLabs Medical 90217 ambulatory blood pressure monitor. *Blood Press. Monit.* **1998**, *3*, 303–307.

39. Sarafidis, P.A.; Lazaridis, A.A.; Imprialos, K.P.; Georgianos, P.I.; Avranas, K.A.; Protogerou, A.D.; Doumas, M.N.; Athyros, V.G.; Karagiannis, A.I. A Comparison Study of Brachial Blood Pressure Recorded with Spacelabs 90217A and Mobil-O-Graph NG Devices Under Static and Ambulatory Conditions. *J. Hum. Hypertens.* **2016**, *30*, 742–749.

40. Abderhalden, L.A.; Meyer, S.; Dawson-Hughes, B.; Orav, E.J.; Meyer, U.; de Godoi Rezende Costa Molino, C.; Theiler, R.; Stahelin, H.B.; Ruschitzka, F.; Egli, A.; et al. Effect of Daily 2000 IU Versus 800 IU Vitamin D on Blood Pressure among Adults Age 60 Years and Older: A Randomized Clinical Trial. *Am. J. Clin. Nutr.* **2020**, *112*, 527–537.

41. Sluyter, J.D.; Camargo, C.A., Jr.; Stewart, A.W.; Waayer, D.; Lawes, C.M.M.; Toop, L.; Khaw, K.T.; Thom, S.A.M.; Hametner, B.; Wassertheurer, S.; et al. Effect of Monthly, High-Dose, Long-Term Vitamin D Supplementation on Central Blood Pressure Parameters: A Randomized Controlled Trial Substudy. *J. Am. Heart Assoc.* **2017**, *6*, e006802. [CrossRef]

42. Vimaleswaran, K.S.; Cavadino, A.; Berry, D.J.; LifeLines Cohort Study investigators; Jorde, R.; Dieffenbach, A.K.; Lu, C.; Alves, A.C.; Heerspink, H.J.; Tikkanen, E.; et al. Association of Vitamin D Status with Arterial Blood Pressure and Hypertension Risk: A Mendelian Randomisation Study. *Lancet Diabetes Endocrinol.* **2014**, *2*, 719–729. [PubMed]

43. Pilz, S.; Marz, W.; Cashman, K.D.; Kiely, M.E.; Whiting, S.J.; Holick, M.F.; Grant, W.B.; Pludowski, P.; Hiligsmann, M.; Trummer, C.; et al. Rationale and Plan for Vitamin D Food Fortification: A Review and Guidance Paper. *Front. Endocrinol.* **2018**, *9*, 373.

44. Cashman, K.D.; Sheehy, T.; O'Neill, C.M. Is Vitamin D Deficiency a Public Health Concern for Low Middle Income Countries? A Systematic Literature Review. *Eur. J. Nutr.* **2019**, *58*, 433–453. [PubMed]

45. Zhang, D.; Cheng, C.; Wang, Y.; Sun, H.; Yu, S.; Xue, Y.; Liu, Y.; Li, W.; Li, X. Effect of Vitamin D on Blood Pressure and Hypertension in the General Population: An Update Meta-Analysis of Cohort Studies and Randomized Controlled Trials. *Prev. Chronic Dis.* **2020**, *17*, E03.

nutrients

Review

Clinical Practice in the Prevention, Diagnosis and Treatment of Vitamin D Deficiency: A Central and Eastern European Expert Consensus Statement

Pawel Pludowski [1,*], Istvan Takacs [2], Mihail Boyanov [3], Zhanna Belaya [4], Camelia C. Diaconu [5], Tatiana Mokhort [6], Nadiia Zherdova [7], Ingvars Rasa [8], Juraj Payer [9] and Stefan Pilz [10]

1   Department of Biochemistry, Radioimmunology and Experimental Medicine, The Children's Memorial Health Institute, 04-730 Warsaw, Poland
2   Department of Internal Medicine and Oncology, Faculty of Medicine, Semmelweis University, 1085 Budapest, Hungary; takacs.istvan@med.semmelweis-univ.hu
3   Department of Internal Medicine, Clinic of Endocrinology and Metabolism, University Hospital Alexandrovska, Medical University of Sofia, 1, G. Sofiyski Str., 1431 Sofia, Bulgaria; mihailboyanov@yahoo.com
4   The National Medical Research Center for Endocrinology, 117036 Moscow, Russia; jannabelaya@gmail.com
5   Faculty of Medicine, "Carol Davila" University of Medicine and Pharmacy, 050474 Bucharest, Romania; drcameliadiaconu@gmail.com
6   Department of Endocrinology, Belarussian State Medical University, 220116 Minsk, Belarus; tatsianamokhort@gmail.com
7   Department of Diagnostics and Treatment of Metabolic Diseases, Center for Innovation Medical Technology, The National Academy of Science of Ukraine, 01030 Kiev, Ukraine; nadejda05.1977@gmail.com
8   Centre for Continuing Education, Riga Stradins University, LV-1002 Riga, Latvia; dr.irasa@inbox.lv
9   Department of Internal Medicine, Comenius University Faculty of Medicine, University Hospital Bratislava, 826 06 Bratislava, Slovakia; payer@ru.unb.sk
10   Department of Internal Medicine, Division of Endocrinology and Diabetology, Medical University of Graz, Auenbruggerplatz 15, 8036 Graz, Austria; stefan.pilz@medunigraz.at
*   Correspondence: p.pludowski@ipczd.pl

Citation: Pludowski, P.; Takacs, I.; Boyanov, M.; Belaya, Z.; Diaconu, C.C.; Mokhort, T.; Zherdova, N.; Rasa, I.; Payer, J.; Pilz, S. Clinical Practice in the Prevention, Diagnosis and Treatment of Vitamin D Deficiency: A Central and Eastern European Expert Consensus Statement. *Nutrients* **2022**, *14*, 1483. https://doi.org/10.3390/nu14071483

Academic Editor: Bruce W. Hollis

Received: 8 March 2022
Accepted: 21 March 2022
Published: 2 April 2022

Publisher's Note: MDPI stays neutral with regard to jurisdictional claims in published maps and institutional affiliations.

Abstract: Vitamin D deficiency has a high worldwide prevalence, but actions to improve this public health problem are challenged by the heterogeneity of nutritional and clinical vitamin D guidelines, with respect to the diagnosis and treatment of vitamin D deficiency. We aimed to address this issue by providing respective recommendations for adults, developed by a European expert panel, using the Delphi method to reach consensus. Increasing the awareness of vitamin D deficiency and efforts to harmonize vitamin D guidelines should be pursued. We argue against a general screening for vitamin D deficiency but suggest 25-hydroxyvitamin D (25(OH)D) testing in certain risk groups. We recommend a vitamin D supplementation dose of 800 to 2000 international units (IU) per day for adults who want to ensure a sufficient vitamin D status. These doses are also recommended for the treatment of vitamin D deficiency, but higher vitamin D doses (e.g., 6000 IU per day) may be used for the first 4 to 12 weeks of treatment if a rapid correction of vitamin D deficiency is clinically indicated before continuing, with a maintenance dose of 800 to 2000 IU per day. Treatment success may be evaluated after at least 6 to 12 weeks in certain risk groups (e.g., patients with malabsorption syndromes) by measurement of serum 25(OH)D, with the aim to target concentrations of 30 to 50 ng/mL (75 to 125 nmol/L).

Keywords: vitamin D; recommendations; guidelines; supplementation; cholecalciferol; treatment

1. Introduction

Vitamin D is crucial for musculoskeletal health, as it plays an important role in the regulation of bone and mineral metabolism, and it can prevent and cure nutritional rickets and osteomalacia [1,2]. In addition, vitamin D receptor (VDR) expression in almost all human cells suggests, or even documents, a more widespread role of vitamin D for overall health, a notion that is supported by several experimental and epidemiological studies [1,3–8]. While there still exist knowledge gaps and controversy regarding potential extra-skeletal effects of vitamin D, there is a wide consensus that the high worldwide prevalence of vitamin D deficiency is of concern and requires actions to improve this situation [2,7,9]. Addressing this issue has to consider the unique metabolism of vitamin D, which is mainly synthesized in the skin stimulated by ultraviolet-B (UV-B) exposure, whereas nutrition is usually only a minor source of vitamin D [10]. Vitamin D from all different sources is metabolized to 25-hydroxyvitamin D (25(OH)D, calcifediol) in the liver, which is the main circulating vitamin D metabolite that is determined to assess vitamin D status. Further hydroxylation of 25(OH)D in the kidneys or certain extra-renal tissues results in the formation of 1,25-dihydroxyvitamin D (1,25(OH)2D, also called calcitriol), which exerts endocrine, autocrine, and paracrine effects as a steroid hormone [10]. Heterogeneous recommendations, regarding several issues in the practical management of vitamin D deficiency, represent a challenge for clinicians and health authorities on how to deal with this public health problem [11–22]. In this context, systematic evaluations of current vitamin D guidelines did, not only, observe a great heterogeneity of the recommendations, but it also reported a low quality score regarding the methodological processes for the majority of these vitamin D guidelines [17,18]. Tables 1 and 2 provide an overview of selected guideline recommendations, with a focus on Central and Eastern European countries, for prevention and treatment of vitamin D deficiency, respectively.

Table 1. Selected guideline recommendations for prevention of vitamin D deficiency in adults with a focus on Central and Eastern European countries, published since 2010.

Authority and/or Country or Region (Year)	Target Population	Age (Years)	Oral Vitamin D (IU)	Reference
Endocrine Society (2011) USA	General population	19–70	600–2000/day	Holick et al. [14]
		>70	800–2000/day	
	Pregnant and lactating women		600–2000/day	
	Obese individuals/Patients on anticonvulsants, glucocorticoids, antifungals, AIDS medications		2–3 times more	
DACH (2012) Germany/Austria/Switzerland	General population	>18	800/day	DGE [23]
EVIDAS (2013) Central Europe	General population	>18	800–2000/day	Płudowski et al. [21]
	Obese individuals and elderly		1600–4000/day	
	Prevention of pregnancy and fetal development complications	>16	1500–2000/day	
	Night workers and dark skin pigmentation		1000–2000/day	
EFSA (2016) Europe	General population	>18	600/day	EFSA [24]
Russia (2016)	General population	>18	800–1000/day	Pigarova et al. [25]
	Pregnant women		800–2000/day	
Poland (2018)	General population	19–75	800–2000/day	Rusińska, Płudowski et al. [26]
	Obese individuals	19–75	1600–4000/day	
	General population	>75	2000–4000/day	
	Obese individuals	>75	4000–8000/day	
	Pregnant and lactating women		2000/day	
Belarus (2013)	General population	>18	800–2000/day	Rudenko [27]
Hungary (2012)	General population	>18	1500–2000/day	Takács et al. [22]
	Pregnant and lactating women		1500–2000/day	
Bulgaria (2019)	General population	>19	600–2000/day	Borisova et al. [28]
	Pregnant and lactating women		600–2000/day	
	Patients on anticonvulsants, glucocorticoids, antifungals		2–3 times more	
Slovakia (2018)	Postmenopausal osteoporosis patients	>50	800–1000/day	Payer et al. [29]

Table 2. Selected guideline recommendations for treatment of vitamin D deficiency in adults with a focus on Central and Eastern European countries, published since 2010.

Authority and/or Country or Region (Year)	Target Population	Oral Vitamin D for Treatment (IU)	Treatment Duration	25(OH)D Target Concentration nmol/L (ng/mL)	Oral Vitamin D for Maintenance (IU)	Reference
Endocrine Society (2011) USA	General population Obese individuals/Patients on anticonvulsants, glucocorticoids, antifungals, AIDS medications	50,000/week or 6000/day 2–3 times more; at least 6000–10,000/day	8 weeks	75 (30)	1500–2000/day 3000–6000/day	Holick et al. [14]
EVIDAS (2013) Central Europe	General population	50,000/week or 7000–10,000/day	4–12 weeks	75–125 (30–50)	a maintenance dose may be instituted	Płudowski et al. [21]
Italy (2018)	General population	50,000/week or 5000/day	8 weeks	>75 (>30)	50,000 IU twice per month or 1500–2000 IU/day	Cesareo et al. [30]
Russia (2016)		25(OH)D < 50 nmol/L (<20 ng/mL): 50,000/week or 200,000/month or 150,000/month or 6000–8000/day 25(OH)D < 75 nmol/L (30 ng/mL): 50,000/week or 200,000 or 150,000 or 6000–8000/day	8 weeks 2 months 3 months 8 weeks 4 weeks single dose single dose 4 weeks		1000–2000/day or 6000–14,000/week	Pigarova et al. [25]
Poland (2018)	General population	6000/day	12 weeks or until a 25(OH)D concentration of 75 nmol/L (30 ng/mL) is reached	>75–125 (>30–50)	maintenance dose i.e., a prophylactic dose recommended for the general population (see Table 1)	Rusińska, Płudowski et al. [26]
Belarus (2013)	General population	25(OH)D < 25 nmol/L (<10 ng/mL): 2000 to 10,000/day 25(OH)D 25–50 nmol/L (10–20 ng/mL): 800 to 4000/day	4–12 weeks 1 year	75–200 (30–80)	800–2000 IU/day	Rudenko [27]
Hungary (2012)	General population	50,000/week or 30,000/week or 2000/day	4–8 weeks 6–12 weeks 12 weeks	75 (30)	1500–2000/day	Takács et al. [22]
Bulgaria (2019)	General population	To maintain bone health: 1000–2000/day For extra–skeletal effects: 2000–4000/day	/ /	50 (20) 75–110 (30–44)	maintenance dose i.e., a prophylactic dose recommended for the general population (see Table 1)	Borisova et al. [28]

In clinical practice, a great variability and controversy is reported regarding vitamin D testing and supplementation, thus requiring an improved guidance for clinicians [31]. Therefore, we aimed to draft an expert consensus statement covering important aspects of the clinical practice in the prevention, diagnosis, and treatment of vitamin D deficiency of adults. Rather than covering all of the above-mentioned vitamin D issues in great and extensive detail, we aimed to provide a simple, easy-to-follow guidance for clinicians. Respective data and recommendations regarding these issues in children can be found elsewhere [14,21,26,32–34].

2. Consensus Development Process

A European expert panel with 10 members and a focus on physicians from Eastern Europe, who are usually underrepresented in such groups, was established by selecting clinicians and key opinion leaders with expertise in vitamin D and related topics from their respective countries. The first and senior authors (P.P. and S.P.) of this article served as the chairs of this expert panel, who selected and invited the other panel members. The consensus-reaching process itself was conducted by using the Delphi method, which is a widely accepted tool for clinical consensus statements [35,36]. The Delphi method was applied by using SurveyMonkey® for voting on various statements on a 9-point scale with the following numeric (and descriptive) anchors: 1 (strongly disagree), 3 (disagree), 5 (neutral), 7 (agree), and 9 (strongly agree). It was pre-specified that a consensus will be established, in the case that ≥75% of the participants agree with the statement, by a voting scale of 7 or above. In case of failing to reach a consensus according to this criterion, it was planned to repeat these surveys after group discussions and statement modifications, if necessary, until a consensus is reached.

After the alignment of the scope of this consensus document and recruitment of the 10 members of the expert panel, the detailed process of this work started with drafting various questions on practical issues regarding vitamin D, which were exclusively discussed and fine-tuned by the chairs. These questions were subsequently used in a first survey round in September 2021. All panel members were involved in this survey, and their answers, along with the existing literature on vitamin D, served as the basis to formulate respective statements by the chairs. At a hybrid meeting in Darmstadt, Germany on the 20 November 2021, the results of the first survey round, along with the subsequently formulated statements, were presented by the chairs. At the same meeting, each statement was subject to a group discussion and a Delphi voting round to reach consensus, which was finally successful for all statements. The 10 authors of this paper are the 10 experts who participated in the Delphi voting rounds. Each voting round was performed for the whole content of each Table or Figure, respectively. Group discussions before each voting round, however, were used to fine tune the statements according to the opinions and recommendations of the panel members. The work on this consensus document was funded and organized by Wörwag Pharma (Böblingen, Germany). The sponsor provided financial and logistic support but did not actively participate in the scientific discussions and consensus-reaching processes.

3. Consensus Recommendations

The final consensus statements with the level of agreement within the expert group are subsequently presented below. The evidence and considerations underpinning each statement are also outlined. It was our aim to base our work on the totality of available evidence by considering the established hierarchy of evidence levels. The expert panel group was encouraged to perform systematic literature reviews on the topics of this consensus document, but we have to acknowledge that this gathering of information did not follow a pre-specified structured process.

3.1. Current Situation of Vitamin D Deficiency Diagnosis, Prevention and Treatment

Epidemiological studies have documented a high prevalence of vitamin D deficiency worldwide [37]. Data from Europe showed that 25(OH)D concentrations below 20 ng/mL (50 nmol/L) and below 12 ng/mL (30 nmol/L) are observed in 40.4% and 13.0% of the general population, respectively [9]. Therefore, a huge gap exists between the recommendations of nutritional societies regarding dietary reference intakes, as well as target 25(OH)D concentrations, and the actual situation, as documented in large population surveys [38]. Public health actions are, therefore, required to improve the vitamin D status in the general population, but regional differences in vitamin D status, related to factors such as latitude, genetics, lifestyle, body composition, or dietary intake have to be considered [9,38–41]. A major issue to achieve this task is to harmonize the current heterogeneous efforts and guideline recommendations (Table 3).

Table 3. Statement regarding the current situation of vitamin D deficiency diagnosis, prevention, and treatment.

Consensus Statement	Consensus Voting Scale	Level of Agreement
	9 (strongly agree)	80%
To ensure an adequate screening, prevention and treatment of vitamin D	8	0%
deficiency in the clinical practice, it is necessary to increase the awareness	7 (agree)	20%
and improve education in the public and medical community.	6	0%
Moreover, national and international guidelines/recommendations should	5 (neutral)	0%
be precise regarding the risk groups that need to be screened, the adequate	4	0%
dosages for prevention and treatment of vitamin D deficiency, the treatment	3 (disagree)	0%
regimen as well as the follow-up to allow transfer into clinical practice.	2	0%
	1 (strongly disagree)	0%
Overall agreement 100%, consensus endorsed		

3.2. Screening of Vitamin D Deficiency in Adults

No published study evaluated the effects of a screening program for vitamin D deficiency in the general population, so the evidence is insufficient to balance the benefits and harms of such a screening [42,43]. Accordingly, we stress that it is currently not justified to recommend a general screening for vitamin D deficiency by measuring 25(OH)D concentrations in the whole general population. Nevertheless, considering that certain groups of individuals or patients are particularly prone to vitamin D deficiency and/or may particularly benefit from vitamin D treatment, we suggest, in line with the Endocrine Society, that 25(OH)D measurements should be considered in these groups, as listed in Table 4 [14].

Total serum 25(OH)D concentration, i.e., the sum of 25(OH)D$_3$ and 25(OH)D$_2$, is the accepted marker for the assessment of vitamin D status, as it best reflects vitamin D supply by all different sources, i.e., endogenous vitamin D synthesis in the skin, diet, supplements, and mobilization from tissue stores. Previous reports on a relatively high inter-assay and inter-laboratory variability of 25(OH)D measurements underscore the need for assay standardization and laboratory quality assurance [44,45]. In patients with vitamin D deficiency and certain related health issues, e.g., bone diseases, it should be considered to measure additional laboratory parameters, including serum calcium, phosphate, alkaline phosphatase, parathyroid hormone (PTH), creatinine (to calculate the estimated glomerular filtration rate), and magnesium, as these laboratory markers may be useful to guide further diagnostics and treatment of these patients. Measurements of, e.g., serum calcium and creatinine are, however, also advised in patients with 25(OH)D concentrations above 100 ng/mL (250 nmol/L), as vitamin D oversupply/toxicity may lead to hypercalciuria, followed by hypercalcemia, potential acute kidney disease, and vascular calcification. Hypercalcemia does, however, usually not occur at 25(OH)D concentrations below 150 ng/mL (375 nmol/L) [46]. There are hardly any contraindications to correct vitamin D deficiency by vitamin D supplementation (e.g., kidney stones are *per se* no contraindication) except of rare conditions with an increased sensitivity to vitamin D treatment, such as inherited 24-hydroxylase-deficiency [47]. This is a rare genetic disorder in which catabolism of vitamin D metabolites is impaired, leading to hypercalcemia, low PTH concentrations, and relatively high serum 25(OH)D concentrations along with an increased risk of nephrolithiasis [47]. If such a disease is suspected, the measurement of 24,25-dihydroxyvitamin D, in a specialized laboratory, aids in the diagnosis as a high ratio of 25(OH)D to 24,25-dihydroxyvitamin D suggests this disease that is further confirmed by genetic analyses [47].

Classification of vitamin D status and its terminology, according to 25(OH)D concentration, remains a controversial issue in the scientific literature [7,16,21]. Being aware that it is an individual continuum from vitamin D deficiency to a sufficient and optimal vitamin D status, as well as to vitamin D toxicity, we suggest a classification system, as indicated in Table 4. It should be kept in mind that such a general classification of vitamin D status cannot take into account variations in the individual sensitivity to vitamin D effects that may be due to genetic polymorphisms, epigenetic or nutritional factors (e.g., magnesium status), as well as co-morbidities or medications [48–52].

Table 4. Statement regarding screening of vitamin D deficiency in adults.

Consensus Statement	Consensus Voting Scale	Level of Agreement
Screening of vitamin D deficiency should be considered in the following patients/individuals or conditions: Osteoporosis; Osteomalacia; Musculoskeletal pain; Chronic kidney disease; Hepatic failure; Malabsorption syndromes (e.g., cystic fibrosis, inflammatory bowel diseases, bariatric surgery, radiation enteritis); Hyperparathyroidism; Chronic treatment with medications that influence vitamin D metabolism (e.g., antiseizure medications, glucocorticoids, AIDS-medications, antifungal agents, cholestyramine); Chronic autoimmune diseases (e.g., multiple sclerosis, rheumatoid arthritis); Pregnant and lactating women; Institutionalized or hospitalized patients; Older adults (>65 years) in general; Older adults with history of falls or nontraumatic fractures; Granuloma-forming disorders (e.g., sarcoidosis, tuberculosis, histoplasmosis, berylliosis, coccidiomycosis); Obesity (BMI $\geq$ 30kg/m^2); dark skin pigmentation. 25(OH)D is recommended as a laboratory marker for the diagnosis of vitamin D deficiency. In patients with diagnosed vitamin D deficiency (25(OH)D < 20 ng/mL (<50 nmol/L)) and suspected related health issues, serum calcium, phosphate, alkaline phosphatase, parathyroid hormone (PTH), creatinine, and serum magnesium levels should be considered for evaluation; in particular in individuals with a 25(OH)D concentration of <10 ng/mL (<25 nmol/L). A 25(OH)D concentration of <20 ng/mL (<50 nmol/L) is considered as vitamin D deficiency A 25(OH)D concentration of $\geq$20 ng/mL ($\geq$50 nmol/L) and <30 ng/mL (<75 nmol/L) is considered as vitamin D insufficiency A 25(OH)D concentration of 30–50 ng/mL (75–125 nmol/L) is considered as vitamin D sufficiency A 25(OH)D concentration of >50–60 ng/mL (125–150 nmol/L) is considered as safe but not as a target level A 25(OH)D concentration of >60–100 ng/mL (150–250 nmol/L) is considered as area of uncertainty with potential benefits or risks. A 25(OH)D concentration of >100 ng/mL (250 nmol/L) is considered as oversupply/vitamin D toxicity	9 (strongly agree) 8 7 (agree) 6 5 (neutral) 4 3 (disagree) 2 1 (strongly disagree)	50% 20% 30% 0% 0% 0% 0% 0% 0%
Overall agreement 100%, consensus endorsed		

3.3. Prevention of Vitamin D Deficiency in Adults

Most nutritional vitamin D guidelines conclude that vitamin D requirements are met for the vast majority (i.e., 97.5%) of the population when achieving a target 25(OH)D concentration of at least 20 ng/mL (50 nmol/L) [11,12]. Recommended dietary reference intakes for vitamin D usually range from 600 to 800 international units (IU) (40 IU are equal to 1 μg) per day and should ensure a sufficient vitamin D status under conditions of minimal-to-no sunlight exposure [11–13,16,53,54]. These vitamin D intake doses were calculated according to meta-regression analyses of so called "winter RCTs" to estimate the dose-response curve of vitamin D intakes and achieved serum 25(OH)D concentrations without relevant endogenous vitamin D synthesis in the skin [11]. It is a major limitation of most nutritional vitamin D guidelines that they performed meta-regression analyses based on aggregate data because such an approach does not adequately capture between person variability in the treatment response [53,54]. Using individual participant data instead of aggregate data for meta-regression analyses, as a superior methodological approach, results in significantly higher vitamin D intakes to achieve certain target 25(OH)D concentrations [53,54]. Individual participant data meta-regression analyses and single RCTs suggest that an overall vitamin D intake of about 1000 IU of vitamin D per day is required to maintain 25(OH)D concentrations of, at least, 20 ng/mL (50 nmol/L) in 97.5% of the population [54,55]. Therefore, we recommend a vitamin D supplement dose of at least 800 IU per day when targeting a sufficient vitamin D status, i.e., a 25(OH)D concentration of at least 20 ng/mL (50 nmol/L). We can, of course, improve and maintain vitamin D status by consuming natural or fortified food sources, but vitamin D intake by diet is usually in the range of about 100 to 200 IU per day in the general population [37,56].

In detail, we recommend a vitamin D supplementation dose of 800 to 2000 IU per day for adults who want to ensure a sufficient vitamin D status, with up to 4000 IU per day for certain groups, particularly for patients with obesity and malabsorption syndromes, as well as for individuals with a dark skin pigmentation (see Table 5). The relatively wide dose ranges for vitamin D account for various differences in the dose-response relationship for a given supplemental vitamin D dose and the achieved 25(OH)D concentration with higher dose requirements with increasing body weight and vice versa [57–64]. If a clinician is asked by a random individual which vitamin D dose is safe and very likely avoids vitamin D deficiency, a dose of 800 to 1000 IU per day should fulfill these criteria for the vast majority, even if individual characteristics, including the 25(OH)D status, is unknown. It should, however, also be noted that a few health authorities and experts consider a 25(OH)D concentration, of at least 10–12 ng/mL (25 to 30 nmol/L), as a reasonable treatment target that can be achieved by supplementation of 400 IU of vitamin D per day [11,13,16,53].

Table 5. Statement regarding prevention of vitamin D deficiency in adults.

Consensus Statement	Consensus Voting Scale	Level of Agreement
In healthy adults without other risk factors, a supplementation of	9 (strongly agree)	30%
800–2000 IU/day, for those who want to achieve a targeted/measured 25(OH)D	8	20%
concentration, should be considered during wintertime (mainly November-April)	7 (agree)	50%
due to insufficient endogenous dermal vitamin D synthesis and depending on the	6	0%
body weight.	5 (neutral)	0%
Due to decreased skin synthesis in elderly (>65 years), a supplementation of	4	0%
800–2000 IU/day is recommended throughout the year.	3 (disagree)	0%
In hospitalized/institutionalized individuals, a supplementation of	2	0%
800–2000 IU/day is recommended throughout the year.	1 (strongly disagree)	0%
Women planning a pregnancy should start or maintain the vitamin D supplementation as recommended for healthy adults without other risk factors (800–2000 IU/day). The vitamin D supplementation should be continued throughout pregnancy and lactation.		
In certain patients/individuals or conditions 2–3 times higher vitamin D dosages, without using vitamin D doses above the UL of 4000 IU/day, are recommended for prevention compared to healthy adults without other risk factors:		
Malabsorption (e.g., cystic fibrosis, inflammatory bowel diseases, bariatric surgery, radiation enteritis)		
Obesity (BMI $\geq$ 30 kg/m^2)		
Dark skin pigmentation		
As vitamin D metabolites are stored in fat and other tissues and gradually released into the blood circulation, a daily or weekly or monthly supplementation regimen is equally effective and safe, if monthly doses are not exceedingly high, for the prevention of vitamin D deficiency.		
A tailored approach for vitamin D administration, involving the patients' preferences of the supplementation regimen (daily, weekly, monthly) might enhance the adherence to preventive vitamin D supplementation.		
For the prevention of vitamin D deficiency, the supplementation of oral cholecalciferol (vitamin D3) is recommended.		
Overall agreement 100%, consensus endorsed		

Daily, weekly, or monthly vitamin D supplementation, at equivalent doses, lead to similar increases in 25(OH)D serum concentrations, when measured after 2 to 3 months [65–67]. Adherence may be better with intermittent vitamin D dosing, but there are also concerns that high intermittent vitamin D doses may be less beneficial or might even be harmful in certain settings [67–69]. In view of the available evidence from clinical vitamin D trials and some pathophysiological considerations (e.g., altered vitamin D metabolism with high intermittent vitamin D doses), a daily vitamin D dosing schedule should rather be preferred, but when exceedingly high intermittent vitamin D doses are avoided, a weekly or monthly dosing schedule can also be applied [66,67]. The panel members could not reach a clear consensus on a clear cut-off for exceedingly high vitamin D doses, but single doses above about 50,000 IU of vitamin D should rather be avoided. Due to superior evidence regarding clinical benefits and dose-response, we rather prefer vitamin D3 (cholecalciferol) over vitamin D2 (ergocalciferol) for the prevention of vitamin D deficiency [51,70].

3.4. Treatment of Vitamin D Deficiency in Adults

Individuals with a measured 25(OH)D concentration below 20 ng/mL (50 nmol/L) should be treated with vitamin D supplementation, because their vitamin D requirements may not be met [11,12]. There is controversy in the scientific literature whether 25(OH)D concentrations between 20 ng/mL (50 nmol/L) and <30 ng/mL (75 nmol/L) justify vitamin D supplementation [71,72]. The recommended dose range of 800 to 2000 IU per day is a reflection of various considerations underlying such treatment goals (see Table 6). When aiming for a minimum 25(OH)D concentration of at least 20 ng/mL (50 nmol/L), a daily vitamin D supplement dose of about 800 IU per day is sufficient for almost all individuals, even during the winter season, in Europe [54,55]. Data are less clear on which vitamin D doses are required to achieve a 25(OH)D concentration of $\geq$30 ng/mL (75 nmol/L) in almost all patients, but doses may be in the range of about 1500 to 2000 IU per day or even higher [14,21,26–28]. The classic rule of thumb that 100 IU of vitamin D per day increases serum 25(OH)D concentrations by about 1 ng/mL (2.5 nmol/L) seems to be a useful approximation, but several factors modulate the individual treatment response [73,74]. For example, increases in 25(OH)D are relatively high at low vitamin D supplement doses and low baseline 25(OH)D concentrations, whereas the dose-response curve flattens with higher vitamin D supplement doses and higher baseline 25(OH)D concentrations [73,74]. Evaluations of treatment success, by measurements of 25(OH)D, may be considered in certain patients, such as those with e.g., malabsorption or questionable adherence, but this should not be done earlier than 6 to 12 weeks after starting vitamin D supplementation, as this is about the time that it takes to reach a steady-state in serum 25(OH)D concentrations [2]. Although there is, of course, a seasonal variation in serum 25(OH)D concentrations, usually with higher levels during summertime, as a consequence of endogenous vitamin D synthesis in the skin, we do, in general, recommend continuous and, usually, fixed doses of vitamin D supplementation throughout the year. The decrease in serum 25(OH)D during winter season is, in many patients, significant, but it is less than could be expected by the half-life of serum 25(OH)D concentrations of about 2 to 3 weeks because of a mobilization of vitamin D and its metabolites from various tissue stores (e.g., adipose tissue and muscle) [75].

If a rapid correction of vitamin D deficiency is clinically indicated, a regimen with a higher initial vitamin D dose, i.e., 6000 IU per day, and in certain cases, even up to 10,000 IU per day, followed by a maintenance dose with 800 to 2000 IU per day is recommended (Table 6). Such doses of 6000 IU, or even up to 10,000 IU, per day for several weeks are usually safe and ensure a more rapid correction of vitamin D deficiency compared to lower doses [14,76,77]. Daily vitamin D doses are generally preferred over intermittent dosing schedules [67]. The clinical indications for a rapid correction of vitamin D are, beyond osteomalacia, not clearly defined but may, according to our opinion, involve conditions such as extremely low 25(OH)D concentrations, osteoporosis patients with a very high fracture risk, patients with secondary hyperparathyroidism, and/or reduced serum calcium concentrations.

Regarding treatment of vitamin D deficiency and its prevention, we want to emphasize that promoting a healthy lifestyle by preventing or reducing obesity, regular physical activity with moderate (cautious) sunlight exposure, and a healthy balanced diet are also effective measures to improve both vitamin D status and overall health. Promoting such lifestyle measures is, of course, also highly recommended, and it should accompany any vitamin D treatment.

As for the prevention of vitamin D deficiency, we recommend vitamin D3 (cholecalciferol) over vitamin D2 (ergocalciferol) for its treatment. Although parenteral, particularly intramuscular, vitamin D treatment can be considered in patients with malabsorption, e.g., inflammatory bowel disease, we primarily suggest to increase the oral vitamin D dose in such settings [78]. Clinicians have to consider that patients with inflammatory bowel disease frequently require higher vitamin D doses, but with daily oral vitamin D supplementation of about 5000 to 10,000 IU, even these patients usually achieve their 25(OH)D target concentrations [78]. If intermittent intramuscular vitamin D injections (e.g., 100,000 IU all three months) are used, the doses are roughly similar and slightly more efficient for intramuscular compared to oral doses, in terms of raising serum 25(OH)D concentrations, but the increase in serum 25(OH)D is slower for intramuscular versus oral vitamin D supplementation [78–81].

Some experts argue that calcifediol (=25(OH)D3, calcidiol) may also be used to correct vitamin D deficiency in certain conditions. The use of calcifediol seems to be more justified in obese people, people with malabsorption syndromes, people with liver disease, patients suffering from chronic kidney disease (stage 3 or 4), and those in all conditions where rapid correction of vitamin D deficiency is required [82–84]. Furthermore, calcifediol use may also be beneficial in patients taking medications that disrupt the hepatic cytochrome P-450 enzyme system, including those taking glucocorticoids, anticonvulsants, anticancer drugs, or antiretroviral drugs [82–85]. The increase in serum 25(OH)D is markedly reduced in patients with obesity (high BMI) and in patients with malabsorption syndromes treated with cholecalciferol, but with calcifediol, the 25(OH)D increase is not significantly different according to BMI or according to the presence, or absence, of malabsorption syndromes. Moreover, the increase in serum 25(OH)D is faster, and the dose-response curve is more linear with the use of calcifediol versus vitamin D3, and when stopping treatment, the decline in 25(OH)D concentration is faster after calcifediol compared to vitamin D3 [75,82–84]. While accumulating evidence suggests that calcifediol may be an attractive alternative to "native" vitamin D, due to the lack of experience with this molecule in Central and Eastern European countries, at this stage, we continue to recommend vitamin D3 (cholecalciferol) [75]. Cholecalciferol and calcifediol appear, so far, as equal molecules in the fight with vitamin D deficiency. However, RCT data are still missing on the superior benefit of calcifediol versus vitamin D, with reference to hard clinical outcomes, but more data on this topic may be available in the future [75,84,85].

Calcitriol (=1,25(OH)2D) and its analogues are used at much lower doses compared to vitamin D3, have a relatively high risk of hypercalcemia and a relatively narrow therapeutic window, and are not recommended for the treatment of common vitamin D deficiency [64]. Therefore, active vitamin D treatment is only indicated in certain diseases, such as chronic hypoparathyroidism, chronic kidney disease, or mineral and bone disorders (CKD-MBD) [64,86].

Table 6. Statement regarding treatment of vitamin D deficiency in adults.

Consensus Statement	Consensus Voting Scale	Level of Agreement
It is recommended to initiate a vitamin D deficiency treatment at a 25(OH)D concentration of <20 ng/mL (<50 nmol/L). At a concentration of <30 ng/mL (<75 nmol/L) a treatment may be considered.	9 (strongly agree)	60%
	8	10%
Individuals with diagnosed vitamin D deficiency can be initially treated with higher vitamin D dosages compared to the preventive dosages recommended for the general population, if a rapid correction of the 25(OH)D concentration is clinically indicated. As initial dose for the treatment of vitamin D deficiency in patients without other risk factors, a dosage of 6000 IU, equivalent to a daily dosage, is recommended.	7 (agree)	30%
	6	0%
	5 (neutral)	0%
	4	0%
	3 (disagree)	0%
	2	0%
	1 (strongly disagree)	0%

In certain individuals or conditions, higher vitamin D dosages, up to 10,000 IU, equivalent to a daily dosage, are recommended for treatment compared to healthy adults without other risk factors (Endocrine Society recommendation) [14]:
-Malabsorption (e.g., cystic fibrosis, inflammatory bowel diseases, bariatric surgery, radiation enteritis)
-Chronic treatment of medications that influence vitamin D metabolism (e.g., antiseizure medications, glucocorticoids, AIDS-medications, antifungal agents, cholestyramine)
-Obesity (BMI $\geq$ 30 kg/m^2)
A treatment duration of 4–12 weeks is recommended, depending on the severity of vitamin D deficiency.
A tailored approach for vitamin D administration, involving the patients' preferences of the treatment regimen (daily, weekly, monthly) might enhance the adherence to the therapy.
As soon as a 25(OH)D concentration of 30–50 ng/mL (75–125 nmol/L) is achieved, a maintenance dose of 800–2000 IU/day is recommended, that can also be used as an initial treatment dose if there is no requirement for a rapid correction of vitamin D deficiency.
Approx. 6–12 weeks after start of the treatment, the effectiveness may be evaluated by measurement of the 25(OH)D concentration particularly in certain risk groups with e.g., malabsorption syndrome.
For the treatment of vitamin D deficiency in adults, oral cholecalciferol (vitamin D3) is preferred.
Calcifediol may be used instead of vitamin D in certain conditions, including obesity or malabsorption.
Calcitriol and active vitamin D analogues may be considered in special patient groups.
In certain risk groups (e.g., patients with severe malabsorption), parenteral vitamin D treatment can be considered.

Overall agreement 100%, consensus endorsed

3.5. Vitamin D in Musculoskeletal Disorders

While older meta-analyses of vitamin D RCTs showed a significant reduction in fractures by vitamin D supplementation at a daily dose of about 800 to 2000 IU per day, these data have recently been challenged by updated meta-analyses, suggesting no significant effect on fractures and falls [5,87,88]. Nevertheless, major osteoporosis guidelines recommend vitamin D treatment in osteoporosis patients, and some studies indicate that sufficient 25(OH)D concentrations are required for optimal bisphosphonate treatment efficacy [51,89]. While it is beyond the scope of our work to release detailed recommendations regarding calcium supplementation in osteoporosis patients, we wish to point out that a recent RCT with the bisphosphonate zoledronate showed excellent anti-fracture effects in patients using pure vitamin D supplementation without additional calcium supplements but consuming 1g of calcium daily by a usual diet [90]. These data may suggest that, even in osteoporosis patients, vitamin D treatment without additional calcium supplements may be sufficient in the case of adequate dietary calcium intake, but this issue is still not clarified in the scientific community, since it is challenging to disentangle the effects of vitamin D and calcium with regards to bone health [5,88,90–92]. Of note, increasing calcium and protein intake by milk, yoghurt, and cheese significantly reduces the risk of falls and fractures in aged care residents [93]. Calcium supplementation is, however, generally indicated in patients with osteoporosis and an insufficient dietary calcium intake. In this context, it should be stressed that most RCTs on osteoporosis drugs were conducted under the conditions of combined calcium plus vitamin D supplementation, thus supporting the use of calcium supplementation, in addition to vitamin D for osteoporosis treatment. Being aware of the uncertainties regarding vitamin D, in the context of osteoporosis, we nevertheless strongly argue to ensure a sufficient vitamin D status in patients with an increased risk of fractures and falls (Table 7). With reference to falls, some studies suggest that high-dose intermittent vitamin D supplementation may even be detrimental so that we prefer daily doses at the lower end of the dosing range of 800 to 2000 IU per day, in this setting [67,85,94]. Therefore, vitamin D overdosing must particularly be avoided in older and severely ill patients [94,95].

Table 7. Statement regarding vitamin D in musculoskeletal disorders.

Consensus Statement	Consensus Voting Scale	Level of Agreement
In osteoporosis patients, a supplementation of 800–2000 IU/day, with	9 (strongly agree)	30%
oral cholecalciferol (vitamin D3) is recommended in combination with	8	10%
calcium, if indicated.	7 (agree)	60%
Vitamin D deficiency may impair the response to the osteoporosis	6	0%
treatment, thus screening of the 25(OH)D level and correction of vitamin	5 (neutral)	0%
D deficiency before starting the osteoporosis treatment with	4	0%
antiresorptive medications is recommended.	3 (disagree)	0%
In patients with an increased risk of falls or fractures, a supplementation	2	0%
of 800–2000 IU/day is recommended.	1 (strongly disagree)	0%
Overall agreement 100%, consensus endorsed		

3.6. Extra-Skeletal Actions of Vitamin D in Adults

The expression of the VDR and of vitamin D-metabolizing enzymes in almost all human tissues and cells suggests a widespread role of vitamin D for overall human health [1,10]. In line with this, numerous epidemiological studies documented that low 25(OH)D concentrations are associated with an increased risk of all-cause mortality and major acute and chronic diseases, such as cancer, cardiovascular and autoimmune diseases, as well as infections (see Table 8) [1,3,4,6,8,92,96–100]. Confounding and reverse causation may, however, explain parts of these associations so that cause and effect relationships are not yet fully established for several of the above mentioned diseases [101].

Meta-analyses of vitamin D RCTs suggest that vitamin D supplementation may reduce the incidence of acute respiratory infections, cancer mortality, as well as asthma and chronic obstructive pulmonary disease (COPD) exacerbations [5,96,102–108]. Considering the totality of evidence regarding vitamin D and a variety of extra-skeletal diseases, we are of the opinion that it is justified to consider screening and preventive vitamin D supplementation in certain populations at risk (see Table 8). The general dilemma with potential extra-skeletal health benefits of vitamin D in the context of vitamin D guidelines is that vitamin D requirements for skeletal health, such as for the prevention of rickets and osteomalacia, may be met at lower 25(OH)D concentrations than the requirements for certain extra-skeletal health benefits [1,14]. Of note, recent large vitamin D RCTs failed to document significant benefits regarding their primary outcomes, such as mortality, cancer, or cardiovascular diseases, but these trials enrolled populations that were, by the vast majority, not vitamin D deficient [109,110].

Numerous studies have been published on vitamin D, in the context of the Coronavirus Disease 2019 (COVID-19) pandemic, caused by the severe acute respiratory syndrome coronavirus 2 (SARS-CoV-2) [111–113]. While there is accumulating evidence suggesting potential benefits of vitamin D or calcifediol, for the prevention and treatment of COVID-19, this hypothesis still requires confirmation in large clinical trials [111–113]. Fortunately, due to the efficacy of vaccines and natural immunity of individuals who recovered from SARS-CoV-2 infections, the COVID-19 pandemic has already been significantly mitigated in early 2022, with expected further improvements in the near future [114–116].

Table 8. Statement regarding extra-skeletal actions of vitamin D in adults.

Consensus Statement	Consensus Voting Scale	Level of Agreement
Results from observational studies consider a low 25(OH)D concentration as a potential risk marker for several diseases such as cancer incidence and mortality, cardiovascular diseases, diabetes mellitus and its comorbidities, chronic autoimmune diseases, metabolic syndrome, acute respiratory tract infections, neurological diseases and total mortality.	9 (strongly agree)	60%
	8	10%
	7 (agree)	30%
	6	0%
Results from meta-analyses of RCTs suggest that beyond musculoskeletal-effects, vitamin D supplementation may have beneficial extra-skeletal effects regarding acute respiratory tract infections and cancer mortality.	5 (neutral)	0%
	4	0%
	3 (disagree)	0%
In patients with or at risk of different types of cancer, certain cardiovascular diseases, diabetes mellitus and its comorbidities, chronic autoimmune diseases, certain neurological diseases and recurrent acute respiratory tract infections, a screening of vitamin D deficiency should be considered, and preventive vitamin D supplementation may be considered.	2	0%
	1 (strongly disagree)	0%
Overall agreement 100%, consensus endorsed		

3.7. Development of a Vitamin D Deficiency Screening and Treatment Algorithm

Based on our consensus statements, we developed an algorithm for the prevention and treatment of vitamin D deficiency that was also subject to a Delphi voting round, with the following point scale results: 9 points (20% of panel members), 8 points (30%), and 7 points (50%) (see Figure 1).

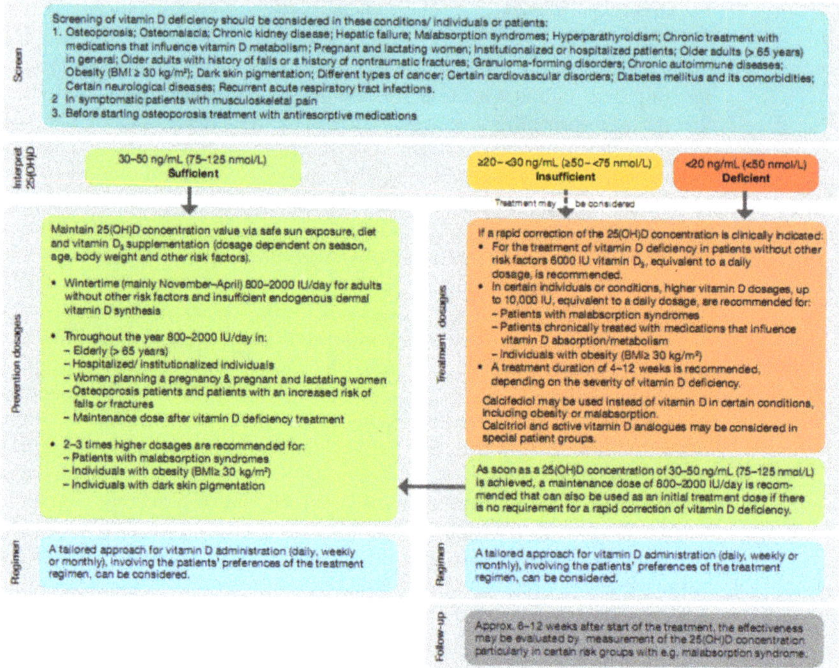

Figure 1. Algorithm for vitamin D deficiency screening and treatment.

4. Strengths and Limitations

As a potential limitation, the financial and logistic support by an industry sponsor (Wörwag Pharma) must be acknowledged. Despite no involvement of the sponsor in the scientific group discussions and consensus reaching processes, we cannot totally exclude some sort of funding bias [17,18]. Moreover, this consensus document was not informed by an a priori structured and pre-registered systematic review of the evidence. A definite strength of our work is the a priori structured process for developing this document, based on the Delphi method for reaching consensus. The number of 10 participants for using the Delphi voting rounds may be considered as too low, but such a group size is not uncommon and, thus, is generally accepted in the scientific literature on this methodology [35]. The involvement of several colleagues from Eastern Europe, who are usually underrepresented in European expert groups, and the well-balanced gender distribution of this panel, may also be regarded as a strength of our work.

5. Conclusions

This consensus statement covers various statements with relevance for the clinical practice in the prevention, diagnosis, and treatment of vitamin D deficiency. We highlighted the relevance of vitamin D for public health and provided guidance regarding this issue by considering the totality of the available scientific evidence, including our personal experience and opinions.

We consider that our work adds to the existing literature by providing a useful and evidence-based guidance for clinicians and health-care workers regarding several relevant and partially controversial topics for the practical management, with reference to vitamin D. In addition, we also addressed various issues with relevance for public health authorities and individuals from the general population that will, hopefully, help to reduce the global health burden of vitamin D deficiency.

Author Contributions: Writing—original draft preparation, P.P. and S.P. Writing a final review and final editing, I.T., M.B., Z.B., C.C.D., T.M., N.Z., I.R., J.P., P.P. and S.P. All authors have read and agreed to the published version of the manuscript.

Funding: This research was supported by Wörwag Pharma.

Institutional Review Board Statement: Not applicable.

Informed Consent Statement: Not applicable.

Data Availability Statement: Not applicable.

Conflicts of Interest: All authors received honoraria from Wörwag Pharma.

References

1. Bouillon, R.; Marcocci, C.; Carmeliet, G.; Bikle, D.; White, J.H.; Dawson-Hughes, B.; Lips, P.; Munns, C.F.; Lazaretti-Castro, M.; Giustina, A.; et al. Skeletal and Extraskeletal Actions of Vitamin D: Current Evidence and Outstanding Questions. *Endocr. Rev.* **2019**, *40*, 1109–1151. [CrossRef] [PubMed]
2. Pilz, S.; Zittermann, A.; Trummer, C.; Theiler-Schwetz, V.; Lerchbaum, E.; Keppel, M.H.; Grubler, M.R.; Marz, W.; Pandis, M. Vitamin D testing and treatment: A narrative review of current evidence. *Endocr. Connect.* **2019**, *8*, R27–R43. [CrossRef]
3. Maretzke, F.; Bechthold, A.; Egert, S.; Ernst, J.B.; Melo van Lent, D.; Pilz, S.; Reichrath, J.; Stangl, G.I.; Stehle, P.; Volkert, D.; et al. Role of Vitamin D in Preventing and Treating Selected Extraskeletal Diseases—An Umbrella Review. *Nutrients* **2020**, *12*, 969. [CrossRef] [PubMed]
4. Zittermann, A.; Trummer, C.; Theiler-Schwetz, V.; Lerchbaum, E.; Marz, W.; Pilz, S. Vitamin D and Cardiovascular Disease: An Updated Narrative Review. *Int. J. Mol. Sci.* **2021**, *22*, 2896. [CrossRef] [PubMed]
5. Bouillon, R.; Manousaki, D.; Rosen, C.; Trajanoska, K.; Rivadeneira, F.; Richards, J.B. The health effects of vitamin D supplementation: Evidence from human studies. *Nat. Rev. Endocrinol.* **2021**, *18*, 96–110. [CrossRef]
6. Pilz, S.; Zittermann, A.; Obeid, R.; Hahn, A.; Pludowski, P.; Trummer, C.; Lerchbaum, E.; Perez-Lopez, F.R.; Karras, S.N.; Marz, W. The Role of Vitamin D in Fertility and during Pregnancy and Lactation: A Review of Clinical Data. *Int. J. Environ. Res. Public Health* **2018**, *15*, 2241. [CrossRef]
7. Holick, M.F. The vitamin D deficiency pandemic: Approaches for diagnosis, treatment and prevention. *Rev. Endocr. Metab. Disord.* **2017**, *18*, 153–165. [CrossRef]

8. Gaksch, M.; Jorde, R.; Grimnes, G.; Joakimsen, R.; Schirmer, H.; Wilsgaard, T.; Mathiesen, E.B.; Njolstad, I.; Lochen, M.L.; Marz, W.; et al. Vitamin D and mortality: Individual participant data meta-analysis of standardized 25-hydroxyvitamin D in 26916 individuals from a European consortium. *PLoS ONE* **2017**, *12*, e0170791. [CrossRef]

9. Cashman, K.D.; Dowling, K.G.; Skrabakova, Z.; Gonzalez-Gross, M.; Valtuena, J.; De Henauw, S.; Moreno, L.; Damsgaard, C.T.; Michaelsen, K.F.; Molgaard, C.; et al. Vitamin D deficiency in Europe: Pandemic? *Am. J. Clin. Nutr.* **2016**, *103*, 1033–1044. [CrossRef]

10. Saponaro, F.; Saba, A.; Zucchi, R. An Update on Vitamin D Metabolism. *Int. J Mol. Sci.* **2020**, *21*, 6573. [CrossRef]

11. Pilz, S.; Trummer, C.; Pandis, M.; Schwetz, V.; Aberer, F.; Grubler, M.; Verheyen, N.; Tomaschitz, A.; Marz, W. Vitamin D: Current Guidelines and Future Outlook. *Anticancer Res.* **2018**, *38*, 1145–1151. [CrossRef]

12. Ross, A.C.; Manson, J.E.; Abrams, S.A.; Aloia, J.F.; Brannon, P.M.; Clinton, S.K.; Durazo-Arvizu, R.A.; Gallagher, J.C.; Gallo, R.L.; Jones, G.; et al. The 2011 report on dietary reference intakes for calcium and vitamin D from the Institute of Medicine: What clinicians need to know. *J. Clin. Endocrinol. Metab.* **2011**, *96*, 53–58. [CrossRef]

13. Buttriss, J.L.; Lanham-New, S.A.; Steenson, S.; Levy, L.; Swan, G.E.; Darling, A.L.; Cashman, K.D.; Allen, R.E.; Durrant, L.R.; Smith, C.P.; et al. Implementation strategies for improving vitamin D status and increasing vitamin D intake in the UK: Current controversies and future perspectives: Proceedings of the 2nd Rank Prize Funds Forum on vitamin D. *Br. J. Nutr.* **2021**, 1–21. [CrossRef]

14. Holick, M.F.; Binkley, N.C.; Bischoff-Ferrari, H.A.; Gordon, C.M.; Hanley, D.A.; Heaney, R.P.; Murad, M.H.; Weaver, C.M.; Endocrine Society. Evaluation, treatment, and prevention of vitamin D deficiency: An Endocrine Society clinical practice guideline. *J. Clin. Endocrinol. Metab.* **2011**, *96*, 1911–1930. [CrossRef]

15. Giustina, A.; Bouillon, R.; Binkley, N.; Sempos, C.; Adler, R.A.; Bollerslev, J.; Dawson-Hughes, B.; Ebeling, P.R.; Feldman, D.; Heijboer, A.; et al. Controversies in Vitamin D: A Statement from the Third International Conference. *JBMR Plus* **2020**, *4*, e10417. [CrossRef]

16. Bouillon, R. Comparative analysis of nutritional guidelines for vitamin D. *Nat. Rev. Endocrinol.* **2017**, *13*, 466–479. [CrossRef]

17. Fraile Navarro, D.; Lopez Garcia-Franco, A.; Nino de Guzman, E.; Rabassa, M.; Zamanillo Campos, R.; Pardo-Hernandez, H.; Ricci-Cabello, I.; Canelo-Aybar, C.; Meneses-Echavez, J.F.; Yepes-Nunez, J.J.; et al. Vitamin D recommendations in clinical guidelines: A systematic review, quality evaluation and analysis of potential predictors. *Int. J. Clin. Pract.* **2021**, *75*, e14805. [CrossRef]

18. Dai, Z.; McKenzie, J.E.; McDonald, S.; Baram, L.; Page, M.J.; Allman-Farinelli, M.; Raubenheimer, D.; Bero, L.A. Assessment of the Methods Used to Develop Vitamin D and Calcium Recommendations—A Systematic Review of Bone Health Guidelines. *Nutrients* **2021**, *13*, 2423. [CrossRef]

19. Rosen, C.J.; Abrams, S.A.; Aloia, J.F.; Brannon, P.M.; Clinton, S.K.; Durazo-Arvizu, R.A.; Gallagher, J.C.; Gallo, R.L.; Jones, G.; Kovacs, C.S.; et al. IOM committee members respond to Endocrine Society vitamin D guideline. *J. Clin. Endocrinol. Metab.* **2012**, *97*, 1146–1152. [CrossRef]

20. Pludowski, P.; Holick, M.F.; Grant, W.B.; Konstantynowicz, J.; Mascarenhas, M.R.; Haq, A.; Povoroznyuk, V.; Balatska, N.; Barbosa, A.P.; Karonova, T.; et al. Vitamin D supplementation guidelines. *J. Steroid Biochem. Mol. Biol.* **2018**, *175*, 125–135. [CrossRef]

21. Pludowski, P.; Karczmarewicz, E.; Bayer, M.; Carter, G.; Chlebna-Sokol, D.; Czech-Kowalska, J.; Debski, R.; Decsi, T.; Dobrzanska, A.; Franek, E.; et al. Practical guidelines for the supplementation of vitamin D and the treatment of deficits in Central Europe—Recommended vitamin D intakes in the general population and groups at risk of vitamin D deficiency. *Endokrynol. Pol.* **2013**, *64*, 319–327. [CrossRef]

22. Takacs, I.; Benko, I.; Toldy, E.; Wikonkal, N.; Szekeres, L.; Bodolay, E.; Kiss, E.; Jambrik, Z.; Szabo, B.; Merkely, B.; et al. Hungarian consensus regarding the role of vitamin D in the prevention and treatment of diseases. *Orv. Hetil.* **2012**, *153* (Suppl. 2), 5–26. [CrossRef]

23. German Nutrition Society (DGE). New reference values for vitamin D. *Ann. Nutr. Metab.* **2012**, *60*, 241–246. [CrossRef]

24. EFSA NDA Panel (EFSA Panel on Dietetic Products, Nutrition and Allergies). Scientific opinion on dietary reference values for vitamin D. *EFSA J.* **2016**, *14*, 4547. [CrossRef]

25. Pigarova, E.A.; Rozhinskaya, L.Y.; Belaya, J.E.; Dzeranova, L.K.; Karonova, T.L.; Ilyin, A.V.; Melnichenko, G.A.; Dedov, I.I. Russian Association of Endocrinologists recommendations for diagnosis, treatment and prevention of vitamin D deficiency in adults. *Probl. Endocrinol.* **2016**, *62*, 60–84. (In Russian) [CrossRef]

26. Rusinska, A.; Pludowski, P.; Walczak, M.; Borszewska-Kornacka, M.K.; Bossowski, A.; Chlebna-Sokol, D.; Czech-Kowalska, J.; Dobrzanska, A.; Franek, E.; Helwich, E.; et al. Vitamin D Supplementation Guidelines for General Population and Groups at Risk of Vitamin D Deficiency in Poland-Recommendations of the Polish Society of Pediatric Endocrinology and Diabetes and the Expert Panel with Participation of National Specialist Consultants and Representatives of Scientific Societies—2018 Update. *Front. Endocrinol.* **2018**, *9*, 246. [CrossRef]

27. Rudenko, E. *National Regulation about Methods of Diagnosis, Prevention and Differentiated Treatment of Vitamin D Deficiency*; Ministry of Health of the Republic of Belarus: Minsk, Belarus, 2013.

28. Borissova, A.-M.; Boyanov, M.A.; Popivanov, P.R.; Kolarov, Z.; Petranova, T.P.; Shinkov, A.D. *Recommendation for Diagnosis, Prevention and Treatment of Vitamin D Deficiency*; Bulgarian Society of Endocrinology: Sofia, Bulgaria, 2019.

29. Payer, J.; Killinger, Z.; Jackuliak, P.; Kužma, M. Postmenopausal osteoporosis: Standard diagnostic and therapeutic practice. *Clin. Osteol.* **2018**, *23*, 18–27.

30. Cesareo, R.; Attanasio, R.; Caputo, M.; Castello, R.; Chiodini, I.; Falchetti, A.; Guglielmi, R.; Papini, E.; Santonati, A.; Scillitani, A.; et al. Italian Association of Clinical Endocrinologists (AME) and Italian Chapter of the American Association of Clinical Endocrinologists (AACE) Position Statement: Clinical Management of Vitamin D Deficiency in Adults. *Nutrients* **2018**, *10*, 546. [CrossRef]

31. Rockwell, M.; Kraak, V.; Hulver, M.; Epling, J. Clinical Management of Low Vitamin D: A Scoping Review of Physicians' Practices. *Nutrients* **2018**, *10*, 493. [CrossRef]

32. Zittermann, A.; Pilz, S.; Berthold, H.K. Serum 25-hydroxyvitamin D response to vitamin D supplementation in infants: A systematic review and meta-analysis of clinical intervention trials. *Eur. J. Nutr.* **2020**, *59*, 359–369. [CrossRef]

33. Munns, C.F.; Shaw, N.; Kiely, M.; Specker, B.L.; Thacher, T.D.; Ozono, K.; Michigami, T.; Tiosano, D.; Mughal, M.Z.; Makitie, O.; et al. Global Consensus Recommendations on Prevention and Management of Nutritional Rickets. *J. Clin. Endocrinol. Metab.* **2016**, *101*, 394–415. [CrossRef] [PubMed]

34. Grossman, Z.; Hadjipanayis, A.; Stiris, T.; Del Torso, S.; Mercier, J.C.; Valiulis, A.; Shamir, R. Vitamin D in European children-statement from the European Academy of Paediatrics (EAP). *Eur. J. Pediatr.* **2017**, *176*, 829–831. [CrossRef] [PubMed]

35. Diamond, I.R.; Grant, R.C.; Feldman, B.M.; Pencharz, P.B.; Ling, S.C.; Moore, A.M.; Wales, P.W. Defining consensus: A systematic review recommends methodologic criteria for reporting of Delphi studies. *J. Clin. Epidemiol.* **2014**, *67*, 401–409. [CrossRef] [PubMed]

36. Rosenfeld, R.M.; Nnacheta, L.C.; Corrigan, M.D. Clinical Consensus Statement Development Manual. *Otolaryngol. Head Neck Surg.* **2015**, *153*, S1–S14. [CrossRef]

37. Cashman, K.D. Global differences in vitamin D status and dietary intake: A review of the data. *Endocr. Connect.* **2022**, *11*, e210282. [CrossRef]

38. Lips, P.; Cashman, K.D.; Lamberg-Allardt, C.; Bischoff-Ferrari, H.A.; Obermayer-Pietsch, B.; Bianchi, M.L.; Stepan, J.; El-Hajj Fuleihan, G.; Bouillon, R. Current vitamin D status in European and Middle East countries and strategies to prevent vitamin D deficiency: A position statement of the European Calcified Tissue Society. *Eur. J. Endocrinol.* **2019**, *180*, P23–P54. [CrossRef]

39. Płudowski, P.; Ducki, C.; Konstantynowicz, J.; Jaworski, M. Vitamin D status in Poland. *Pol. Arch. Med. Wewn.* **2016**, *126*, 530–539. [CrossRef]

40. Avdeeva, V.A.; Suplotova, L.A.; Pigarova, E.A.; Rozhinskaya, L.Y.; Troshina, E.A. Vitamin D deficiency in Russia: The first results of a registered, non-interventional study of the frequency of vitamin D deficiency and insufficiency in various geographic regions of the country. *Probl. Endokrinol.* **2021**, *67*, 84–92. [CrossRef]

41. Nikolova, M.G.; Boyanov, M.A.; Tsakova, A.D. Correlations of Serum Vitamin D with Metabolic Parameters in Adult Outpatients with Different Degrees of Overweight/Obesity Coming from an Urban Community. *Acta Endocrinol.* **2018**, *14*, 375–383. [CrossRef]

42. Burnett-Bowie, S.M.; Cappola, A.R. The USPSTF 2021 Recommendations on Screening for Asymptomatic Vitamin D Deficiency in Adults: The Challenge for Clinicians Continues. *JAMA* **2021**, *325*, 1401–1402. [CrossRef]

43. US Preventive Services Task Force; Krist, A.H.; Davidson, K.W.; Mangione, C.M.; Cabana, M.; Caughey, A.B.; Davis, E.M.; Donahue, K.E.; Doubeni, C.A.; Epling, J.W., Jr.; et al. Screening for Vitamin D Deficiency in Adults: US Preventive Services Task Force Recommendation Statement. *JAMA* **2021**, *325*, 1436–1442. [CrossRef]

44. Macova, L.; Bicikova, M. Vitamin D: Current Challenges between the Laboratory and Clinical Practice. *Nutrients* **2021**, *13*, 1758. [CrossRef]

45. Binkley, N.; Dawson-Hughes, B.; Durazo-Arvizu, R.; Thamm, M.; Tian, L.; Merkel, J.M.; Jones, J.C.; Carter, G.D.; Sempos, C.T. Vitamin D measurement standardization: The way out of the chaos. *J. Steroid Biochem. Mol. Biol.* **2017**, *173*, 117–121. [CrossRef]

46. Marcinowska-Suchowierska, E.; Kupisz-Urbanska, M.; Lukaszkiewicz, J.; Pludowski, P.; Jones, G. Vitamin D Toxicity—A Clinical Perspective. *Front. Endocrinol.* **2018**, *9*, 550. [CrossRef]

47. Cappellani, D.; Brancatella, A.; Morganti, R.; Borsari, S.; Baldinotti, F.; Caligo, M.A.; Kaufmann, M.; Jones, G.; Marcocci, C.; Cetani, F. Hypercalcemia due to CYP24A1 mutations: A systematic descriptive review. *Eur. J. Endocrinol.* **2021**, *186*, 137–149. [CrossRef]

48. Carlberg, C. Vitamin D: A Micronutrient Regulating Genes. *Curr. Pharm. Des.* **2019**, *25*, 1740–1746. [CrossRef]

49. Hanel, A.; Neme, A.; Malinen, M.; Hamalainen, E.; Malmberg, H.R.; Etheve, S.; Tuomainen, T.P.; Virtanen, J.K.; Bendik, I.; Carlberg, C. Common and personal target genes of the micronutrient vitamin D in primary immune cells from human peripheral blood. *Sci. Rep.* **2020**, *10*, 21051. [CrossRef]

50. Krasniqi, E.; Boshnjaku, A.; Wagner, K.H.; Wessner, B. Association between Polymorphisms in Vitamin D Pathway-Related Genes, Vitamin D Status, Muscle Mass and Function: A Systematic Review. *Nutrients* **2021**, *13*, 3109. [CrossRef]

51. Bilezikian, J.P.; Formenti, A.M.; Adler, R.A.; Binkley, N.; Bouillon, R.; Lazaretti-Castro, M.; Marcocci, C.; Napoli, N.; Rizzoli, R.; Giustina, A. Vitamin D: Dosing, levels, form, and route of administration: Does one approach fit all? *Rev. Endocr. Metab. Disord.* **2021**, *22*, 1201–1218. [CrossRef]

52. Kupisz-Urbanska, M.; Pludowski, P.; Marcinowska-Suchowierska, E. Vitamin D Deficiency in Older Patients—Problems of Sarcopenia, Drug Interactions, Management in Deficiency. *Nutrients* **2021**, *13*, 1247. [CrossRef]

53. Cashman, K.D. Vitamin D Requirements for the Future-Lessons Learned and Charting a Path Forward. *Nutrients* **2018**, *10*, 533. [CrossRef]

54. Cashman, K.D.; Ritz, C.; Kiely, M.; Odin, C. Improved Dietary Guidelines for Vitamin D: Application of Individual Participant Data (IPD)-Level Meta-Regression Analyses. *Nutrients* **2017**, *9*, 469. [CrossRef]

55. Pilz, S.; Hahn, A.; Schon, C.; Wilhelm, M.; Obeid, R. Effect of Two Different Multimicronutrient Supplements on Vitamin D Status in Women of Childbearing Age: A Randomized Trial. *Nutrients* **2017**, *9*, 30. [CrossRef]

56. Calvo, M.S.; Whiting, S.J. Public health strategies to overcome barriers to optimal vitamin D status in populations with special needs. *J. Nutr.* **2006**, *136*, 1135–1139. [CrossRef]

57. Mo, M.; Wang, S.; Chen, Z.; Muyiduli, X.; Wang, S.; Shen, Y.; Shao, B.; Li, M.; Chen, D.; Chen, Z.; et al. A systematic review and meta-analysis of the response of serum 25-hydroxyvitamin D concentration to vitamin D supplementation from RCTs from around the globe. *Eur. J. Clin. Nutr.* **2019**, *73*, 816–834. [CrossRef]

58. Bacha, D.S.; Rahme, M.; Al-Shaar, L.; Baddoura, R.; Halaby, G.; Singh, R.J.; Mahfoud, Z.R.; Habib, R.; Arabi, A.; El-Hajj Fuleihan, G. Vitamin D3 Dose Requirement That Raises 25-Hydroxyvitamin D to Desirable Level in Overweight and Obese Elderly. *J. Clin. Endocrinol. Metab.* **2021**, *106*, e3644–e3654. [CrossRef]

59. Zittermann, A.; Ernst, J.B.; Gummert, J.F.; Borgermann, J. Vitamin D supplementation, body weight and human serum 25-hydroxyvitamin D response: A systematic review. *Eur. J. Nutr.* **2014**, *53*, 367–374. [CrossRef] [PubMed]

60. Autier, P.; Gandini, S.; Mullie, P. A systematic review: Influence of vitamin D supplementation on serum 25-hydroxyvitamin D concentration. *J. Clin. Endocrinol. Metab.* **2012**, *97*, 2606–2613. [CrossRef] [PubMed]

61. Cashman, K.D.; Ritz, C.; Adebayo, F.A.; Dowling, K.G.; Itkonen, S.T.; Ohman, T.; Skaffari, E.; Saarnio, E.M.; Kiely, M.; Lamberg-Allardt, C. Differences in the dietary requirement for vitamin D among Caucasian and East African women at Northern latitude. *Eur. J. Nutr.* **2019**, *58*, 2281–2291. [CrossRef] [PubMed]

62. Cashman, K.D.; Kiely, M.E.; Andersen, R.; Gronborg, I.M.; Tetens, I.; Tripkovic, L.; Lanham-New, S.A.; Lamberg-Allardt, C.; Adebayo, F.A.; Gallagher, J.C.; et al. Individual participant data (IPD)-level meta-analysis of randomised controlled trials to estimate the vitamin D dietary requirements in dark-skinned individuals resident at high latitude. *Eur. J. Nutr.* **2021**, *61*, 1015–1034. [CrossRef] [PubMed]

63. Mazahery, H.; von Hurst, P.R. Factors Affecting 25-Hydroxyvitamin D Concentration in Response to Vitamin D Supplementation. *Nutrients* **2015**, *7*, 5111–5142. [CrossRef]

64. Ramasamy, I. Vitamin D Metabolism and Guidelines for Vitamin D Supplementation. *Clin. Biochem. Rev.* **2020**, *41*, 103–126. [CrossRef]

65. Ish-Shalom, S.; Segal, E.; Salganik, T.; Raz, B.; Bromberg, I.L.; Vieth, R. Comparison of daily, weekly, and monthly vitamin D3 in ethanol dosing protocols for two months in elderly hip fracture patients. *J. Clin. Endocrinol. Metab.* **2008**, *93*, 3430–3435. [CrossRef]

66. Takacs, I.; Toth, B.E.; Szekeres, L.; Szabo, B.; Bakos, B.; Lakatos, P. Randomized clinical trial to comparing efficacy of daily, weekly and monthly administration of vitamin D3. *Endocrine* **2017**, *55*, 60–65. [CrossRef]

67. Mazess, R.B.; Bischoff-Ferrari, H.A.; Dawson-Hughes, B. Vitamin D: Bolus Is Bogus—A Narrative Review. *JBMR Plus* **2021**, *5*, e10567. [CrossRef]

68. Jorde, R.; Grimnes, G. Serum cholecalciferol may be a better marker of vitamin D status than 25-hydroxyvitamin D. *Med. Hypotheses* **2018**, *111*, 61–65. [CrossRef]

69. Rothen, J.P.; Rutishauser, J.; Walter, P.N.; Hersberger, K.E.; Arnet, I. Oral intermittent vitamin D substitution: Influence of pharmaceutical form and dosage frequency on medication adherence: A randomized clinical trial. *BMC Pharmacol. Toxicol.* **2020**, *21*, 51. [CrossRef]

70. Tripkovic, L.; Lambert, H.; Hart, K.; Smith, C.P.; Bucca, G.; Penson, S.; Chope, G.; Hypponen, E.; Berry, J.; Vieth, R.; et al. Comparison of vitamin D2 and vitamin D3 supplementation in raising serum 25-hydroxyvitamin D status: A systematic review and meta-analysis. *Am. J. Clin. Nutr.* **2012**, *95*, 1357–1364. [CrossRef]

71. Bjelakovic, G.; Gluud, L.L.; Nikolova, D.; Whitfield, K.; Wetterslev, J.; Simonetti, R.G.; Bjelakovic, M.; Gluud, C. Vitamin D supplementation for prevention of mortality in adults. *Cochrane Database Syst. Rev.* **2008**, CD007470. [CrossRef]

72. Holick, M.F.; Binkley, N.C.; Bischoff-Ferrari, H.A.; Gordon, C.M.; Hanley, D.A.; Heaney, R.P.; Murad, M.H.; Weaver, C.M. Guidelines for preventing and treating vitamin D deficiency and insufficiency revisited. *J. Clin. Endocrinol. Metab.* **2012**, *97*, 1153–1158. [CrossRef]

73. Heaney, R.P.; Armas, L.A. Quantifying the vitamin D economy. *Nutr. Rev.* **2015**, *73*, 51–67. [CrossRef] [PubMed]

74. Pludowski, P.; Socha, P.; Karczmarewicz, E.; Zagorecka, E.; Lukaszkiewicz, J.; Stolarczyk, A.; Piotrowska-Jastrzebska, J.; Kryskiewicz, E.; Lorenc, R.S.; Socha, J. Vitamin D supplementation and status in infants: A prospective cohort observational study. *J. Pediatr. Gastroenterol. Nutr.* **2011**, *53*, 93–99. [CrossRef] [PubMed]

75. Vieth, R. Vitamin D supplementation: Cholecalciferol, calcifediol, and calcitriol. *Eur. J. Clin. Nutr.* **2020**, *74*, 1493–1497. [CrossRef] [PubMed]

76. Billington, E.O.; Burt, L.A.; Rose, M.S.; Davison, E.M.; Gaudet, S.; Kan, M.; Boyd, S.K.; Hanley, D.A. Safety of High-Dose Vitamin D Supplementation: Secondary Analysis of a Randomized Controlled Trial. *J. Clin. Endocrinol. Metab.* **2020**, *105*, 1261–1273. [CrossRef]

77. Vieth, R. Vitamin D and cancer mini-symposium: The risk of additional vitamin D. *Ann. Epidemiol.* **2009**, *19*, 441–445. [CrossRef]

78. Nielsen, O.H.; Hansen, T.I.; Gubatan, J.M.; Jensen, K.B.; Rejnmark, L. Managing vitamin D deficiency in inflammatory bowel disease. *Frontline Gastroenterol.* **2019**, *10*, 394–400. [CrossRef]

79. Masood, M.Q.; Khan, A.; Awan, S.; Dar, F.; Naz, S.; Naureen, G.; Saghir, S.; Jabbar, A. Comparison of Vitamin D Replacement Strategies with High-Dose Intramuscular or Oral Cholecalciferol: A Prospective Intervention Study. *Endocr. Pract.* **2015**, *21*, 1125–1133. [CrossRef]

80. Wylon, K.; Drozdenko, G.; Krannich, A.; Heine, G.; Dolle, S.; Worm, M. Pharmacokinetic Evaluation of a Single Intramuscular High Dose versus an Oral Long-Term Supplementation of Cholecalciferol. *PLoS ONE* **2017**, *12*, e0169620. [CrossRef]

81. Gupta, N.; Farooqui, K.J.; Batra, C.M.; Marwaha, R.K.; Mithal, A. Effect of oral versus intramuscular Vitamin D replacement in apparently healthy adults with Vitamin D deficiency. *Indian J. Endocrinol. Metab.* **2017**, *21*, 131–136. [CrossRef]

82. Sosa Henriquez, M.; Gomez de Tejada Romero, M.J. Cholecalciferol or Calcifediol in the Management of Vitamin D Deficiency. *Nutrients* **2020**, *12*, 1617. [CrossRef]

83. Quesada-Gomez, J.M.; Bouillon, R. Is calcifediol better than cholecalciferol for vitamin D supplementation? *Osteoporos. Int.* **2018**, *29*, 1697–1711. [CrossRef]

84. Perez-Castrillon, J.L.; Duenas-Laita, A.; Brandi, M.L.; Jodar, E.; Del Pino-Montes, J.; Quesada-Gomez, J.M.; Cereto Castro, F.; Gomez-Alonso, C.; Gallego Lopez, L.; Olmos Martinez, J.M.; et al. Calcifediol is superior to cholecalciferol in improving vitamin D status in postmenopausal women: A randomized trial. *J. Bone Miner. Res.* **2021**, *36*, 1967–1978. [CrossRef]

85. Bischoff-Ferrari, H.A.; Dawson-Hughes, B.; Orav, E.J.; Staehelin, H.B.; Meyer, O.W.; Theiler, R.; Dick, W.; Willett, W.C.; Egli, A. Monthly High-Dose Vitamin D Treatment for the Prevention of Functional Decline: A Randomized Clinical Trial. *JAMA Intern. Med.* **2016**, *176*, 175–183. [CrossRef]

86. Bollerslev, J.; Rejnmark, L.; Zahn, A.; Heck, A.; Appelman-Dijkstra, N.M.; Cardoso, L.; Hannan, F.M.; Cetani, F.; Sikjaer, T.; Formenti, A.M.; et al. European Expert Consensus on Practical Management of Specific Aspects of Parathyroid Disorders in Adults and in Pregnancy: Recommendations of the ESE Educational Program of Parathyroid Disorders. *Eur. J. Endocrinol.* **2022**, *186*, R33–R63. [CrossRef]

87. Bischoff-Ferrari, H.A.; Willett, W.C.; Orav, E.J.; Lips, P.; Meunier, P.J.; Lyons, R.A.; Flicker, L.; Wark, J.; Jackson, R.D.; Cauley, J.A.; et al. A pooled analysis of vitamin D dose requirements for fracture prevention. *N. Engl. J. Med.* **2012**, *367*, 40–49. [CrossRef]

88. Bolland, M.J.; Grey, A.; Avenell, A. Effects of vitamin D supplementation on musculoskeletal health: A systematic review, meta-analysis, and trial sequential analysis. *Lancet Diabetes Endocrinol.* **2018**, *6*, 847–858. [CrossRef]

89. Carmel, A.S.; Bockman, R.S. Vitamin D and bisphosphonate response. *Osteoporos. Int.* **2014**, *25*, 2155. [CrossRef]

90. Reid, I.R.; Horne, A.M.; Mihov, B.; Stewart, A.; Garratt, E.; Wong, S.; Wiessing, K.R.; Bolland, M.J.; Bastin, S.; Gamble, G.D. Fracture Prevention with Zoledronate in Older Women with Osteopenia. *N. Engl. J. Med.* **2018**, *379*, 2407–2416. [CrossRef]

91. Li, S.; Xi, C.; Li, L.; Long, Z.; Zhang, N.; Yin, H.; Xie, K.; Wu, Z.; Tian, J.; Wang, F.; et al. Comparisons of different vitamin D supplementation for prevention of osteoporotic fractures: A Bayesian network meta-analysis and meta-regression of randomised controlled trials. *Int. J. Food Sci. Nutr.* **2021**, *72*, 518–528. [CrossRef]

92. Ganmaa, D.; Enkhmaa, D.; Nasantogtokh, E.; Sukhbaatar, S.; Tumur-Ochir, K.E.; Manson, J.E. Vitamin D, respiratory infections, and chronic disease: Review of meta-analyses and randomized clinical trials. *J. Intern. Med.* **2021**. [CrossRef]

93. Iuliano, S.; Poon, S.; Robbins, J.; Bui, M.; Wang, X.; De Groot, L.; Van Loan, M.; Zadeh, A.G.; Nguyen, T.; Seeman, E. Effect of dietary sources of calcium and protein on hip fractures and falls in older adults in residential care: Cluster randomised controlled trial. *BMJ* **2021**, *375*, n2364. [CrossRef]

94. Appel, L.J.; Michos, E.D.; Mitchell, C.M.; Blackford, A.L.; Sternberg, A.L.; Miller, E.R., 3rd; Juraschek, S.P.; Schrack, J.A.; Szanton, S.L.; Charleston, J.; et al. The Effects of Four Doses of Vitamin D Supplements on Falls in Older Adults: A Response-Adaptive, Randomized Clinical Trial. *Ann. Intern. Med.* **2021**, *174*, 145–156. [CrossRef]

95. Zittermann, A.; Ernst, J.B.; Prokop, S.; Fuchs, U.; Dreier, J.; Kuhn, J.; Knabbe, C.; Birschmann, I.; Schulz, U.; Berthold, H.K.; et al. Effect of vitamin D on all-cause mortality in heart failure (EVITA): A 3-year randomized clinical trial with 4000 IU vitamin D daily. *Eur. Heart J.* **2017**, *38*, 2279–2286. [CrossRef]

96. Gnagnarella, P.; Muzio, V.; Caini, S.; Raimondi, S.; Martinoli, C.; Chiocca, S.; Miccolo, C.; Bossi, P.; Cortinovis, D.; Chiaradonna, F.; et al. Vitamin D Supplementation and Cancer Mortality: Narrative Review of Observational Studies and Clinical Trials. *Nutrients* **2021**, *13*, 3285. [CrossRef]

97. Charoenngam, N.; Holick, M.F. Immunologic Effects of Vitamin D on Human Health and Disease. *Nutrients* **2020**, *12*, 2097. [CrossRef]

98. Hajhashemy, Z.; Shahdadian, F.; Moslemi, E.; Mirenayat, F.S.; Saneei, P. Serum vitamin D levels in relation to metabolic syndrome: A systematic review and dose-response meta-analysis of epidemiologic studies. *Obes. Rev.* **2021**, *22*, e13223. [CrossRef]

99. Han, J.; Guo, X.; Yu, X.; Liu, S.; Cui, X.; Zhang, B.; Liang, H. 25-Hydroxyvitamin D and Total Cancer Incidence and Mortality: A Meta-Analysis of Prospective Cohort Studies. *Nutrients* **2019**, *11*, 2295. [CrossRef]

100. Hahn, J.; Cook, N.R.; Alexander, E.K.; Friedman, S.; Walter, J.; Bubes, V.; Kotler, G.; Lee, I.M.; Manson, J.E.; Costenbader, K.H. Vitamin D and marine omega 3 fatty acid supplementation and incident autoimmune disease: VITAL randomized controlled trial. *BMJ* **2022**, *376*, e066452. [CrossRef]

101. Autier, P.; Boniol, M.; Pizot, C.; Mullie, P. Vitamin D status and ill health: A systematic review. *Lancet Diabetes Endocrinol.* **2014**, *2*, 76–89. [CrossRef]

102. Jolliffe, D.A.; Camargo, C.A., Jr.; Sluyter, J.D.; Aglipay, M.; Aloia, J.F.; Ganmaa, D.; Bergman, P.; Bischoff-Ferrari, H.A.; Borzutzky, A.; Damsgaard, C.T.; et al. Vitamin D supplementation to prevent acute respiratory infections: A systematic review and meta-analysis of aggregate data from randomised controlled trials. *Lancet Diabetes Endocrinol.* **2021**, *9*, 276–292. [CrossRef]

103. Sluyter, J.D.; Manson, J.E.; Scragg, R. Vitamin D and Clinical Cancer Outcomes: A Review of Meta-Analyses. *JBMR Plus* **2021**, *5*, e10420. [CrossRef] [PubMed]

104. Scragg, R.; Sluyter, J.D. Is There Proof of Extraskeletal Benefits from Vitamin D Supplementation from Recent Mega Trials of Vitamin D? *JBMR Plus* **2021**, *5*, e10459. [CrossRef] [PubMed]

105. Jolliffe, D.A.; Greenberg, L.; Hooper, R.L.; Mathyssen, C.; Rafiq, R.; de Jongh, R.T.; Camargo, C.A.; Griffiths, C.J.; Janssens, W.; Martineau, A.R. Vitamin D to prevent exacerbations of COPD: Systematic review and meta-analysis of individual participant data from randomised controlled trials. *Thorax* **2019**, *74*, 337–345. [CrossRef]

106. Li, X.; He, J.; Yu, M.; Sun, J. The efficacy of vitamin D therapy for patients with COPD: A meta-analysis of randomized controlled trials. *Ann. Palliat. Med.* **2020**, *9*, 286–297. [CrossRef] [PubMed]

107. Forno, E.; Bacharier, L.B.; Phipatanakul, W.; Guilbert, T.W.; Cabana, M.D.; Ross, K.; Covar, R.; Gern, J.E.; Rosser, F.J.; Blatter, J.; et al. Effect of Vitamin D3 Supplementation on Severe Asthma Exacerbations in Children with Asthma and Low Vitamin D Levels: The VDKA Randomized Clinical Trial. *JAMA* **2020**, *324*, 752–760. [CrossRef]

108. Zittermann, A.; Pilz, S.; Hoffmann, H.; Marz, W. Vitamin D and airway infections: A European perspective. *Eur. J. Med. Res.* **2016**, *21*, 14. [CrossRef] [PubMed]
109. Pilz, S.; Trummer, C.; Theiler-Schwetz, V.; Grubler, M.R.; Verheyen, N.D.; Odler, B.; Karras, S.N.; Zittermann, A.; Marz, W. Critical Appraisal of Large Vitamin D Randomized Controlled Trials. *Nutrients* **2022**, *14*, 303. [CrossRef]
110. Neale, R.E.; Baxter, C.; Romero, B.D.; McLeod, D.S.A.; English, D.R.; Armstrong, B.K.; Ebeling, P.R.; Hartel, G.; Kimlin, M.G.; O'Connell, R.; et al. The D-Health Trial: A randomised controlled trial of the effect of vitamin D on mortality. *Lancet Diabetes Endocrinol.* **2022**, *10*, 120–128. [CrossRef]
111. Zemb, P.; Bergman, P.; Camargo, C.A., Jr.; Cavalier, E.; Cormier, C.; Courbebaisse, M.; Hollis, B.; Joulia, F.; Minisola, S.; Pilz, S.; et al. Vitamin D deficiency and the COVID-19 pandemic. *J. Glob. Antimicrob. Resist.* **2020**, *22*, 133–134. [CrossRef]
112. Smaha, J.; Kuzma, M.; Jackuliak, P.; Payer, J. Vitamin D supplementation as an important factor in COVID-19 prevention and treatment: What evidence do we have? *Vnitr. Lek.* **2020**, *66*, 494–500. [CrossRef]
113. Stroehlein, J.K.; Wallqvist, J.; Iannizzi, C.; Mikolajewska, A.; Metzendorf, M.I.; Benstoem, C.; Meybohm, P.; Becker, M.; Skoetz, N.; Stegemann, M.; et al. Vitamin D supplementation for the treatment of COVID-19: A living systematic review. *Cochrane Database Syst. Rev.* **2021**, *5*, CD015043. [CrossRef]
114. Liu, Q.; Qin, C.; Liu, M.; Liu, J. Effectiveness and safety of SARS-CoV-2 vaccine in real-world studies: A systematic review and meta-analysis. *Infect. Dis. Poverty* **2021**, *10*, 132. [CrossRef]
115. Pilz, S.; Chakeri, A.; Ioannidis, J.P.; Richter, L.; Theiler-Schwetz, V.; Trummer, C.; Krause, R.; Allerberger, F. SARS-CoV-2 re-infection risk in Austria. *Eur. J. Clin. Investig.* **2021**, *51*, e13520. [CrossRef]
116. Pilz, S.; Theiler-Schwetz, V.; Trummer, C.; Krause, R.; Ioannidis, J.P.A. SARS-CoV-2 reinfections: Overview of efficacy and duration of natural and hybrid immunity. *Environ. Res.* **2022**, *209*, 112911. [CrossRef]

Article

Prevalence of Vitamin D Deficiency in Patients Treated for Juvenile Idiopathic Arthritis and Potential Role of Methotrexate: A Preliminary Study

Maciej K. Stawicki [1], Paweł Abramowicz [1,*], Adrian Góralczyk [2], Justyna Młyńczyk [1], Anna Kondratiuk [1] and Jerzy Konstantynowicz [1]

[1] Department of Pediatrics, Rheumatology, Immunology, and Metabolic Bone Diseases, Medical University of Bialystok, Waszyngtona Street 17, 15274 Bialystok, Poland; maciej.stawicki@umb.edu.pl (M.K.S.); justyna.mlynczyk@umb.edu.pl (J.M.); an.kondratiuk@gmail.com (A.K.); jurekonstant@o2.pl (J.K.)

[2] Department of Orthopaedics and Traumatology, Hospital of Ministry of Administration and Internal Affairs in Bialystok, Fabryczna Street 27, 15471 Bialystok, Poland; adriangoralczyk1@yahoo.com

* Correspondence: pawel.abramowicz@umb.edu.pl; Tel.: +48-857-450-622; Fax: +48-857-450-644

Citation: Stawicki, M.K.; Abramowicz, P.; Góralczyk, A.; Młyńczyk, J.; Kondratiuk, A.; Konstantynowicz, J. Prevalence of Vitamin D Deficiency in Patients Treated for Juvenile Idiopathic Arthritis and Potential Role of Methotrexate: A Preliminary Study. Nutrients 2022, 14, 1645. https://doi.org/10.3390/nu14081645

Academic Editors: Tyler Barker and Connie Weaver

Received: 2 March 2022
Accepted: 13 April 2022
Published: 14 April 2022

Abstract: Background: Vitamin D deficiency is reported in rheumatological diseases in adults. The aim was to evaluate the prevalence of vitamin D deficiency in children with juvenile idiopathic arthritis (JIA) and to investigate potential correlations between vitamin D status and clinical factors, laboratory traits, and medical treatment, including methotrexate (MTX) and glucocorticoids (GCs). Methods: In 189 patients aged 3–17.7 years, with JIA in the stable stage of the disease, anthropometry, clinical status, serum 25-hydroxyvitamin D [25(OH)D], calcium (Ca), phosphate (PO_4), total alkaline phosphatase (ALP), erythrocyte sedimentation rate (ESR), and C-reactive protein (CRP) were assessed. Results: Median 25(OH)D level was 15.00 ng/mL, interquartile range (IQR) 12.00 ng/mL. Vitamin D deficiency was found in 67.2% and was independent of sex, disease manifestation, and CRP, ESR, ALP, or PO_4 levels. Higher doses of MTX corresponded with lower 25(OH)D levels using both univariate and multivariate models ($p < 0.05$). No such trend was found for GCs treatment. Serum Ca was lower in patients treated with GCs ($p = 0.004$), MTX ($p = 0.03$), and combined GCs/MTX ($p = 0.034$). Conclusions: JIA patients are vitamin D depleted independently of disease activity or inflammatory markers. MTX therapy may be an iatrogenic factor leading to inadequate 25(OH)D levels. Vitamin D supplementation should be considered in all children with JIA, particularly those receiving long-term MTX therapy.

Keywords: juvenile idiopathic arthritis; vitamin D; calcium/phosphate metabolism; methotrexate; children

1. Introduction

Juvenile idiopathic arthritis (JIA) is a heterogenic group of chronic autoimmune disorders with a large spectrum of clinical manifestations and varying severity [1]. The causes of the disease are yet to be discovered. It is suggested that immunogenic mechanisms secondary to genetic and environmental factors are the background of the disease, while infections together with stress and trauma are suspected to be the most possible etiological agents for JIA [2].

As the incidence of JIA has been reported to increase worldwide, the management of this condition has become an important issue in pediatric care [3,4]. The complex autoimmune, inflammatory, and destructive processes, which are key pathogenic mechanisms in the disease, may lead to disability early in childhood and adolescence and may either persist to adulthood or confer a risk of later significant rheumatologic conditions including rheumatoid arthritis (RA).

Vitamin D is a fat-soluble vitamin and an important hormone involved in many physiological processes in the human body, such as bone mineralization, insulin regulation, and immune regulation [5–7]. Vitamin D is affecting the tissues by specific vitamin D receptors (VDR) inducing its biological activities. VDR is widely expressed in different cells, i.e., immune cells. There are many polymorphic variants of the VDR gene that possibly affect the functionality of the receptor [8].

Recent research focused on vitamin D in relation to a variety of inflammatory disorders revealed several controversial results. Furthermore, a large body of evidence was published within the last decade to demonstrate a vast range of vitamin D deficits in general, otherwise healthy, populations worldwide [9–14]. The role of vitamin D, being mainly a well-known regulator of calcium/phosphate metabolism, has been extensively investigated in adult rheumatoid diseases, demonstrating a potential beneficial effect on the disease course and activity; however, some studies did not support such associations [15].

Available reports show that vitamin D may have a significant influence on pathogenesis [16] and the outcome of JIA, i.e., a lower level of 25-hydroxyvitamin D [25(OH)D] was found in JIA patients compared with healthy children [17,18]. At present, there is a strongly held general view, based also on prospective studies, that vitamin D has pleiotropic multidimensional effects on human metabolism and may interact in situ with specific tissues and, therefore, demonstrates some preventive potential including anti-proliferative, anti-inflammatory, and immunomodulatory actions. On the other hand, the long-lasting vitamin D deficiency, reflected by decreased 25(OH)D concentrations, can deteriorate immune-mediated mechanisms or even exacerbate the course of the disease [19].

Assuming that vitamin D in pediatric rheumatoid diseases may be of importance and that several questions regarding vitamin D deficit remain unanswered, we attempted to investigate correlations between vitamin D status, clinical manifestation, and medical treatment of JIA. This study aimed to determine the prevalence of vitamin D deficiency and to evaluate potential risk factors of decreased serum 25-hydroxyvitamin D levels in children diagnosed with JIA.

2. Material and Methods

In this cross-sectional study, 189 Caucasian individuals (both in- and outpatients) treated for juvenile idiopathic arthritis were examined. The diagnosis of JIA was ascertained using standard classification criteria [20]. Clinical assessment, anthropometric measurements using standardized methods, and laboratory tests were performed. Blood samples were collected at the beginning of hospitalization. Clinical assessment, based on physical examination and functional tests, was performed during scheduled hospital admission. Juvenile arthritis disease activity score (JADAS27) was used to determine disease activity status [21]. Anthropometric measurements were carried out with standardized methods, compliant with WHO guidelines [22], and included body weight using an electronic scale (Seca^{TM} 799, Hamburg, Germany) and standing height obtained with a wall-mounted stadiometer (Seca^{TM} 216, Hamburg, Germany). Body Mass Index (BMI) was calculated with a standard formula.

Vitamin D status was determined by measuring serum 25(OH)D concentration using the automatic immunoenzyme method using Immulite$^{®}$2000 Immunoassay System (Siemens AG, Munich, Germany). Vitamin D deficiency was defined as serum 25(OH)D level < 20 ng/mL, consistently with the Institute of Medicine recommendations and the updated guidelines for Central Europe [5,10]. To assess inflammation activity, erythrocyte sedimentation rate (ESR), and C-reactive protein (CRP) concentration were measured. Calcium (Ca) and phosphate (PO$_4$) serum concentrations and total alkaline phosphatase (ALP) activity were also tested and referred to age-specific normative values to check basic bone mineral metabolism.

All procedures were approved by the Ethical Committee of the Medical University of Bialystok upon informed consent obtained from all participants and/or their legal guardians according to the Declaration of Helsinki and its later amendments.

The statistical analyses were performed with the STATISTICA software (version 13.3, Tibco Software Inc., Palo Alto, CA, USA) and statsmodels.org (version 0.13.2). To evaluate the normality of data distribution, Shapiro–Wilk test was used. Variables distributed normally were expressed as mean and standard deviation, whereas for those with skewed distribution, median and IQR were used as a method of result presentation. According to the data distribution, Student's *t*-test or, the Mann–Whitney *U* test was applied. Subsequently, Spearman rank correlation was used to test the relation between pairs of factors. Furthermore, multinomial logistic regression was used to investigate associations between 25(OH)D concentration and body weight, BMI, MTX dose, GCs dose, CRP, ESR, Ca, P, and ALP, which were incorporated as covariates in the models.

3. Results

A total of 189 children and adolescents (113 girls and 76 boys), aged 3–17.7 years (median 13.12, IQR 6.23) were included; all were Caucasian, living at a similar latitude, none of the participants had been diagnosed with comorbidities potentially affecting vitamin D or bone metabolism, and none had been supplemented with vitamin D at the time of the recruitment to the study.

Among the whole studied group, 49% had oligoarticular manifestation, 44% presented polyarticular manifestation, and 7% had systemic-onset JIA (Table 1). All of them were in a stable stage of the disease (remission or minimal disease activity) according to the JADAS27 scoring [21].

Table 1. Basic characteristics of the study group (* Median and IQR value are shown when applicable).

	Total (n = 189)
Age (years) *	13.12 (6.23)
Male-to-female	76/113
Weight (kg) *	48.50 (24.00)
Height (cm) *	155.00 (28.00)
BMI (kg/m^2) *	19.58 (5.26)
Polyarticular JIA (n; %)	83 (43.90%)
Oligoarticular JIA (n; %)	93 (49.20%)
Systemic-onset JIA (n; %)	13 (6.90%)
Treated with GCs (n; %)	73 (38.60%)
Treated with MTX (n; %)	84 (44.40%)

Methotrexate at a weekly dose of 10–20 mg per m^2, administered orally or subcutaneously, was the only conventional synthetic disease-modifying anti-rheumatic drug (csDMARD) used in these patients, whereas no other DMARDs combination was introduced. In addition, some patients required interim GCs as a bridging therapy, in rare cases given by intra-articular injections.

The median 25(OH)D serum concentration was 15.00 ng/mL, and the IQR was 12.00 in the whole studied group. Vitamin D deficiency was found in 127 patients of both sexes (67.2% of the examined population). Comparisons between the groups in relation to serum 25(OH)D concentration based on the cut-off level of 20 ng/mL are shown in Table 2.

Vitamin D status in children with JIA was independent of sex, age, clinical manifestation, disease activity, or inflammatory markers. Serum 25(OH)D was inversely associated with BMI (r = −0.19), i.e., overweight JIA patients had lower vitamin D levels. Additionally, a weak yet significant correlation between 25(OH)D and serum Ca (r = 0.19) was found. An association was observed between vitamin D status and pharmacological therapy used in children with JIA. Furthermore, the treatment option affected calcium/phosphate metabolism in both boys and girls. Weekly MTX dose was found to be significantly higher in patients with vitamin D deficiency than in those with serum 25(OH)D > 20 ng/mL (Table 2). This association was consistent with Pearson's correlation coefficient, indicating an inverse relationship between the dose of MTX and 25(OH)D concentration (r = −0.34, $p < 0.05$). There was a significant negative correlation between MTX dose and Ca and PO_4 levels but not with serum ALP. Furthermore, the GCs dose had no significant effect on serum 25(OH)D concentration in children with JIA, whereas the daily dose of GCs was inversely associated with ALP activity, Ca, and PO_4 levels. Significant correlations between the two treatment options are shown in Table 3.

Table 2. Characteristics of the patients with JIA in relation to their serum 25(OH)D concentration below and above 20 ng/mL (* mean ± SD or ** median and IQR are given).

	Low 25(OH)D Level <20 ng/mL	Normal 25(OH)D Level ≥20 ng/mL	*p* Value
Patients (N; %)	127 (67.2%)	62 (32.8%)	
Age (years) **	13.34; (5.18)	11.84 (8.32)	0.19
Weight (kg) **	50.00 (22.00)	45.35 (33.00)	0.14
Height (kg) **	156.50 (25.00)	148.00 (37.50)	0.15
BMI (kg/m^2) **	19.81 (4.88)	19.06 (6.10)	0.17
GCs (mg) ** *daily dose*	5.00 (5.00)	5.00 (5.00)	0.29
MTX (mg) ** *weekly dose per m^2*	15.00 (7.50)	12.5 (7.50)	0.02
CRP (mg/L) **	1.00 (3.70)	1.60 (12.0)	0.23
25(OH)D (ng/mL) **	12.00 (8.00)	25.5 (6.0)	<0.001
Ca (mmol/L) *	2.48 ± 0.09	2.52 ± 0.12	0.01
P (mg/dL) *	4.50 ± 0.63	4.54 ± 0.72	0.76
ALP (U/L) **	165.00 (138.00)	172.00 (124.00)	0.70
ESR (mm/h) **	11.50 (22.00)	19.00 (30.00)	0.05
JADAS27 score **	1.50 (0.70)	1.50 (0.50)	0.64

Table 3. Univariate correlations for methotrexate (MTX), glucocorticoids (GCs), and serum calcium/phosphate parameters in children with JIA.

	MTX *weekly dose per m^2*	GCs *daily dose*
25(OH)D	r = −0.33; p = 0.003	r = −0.08; p = 0.26
Calcium	r = −0.31; p = 0.01	r = −0.23; p = 0.01
Phosphate	r = −0.42; p = 0.03	r = −0.27; p = 0.004
ALP	r = −0.14; p = 0.17	r = −0.79; p = 0.004

Multinomial logistic regression analysis showed that, out of all factors introduced in the model, only the weekly MTX dose per m^2 was inversely associated with serum 25(OH)D concentration. The calculated coefficient was 1.79 for MTX/week/m^2 (95% confidence interval [CI]: 0.33–3.24), whereas no other variables were significantly associated with vitamin D concentration. The results of the multivariate analysis are presented in Table 4.

Table 4. Results of multinomial logistic regression performed to investigate multivariate analysis of factors associated with 25(OH)D concentration.

	Coefficient	Standard Error	95% CI		*p* Value
Body weight	0.09	0.40	−0.69	0.87	0.82
BMI	−0.08	0.32	−0.70	0.54	0.80
GCs *daily dose*	0.18	0.21	−0.24	0.60	0.41
MTX *weekly dose per m^2*	1.79	0.74	0.33	3.24	0.02
CRP (mg/L)	−0.29	0.23	−0.74	0.17	0.21
ESR (mm/h)	−0.72	0.23	−0.51	0.37	0.75
Ca (mmol/L)	−0.13	0.19	−0.51	0.24	0.48
P (mg/dL)	0.32	0.21	−0.10	0.73	0.13
ALP (U/L)	0.01	0.20	−0.37	0.40	0.94

4. Discussion

Vitamin D deficiency is common in the general population during growth, according to the available supportive evidence [9–11]. This observation has been extended by the present study demonstrating a disease-specific deficiency in patients with juvenile idiopathic arthritis. Several studies have reported suboptimal vitamin D status in children with rheumatic diseases resulting from multifactorial mechanisms associated with the autoimmunity and/or iatrogenic effects [4,15]. The main finding of our study was an association between long-term methotrexate therapy in children with JIA and a deteriorated vitamin D status. Due to the cross-sectional design of this study, causal pathways may not be clearly elucidated; however, the unfavorable effect of the MTX therapy on 25-hydroxyvitamin D may indicate a role of this particular medication in an increased risk of deficiency.

Presumably, the above-mentioned treatment essentially affected the vitamin D status and calcium/phosphate metabolism, as it was found to interfere at most with vitamin D deficiency, among other variables analyzed in this study. Noteworthy, the strategies of therapeutic management are similar in RA and JIA, excluding current therapies with a single drug, e.g., biologics. Effective recommendations include the proposal of subsuming a sequential application of non-steroidal anti-inflammatory drugs (NSAIDs), GCs, and non-biological/biological DMARDs in the treatment of RA depending on disease activity and severity [23]. Similarly, the currently binding approach to the complex medication of children with juvenile arthritis is based on analogous recommendations [24]. Methotrexate—out of all non-biological DMARDs—appears the most effective and preferably applicable agent due to its well-known effectiveness for restricting autoimmune and inflammatory processes. Moreover, recent studies have shown that MTX is also an inhibitor of osteoclastogenesis by impeding RANKL-induced calcium influx into osteoclast progenitor cells [25]. Kanagawa et al. postulated that MTX would have a protective role against osteoporosis and joint destruction via some of the above-mentioned specific mechanisms [25]. In the light of the multivariate approach, the results of our study suggest a different view, showing that MTX use may be associated with a decreased 25(OH)D level. Initial univariate analyses also showed a dose-dependent effect, i.e., the weekly dose of MTX is negatively associated with serum calcium and phosphate, although multivariate analyses failed to support those results. The influence of MTX on calcium/phosphate metabolism can be direct or indirect—just by affecting vitamin D metabolism. Assuming these causal effects may be true, a question arises: During which transformation phase of vitamin D precursors does this drug interfere? Possible interaction may occur at intestinal absorption, during which methotrexate may deteriorate vitamin D bioavailability from nutrients. Furthermore, it can also considerably downregulate hepatic hydroxylation of calciferol (due to its fully understood liver toxicity), or it can even affect skin synthesis of vitamin D.

These associations have not yet been probably reported in the literature concerning rheumatoid diseases, i.e., JIA or RA. Methotrexate may be regarded as a risk factor for secondary osteoporosis in adults, even if the available data are inconsistent [26,27]. Nevertheless, there is a need for further relevant investigation to determine if long-term MTX use is an independent factor of bone loss in children with rheumatoid conditions. Moreover, a workout of a molecular mechanism through which MTX affects vitamin D metabolism is necessary to prevent the negative effects of MTX treatment on the growing skeleton.

Surprisingly, the prolonged use of GCs in our patients was not associated with a decrease in 25(OH)D concentration despite an evident inverse relationship between GCs and calcium or phosphate metabolism. Some studies support our results by demonstrating clearly that the use of systemic steroids does not influence 25(OH)D levels [28]. Other reports show, by contrast, that GCs may have a specific regulatory effect on vitamin D metabolism [19,29]. Our finding seems even more interesting considering an insight into the molecular mechanism of actions of GCs, reflecting "anti-vitamin D effects". Possible explanations include that glucocorticoids increase calcium and phosphate renal excretion, being antagonists of $1,25(OH)_2D$, while not influencing its serum concentration. Furthermore, there was a strong negative association between the GCs dose and total alkaline phosphatase activity. It has been reported that the decreased serum Ca and PO_4 levels, as well as reduced ALP activity, may be a compelling contribution to reduced bone mineral apparent density (BMAD) concomitant with long-term GCs treatment in children with JIA [30]. The causal relationship between exposure to GCs and suboptimal bone mineral acquisition during growth, including glucocorticoid-induced osteoporosis, has been widely documented, although not all mechanisms have been clarified.

Disease activity, duration, and active inflammation play an important role in the deterioration of mineral and bone metabolism in the course of chronic rheumatologic conditions [15]. We point out that the inflammatory process is supposed to be another risk factor for vitamin D deficiency in children with confirmed JIA. Several published reports are attempting to elucidate this association, but the results are inconsistent and provide ambiguous information [31]. To optimize the usefulness of our study, the JADAS27 scale was applied for disease activity assessment. All studied patients were in remission or had minimal activity of JIA based on the JADAS criteria [21]. These characteristics allowed us to minimize the effect of the confounder, i.e., the impact of disease activity and severity on the results. Based on large cohort studies, there was an inverse relationship between vitamin D intake and RA disease risk [32]. However, the data are different when comparing RA and JIA, while results vary across published studies. More recent reports showed that vitamin D level was significantly reduced in patients with active RA [33–36]. In those studies, the scores DAS28, JADAS-27, and inflammatory markers (ESR, CRP, fibrinogen serum concentration) were used to assess disease activity. Some investigators reported that the prevalence of vitamin D deficiency or insufficiency was high in juvenile idiopathic arthritis; however, it was unassociated with either intensity of inflammation [37–39] or the genotypes of the vitamin D receptor [8]. Other studies revealed the relationship between the inflammatory process and reduced 25(OH)D [40–42]. Active forms of vitamin D have been shown to diminish the inflammatory process through the inhibition of interleukin-6 which is a key cytokine involved in joint destruction [43]. Finally, more profound and corrected analyses may detect true associations. There are published studies in which univariate analyses indicating significant correlation have been essentially altered by multivariate analyses [44]. Our study supports the view that 25(OH)D is independent of disease activity, despite a slight negative correlation between ESR and vitamin D levels. We believe that inflammatory markers alone may not be specific enough to assess disease activity categorically, as both clinical manifestation and the severity of JIA are determined by a multiplicity of factors. If so, it was difficult to find or confirm associations between biochemical markers of inflammation and 25(OH)D serum concentration in this study. Some reports showed an association between a clinical manifestation of RA or JIA and the overall disturbance of the metabolism [45]. Interestingly, in the present study, no differences were observed in vitamin D status between individuals with polyarthritis and systemic-onset JIA even after adjustment for age and sex as possible confounding effects.

In summary, our preliminary study showed that methotrexate may have a general negative influence on vitamin D status in children with JIA. This confers a possible risk of deteriorated bone density and impaired skeletal accrual during growth. There is always a need for supplementation of vitamin D and calcium to be considered in these patients accordingly to the general guidelines. The maintenance of the optimal vitamin D status can be useful in reducing pain symptoms and improving the quality of life in these children, as it was proven in patients with RA [46], considering particular pleiotropic effects of calcitriol, including anti-inflammatory, immunomodulatory, and cell-protecting features. Though there are still some controversial findings in the literature suggesting that vitamin D supply, although effective, does not sufficiently enhance bone health. For example, Hillman et al. proved that vitamin D_3 treatment with 2000 IU/day plus calcium increased 25(OH)D concentration and allowed maintaining the 1,25(OH)$_2$D level but did not improve BMD accretion [47].

We are aware that our study has some relevant limitations, and the cross-sectional design does not allow establishing firm conclusions concerning causal effects. Accessibility to relevant solid data was limited, specifically related to the duration of GCs, and MTX therapy was ineffective. The difficulty resulted from the study design, as the treatment courses were established individually for each patient and were subsequently adjusted depending on clinical manifestation and course of the disease. Because of essential technical issues, it was not possible to assess skeletal status by measuring bone mineral density.

5. Conclusions

The majority of children with juvenile idiopathic arthritis have significantly decreased 25-hydroxyvitamin D serum concentrations independent of clinical manifestations, disease activity, age, sex, or inflammatory markers. Iatrogenic factors play an important role in the development of vitamin D deficiency in JIA. According to our study, long-term methotrexate therapy appears to be the factor associated with reduced 25(OH)D levels. Although glucocorticoids used in JIA essentially affect calcium/phosphate metabolism, indicators of their influence on vitamin D status were not found. Our findings suggest the necessity of extensive vitamin D supplementation in children with JIA, particularly those treated with methotrexate. There is a need for further studies on the effects of methotrexate on vitamin D status in this population with special regard to the underlying molecular mechanism.

Author Contributions: Conceptualization, M.K.S., P.A., A.G. and J.K.; methodology, P.A., A.G. and M.K.S.; validation, J.K., P.A. and M.K.S.; formal analysis, J.K. and P.A.; investigation, A.K., A.G., J.K., M.K.S., J.M. and P.A.; resources, A.K., A.G., M.K.S., J.M. and P.A.; data curation, M.K.S. and A.G.; writing—original draft preparation, M.K.S., A.G., J.K. and P.A.; writing—review and editing, J.K., M.K.S., A.G. and J.K.; visualization, P.A., M.K.S. and A.G.; supervision, P.A. and J.K.; project administration, J.K.; funding acquisition, J.K. All authors have read and agreed to the published version of the manuscript.

Funding: The study was supported by the Medical University of Bialystok grants No: 123-26787L and 113-26839L.

Institutional Review Board Statement: The study was conducted according to the guidelines of the Declaration of Helsinki and approved by the Ethics Committee of the Medical University of Bialystok (protocol code R-I-002/373/2015 and date of approval: 29 October 2015).

Informed Consent Statement: Informed consent was obtained from all subjects involved in the study and/or their legal guardians.

Data Availability Statement: Data are available on request.

Conflicts of Interest: The authors declare no conflict of interest.

References

1. Prakken, B.; Albani, S.; Martini, A. Juvenile idiopathic arthritis. *Lancet* **2011**, *377*, 2138–2149. [CrossRef]
2. Weiss, J.E.; Ilowite, N.T. Juvenile idiopathic arthritis. *Pediatr. Clin. N. Am.* **2005**, *52*, 413–442. [CrossRef]
3. Soybiligic, A.; Tesher, M.; Wagner-Weiner, L.; Onel, K.B. A survey of steroid-related osteoporosis diagnosis, prevention and treatment practices of pediatric rheumatologists in North America. *Pediatr. Rheumatol. Online J.* **2014**, *12*, 24. [CrossRef]
4. Bardare, M.; Bianchi, M.L.; Furia, M.; Gandolini, G.G.; Cohen, E.; Montesano, A. Bone metabolism in juvenile chronic arthritis: The influence of steroids. *Clin. Exp. Rheumatol.* **1991**, *9* (Suppl. S6), 29–31.
5. Pludowski, P.; Holick, M.F.; Pilz, S.; Wagner, C.L.; Hollis, B.W.; Grant, W.B.; Shoenfeld, Y.; Lerchbaum, E.; Llewellyn, D.J.; Kienreich, K.; et al. Vitamin D effects on musculoskeletal health, immunity, autoimmunity, cardiovascular disease, cancer, fertility, pregnancy, dementia and mortality—A review of recent evidence. *Autoimmun. Rev.* **2013**, *12*, 976–989. [CrossRef]
6. Holick, M.F. Vitamin D deficiency. *N. Engl. J. Med.* **2007**, *357*, 266–281. [CrossRef]
7. DeLuca, H.F. Overview of general physiologic features and functions of vitamin D. *Am. J. Clin. Nutr.* **2004**, *80* (Suppl. S6), 1689–1696. [CrossRef]
8. Marini, F.; Falcini, F.; Stagi, S.; Fabbri, S.; Ciuffi, S.; Rigante, D.; Cerinic, M.M.; Brandi, M.L. Study of vitamin D status and vitamin D receptor polymorphisms in a cohort of Italian patients with juvenile idiopathic arthritis. *Sci. Rep.* **2020**, *10*, 17550. [CrossRef]
9. Holick, M.F.; Binkley, N.C.; Bischoff-Ferrari, H.A.; Gordon, C.M.; Hanley, D.A.; Heaney, R.P.; Murad, M.H.; Weaver, C.M.; Endocrine Society. Evaluation, treatment, and prevention of vitamin D deficiency: An Endocrine Society clinical practice guideline. *J. Clin. Endocrinol. Metab.* **2011**, *96*, 1911–1930. [CrossRef]

10. Pludowski, P.; Grant, W.B.; Bhattoa, H.P.; Bayer, M.; Povoroznyuk, V.; Rudenka, E.; Ramanau, H.; Varbiro, S.; Rudenka, A.; Karczmarewicz, E.; et al. Vitamin D status in Central Europe. *Int. J. Endocrinol.* **2014**, *2014*, 589587. [CrossRef]

11. Botros, R.M.; Sabry, I.M.; Abdelbaky, R.S.; Eid, Y.M.; Nasr, M.S.; Hendawy, L.M. Vitamin D deficiency among healthy Egyptian females. *Endocrinol. Nutr.* **2015**, *62*, 314–321. [CrossRef]

12. Hoge, A.; Donneau, A.F.; Streel, S.; Kolh, P.; Chapelle, J.P.; Albert, A.; Cavalier, E.; Guillaume, M. Vitamin D deficiency is common among adults in Wallonia (Belgium, 51°30′ North): Findings from the Nutrition, Environment and Cardio-Vascular Health study. *Nutr. Res.* **2015**, *35*, 716–725. [CrossRef]

13. Ning, Z.; Song, S.; Miao, L.; Zhang, P.; Wang, X.; Liu, J.; Hu, Y.; Xu, Y.; Zhao, T.; Liang, Y.; et al. High prevalence of vitamin D deficiency in urban health checkup population. *Clin. Nutr.* **2015**, *35*, 859–863. [CrossRef]

14. Płudowski, P.; Karczmarewicz, E.; Bayer, M.; Carter, G.; Chlebna-Sokół, D.; Czech-Kowalska, J.; Dębski, R.; Decsi, T.; Dobrzańska, A.; Franek, E.; et al. Practical guidelines for the supplementation of vitamin D and the treatment of deficits in Central Europe-recommended vitamin D intakes in the general population and groups at risk of vitamin D deficiency. *Endokrynol. Pol.* **2013**, *64*, 319–327. [CrossRef]

15. Lin, Z.; Li, W. The roles of vitamin D and its analogs in inflammatory diseases. *Curr. Top. Med. Chem.* **2016**, *16*, 1242–1261. [CrossRef]

16. Pelajo, C.F.; Lopez-Benitez, J.M.; Miller, L.C. Vitamin D and autoimmune rheumatologic disorders. *Autoimmun. Rev.* **2009**, *9*, 507–510. [CrossRef]

17. Pelajo, C.F.; Lopez-Benitez, J.M.; Miller, L.C. 25-hydroxyvitamin D levels and vitamin D deficiency in children with rheumatologic disorders and controls. *J. Rheumatol.* **2011**, *38*, 2000–2004. [CrossRef]

18. Zou, J.; Thornton, C.; Chambers, E.S.; Rosser, E.C.; Ciurtin, C. Exploring the Evidence for an Immunomodulatory Role of Vitamin D in Juvenile and Adult Rheumatic Disease. *Front. Immunol.* **2021**, *18*, 616483. [CrossRef]

19. Vojinovic, J.; Cimaz, R. Vitamin D update for the pediatric rheumatologists. *Pediatr. Rheumatol. Online J.* **2015**, *13*, 18. [CrossRef]

20. Petty, R.E.; Southwood, T.R.; Manners, P.; Baum, J.; Glass, D.N.; Goldenberg, J.; He, X.; Maldonado-Cocco, J.; Orozco-Alcala, J.; Prieur, A.M.; et al. International League of Associations for Rheumatology classification of juvenile idiopathic arthritis: Second revision, Edmonton, 2001. *J. Rheumatol.* **2004**, *31*, 390–392.

21. Consolaro, A.; Bracciolini, G.; Ruperto, N.; Pistorio, A.; Magni-Manzoni, S.; Malattia, C.; Pederzoli, S.; Davi, S.; Martini, A.; Ravelli, A. Remission, minimal disease activity, and acceptable symptom state in juvenile idiopathic arthritis: Defining criteria based on the juvenile arthritis disease activity score. *Arthritis Rheum* **2012**, *64*, 2366–2374. [CrossRef] [PubMed]

22. Physical status: The use and interpretation of anthropometry. Report of a WHO Expert Committee World Health Organ. *Tech. Rep. Ser.* **1995**, *854*, 1–452.

23. Bathon, J.M.; Cohen, S.B. The 2008 American College of Rheumatology recommendations for the use of nonbiologic and biologic disease-modifying antirheumatic drugs in rheumatoid arthritis: Where the rubber meets the road. *Arthritis Rheum* **2008**, *59*, 757–759. [CrossRef] [PubMed]

24. Ringold, S.; Weiss, P.F.; Beukelman, T.; DeWitt, E.M.; Ilowite, N.T.; Kimura, Y.; Laxer, R.M.; Lovell, D.J.; Nigrovic, P.A.; Robinson, A.B.; et al. 2013 update of the 2011 American College of Rheumatology recommendations for the treatment of juvenile idiopathic arthritis: Recommendations for the medical therapy of children with systemic juvenile idiopathic arthritis and tuberculosis screening among children receiving biologic medications. *Arthritis Rheum* **2013**, *65*, 2499–2512. [CrossRef]

25. Kanagawa, H.; Masuyama, R.; Morita, M.; Sato, Y.; Niki, Y.; Kobayashi, T.; Katsuyama, E.; Fujie, A.; Hao, W.; Tando, T.; et al. Methotrexate inhibits osteoclastogenesis by decreasing RANKL-induced calcium influx into osteoclasts progenitors. *J. Bone Miner. Metab.* **2016**, *34*, 526–531. [CrossRef]

26. Uehara, R.; Suzuki, Y.; Ichikawa, Y. Methotrexate (MTX) inhibits osteoblastic differentiation in vitro: Possible mechanism of MTX osteopathy. *J. Rheumatol.* **2001**, *28*, 251–256.

27. Minaur, N.J.; Kounali, D.; Vedi, S.; Compston, J.E.; Beresford, J.N.; Bhalla, A.K. Methotrexate in the treatment of rheumatoid arthritis. II. In vivo effects on bone mineral density. *Rheumatology* **2002**, *41*, 741–749. [CrossRef]

28. De Sousa Studart, S.A.; Leite, A.C.; Marinho, A.L.; Pinto, A.C.; Rabelo Júnior, C.N.; de Melo Nunes, R.; Rocha, H.A.; Rocha, F.A. Vitamin D levels in juvenile idiopathic arthritis from an equatorial region. *Rheumatol. Int.* **2015**, *35*, 1717–1723. [CrossRef]

29. Skversky, A.L.; Kumar, J.; Abramowitz, M.K.; Kaskel, F.J.; Melamed, M.L. Association of glucocorticoid use and low 25-hydroxyvitamin D levels: Results from the National Health and Nutrition Examination Survey (NHANES): 2001–2006. *J. Clin. Endocrinol. Metab.* **2011**, *96*, 3838–3845. [CrossRef]

30. Stagi, S.; Masi, L.; Capannini, S.; Cimaz, R.; Tonini, G.; Matucci-Cerinic, M.; de Martino, M.; Falcini, F. Cross-sectional and longitudinal evaluation of bone mass in children and young adults with juvenile idiopathic arthritis: The role of bone mass determinants in a large cohort of patients. *J. Rheumatol.* **2010**, *37*, 1935–1943. [CrossRef]

31. Finch, S.L.; Rosenberg, A.M.; Vatanparast, H. Vitamin D and juvenile idiopathic arthritis. *Pediatr. Rheumatol. Online J.* **2018**, *16*, 34. [CrossRef] [PubMed]

32. Merlino, L.A.; Curtis, J.; Mikuls, T.R.; Cerhan, J.R.; Criswell, L.A.; Saag, K.G. Iowa Women's Health Study. Vitamin D intake is inversely associated with rheumatoid arthritis: Results from the Iowa Women's Health Study. *Arthritis. Rheum.* **2004**, *50*, 72–77. [CrossRef] [PubMed]

33. Lo Gullo, A.; Mandraffino, G.; Bagnato, G.; Aragona, C.O.; Imbalzano, E.; D'Ascola, A.; Rotondo, F.; Cinquegrani, A.; Mormina, E.; Saitta, C.; et al. Vitamin D Status in Rheumatoid Arthritis: Inflammation, Arterial Stiffness and Circulating Progenitor Cell Number. *PLoS ONE* **2010**, *10*, e0134602. [CrossRef] [PubMed]

34. Sharma, R.; Saigal, R.; Goyal, L.; Mital, P.; Yadav, R.N.; Meena, P.D.; Agrawal, A. Estimation of vitamin D levels in rheumatoid arthritis patients and its correlation with the disease activity. *J. Assoc. Physicians India* **2014**, *62*, 678–681.

35. Baker, J.F.; Baker, D.G.; Toedter, G.; Shults, J.; Von Feldt, J.M.; Leonard, M.B. Associations between vitamin D, disease activity, and clinical response to therapy in rheumatoid arthritis. *Clin. Exp. Rheumatol.* **2012**, *30*, 658–664.

36. Lee, Y.H.; Bae, S.C. Vitamin D level in rheumatoid arthritis and its correlation with the disease activity: A meta-analysis. *Clin. Exp. Rheumatol.* **2016**, *34*, 827–833.

37. Pelajo, C.F.; Lopez- Benitez, J.M.; Kent, D.M.; Price, L.L.; Miller, L.C.; Dawson-Hughes, B. 25-hydroxyvitamin D levels and juvenile idiopathic arthritis: Is there an association with disease activity? *Rheumatol. Int.* **2012**, *32*, 3923–3929. [CrossRef]

38. Munekata, R.V.; Terreri, M.T.; Peracchi, O.A.; Len, C.; Lazaretti-Castro, M.; Sarni, R.O.; Hilário, M.O. Serum 25-hydroxyvitamin D and biochemical markers of bone metabolism in patients with juvenile idiopathic arthritis. *Braz. J. Med. Biol. Res.* **2013**, *46*, 98–102. [CrossRef]

39. Dağdeviren-Çakır, A.; Arvas, A.; Barut, K.; Gür, E.; Kasapçopur, Ö. Serum vitamin D levels during activation and remission periods of patients with juvenile idiopathic arthritis and familial Mediterranean fever. *Turk. J. Pediatr.* **2016**, *58*, 125–131. [CrossRef]

40. Stagi, S.; Bertini, F.; Cavalli, L.; Matuci-Cerinic, M.; Brandi, M.L.; Falcini, F. Determinants of vitamin D levels in children, adolescents, and young adults with juvenile idiopathic arthritis. *J. Rheumatol.* **2014**, *41*, 1884–1892. [CrossRef]

41. Nandi, M.; Mullick, M.A.S.; Nandy, A.; Samanta, M.; Sarkar, S.; Sabui, T.K. Evaluation of vitamin D profile in juvenile idiopathic arthritis. *Mod. Rheumatol.* **2021**. *Epub ahead of print*. [CrossRef] [PubMed]

42. Sengler, C.; Zink, J.; Klotsche, J.; Niewerth, M.; Liedmann, I.; Horneff, G.; Kessel, C.; Ganser, G.; Thon, A.; Haas, J.P.; et al. Vitamin D deficiency is associated with higher disease activity and the risk for uveitis in juvenile idiopathic arthritis-data from a German inception cohort. *Arthritis Res. Ther.* **2018**, *13*, 276. [CrossRef] [PubMed]

43. Cutolo, M.; Otsa, K.; Uprus, M.; Paolino, S.; Seriolo, B. Vitamin D in rheumatoid arthritis. *Autoimmun. Rev.* **2007**, *7*, 59–64. [CrossRef] [PubMed]

44. Bouaddi, I.; Rostom, S.; El Badri, D.; Hassani, A.; Chkirate, B.; Abouqal, R.; Amine, B.; Hajjaj-Hassouni, N. Vitamin D concentrations and disease activity in Moroccan children with juvenile idiopathic arthritis. *BMC Musculoskelet. Disord.* **2014**, *15*, 115. [CrossRef]

45. Bianchi, M.L.; Bardare, M.; Caraceni, M.P.; Cohen, E.; Falvella, S.; Borzani, M.; DeGaspari, M.G. Bone metabolism in juvenile rheumatoid arthritis. *Bone Miner.* **1990**, *9*, 153–162. [CrossRef]

46. Charoenngam, N. Vitamin D and Rheumatic Diseases: A Review of Clinical Evidence. *Int. J. Mol. Sci.* **2021**, *1*, 10659. [CrossRef]

47. Hillman, L.S.; Cassidy, J.T.; Chanetsa, F.; Hewett, J.E.; Higgins, B.J.; Robertson, J.D. Percent true calcium absorption, mineral metabolism, and bone mass in children with arthritis: Effect of supplementation with vitamin D3 and calcium. *Arthritis Rheum* **2008**, *58*, 3255–3263. [CrossRef]

Article

Serum 25-hydroxyvitamin D Concentration Significantly Decreases in Patients with COVID-19 Pneumonia during the First 48 Hours after Hospital Admission

Juraj Smaha [1,*], Martin Kužma [1], Peter Jackuliak [1], Samuel Nachtmann [1], Filip Max [2], Elena Tibenská [1,3], Neil Binkley [4] and Juraj Payer [1]

[1] 5th Department of Internal Medicine, Comenius University Faculty of Medicine, University Hospital, Ruzinovska 6, 826 06 Bratislava, Slovakia; martin.kuzma@fmed.uniba.sk (M.K.); peter.jackuliak@fmed.uniba.sk (P.J.); samuel.nachtmann@gmail.com (S.N.); elena.tibenska@medirex.sk (E.T.); payer@ru.unb.sk (J.P.)

[2] Department of Pharmacology and Toxicology, Comenius University Faculty of Pharmacy, Odbojarov 10, 832 32 Bratislava, Slovakia; max1@uniba.sk

[3] Medirex, a.s., Galvaniho 17/C, 820 16 Bratislava, Slovakia

[4] Department of Medicine, Geriatrics Faculty, Medical Sciences Center, University of Wisconsin, 1300 University Ave, Madison, WI 53706-1510, USA; nbinkley@wisc.edu

* Correspondence: smaha1@uniba.sk

Citation: Smaha, J.; Kužma, M.; Jackuliak, P.; Nachtmann, S.; Max, F.; Tibenská, E.; Binkley, N.; Payer, J. Serum 25-hydroxyvitamin D Concentration Significantly Decreases in Patients with COVID-19 Pneumonia during the First 48 Hours after Hospital Admission. *Nutrients* **2022**, *14*, 2362. https://doi.org/10.3390/nu14122362

Academic Editor: Carsten Carlberg

Received: 18 May 2022
Accepted: 5 June 2022
Published: 7 June 2022

Publisher's Note: MDPI stays neutral with regard to jurisdictional claims in published maps and institutional affiliations.

Abstract: It is unclear how ongoing inflammation in Coronavirus Disease 2019 (COVID-19) affects 25-hydroxyvitamin D (25[OH]D) concentration. The objective of our study was to examine serum 25(OH)D levels during COVID-19 pneumonia. Patients were admitted between 1 November and 31 December 2021. Blood samples were taken on admission (day 0) and every 24 h for the subsequent four days (day 1–4). On admission, 59% of patients were 25(OH)D sufficient (>30 ng/mL), and 41% had 25(OH)D inadequacy (<30 ng/mL). A significant fall in mean 25(OH)D concentration from admission to day 2 (first 48 h) was observed (30.7 ng/mL vs. 26.4 ng/mL; $p < 0.0001$). No subsequent significant change in 25(OH)D concentration was observed between day 2 and 3 (26.4 ng/mL vs. 25.9 ng/mL; $p = 0.230$) and day 3 and day 4 (25.8 ng/mL vs. 25.9 ng/mL; $p = 0.703$). The absolute 25(OH)D change between hospital admission and day 4 was 16% (4.8 ng/mL; $p < 0.0001$). On day 4, the number of patients with 25(OH)D inadequacy increased by 18% ($p = 0.018$). Therefore, serum 25(OH)D concentration after hospital admission in acutely ill COVID-19 patients should be interpreted with caution. Whether low 25(OH)D in COVID-19 reflects tissue level vitamin D deficiency or represents only a laboratory phenomenon remains to be elucidated in further prospective trials of vitamin D supplementation.

Keywords: vitamin D; 25-hydroxyvitamin D; COVID-19; inflammation; pneumonia; SARS-CoV-2

1. Introduction

The role of vitamin D metabolites as a potentially modifiable risk factor in Coronavirus Disease 19 (COVID-19) infection was suggested early in the pandemic [1]. Since then, many studies relating vitamin D status with COVID-19 severity and outcome have been published. Regarding COVID-19, low 25-hydroxyvitamin D (25[OH]D) levels have been associated with a higher risk of testing positive [2], the severity of infection [3], higher need for invasive ventilation [4], and higher mortality [5]. Our previous study found that serum 25(OH)D concentration measured at admission in patients with severe COVID-19 is an independent risk factor for mortality. Moreover, patients with severe vitamin D deficiency, (i.e., 25[OH]D < 12 ng/mL) had a higher viral load, higher Charlson comorbidity index, and an 11% increase in mortality rate than patients with a serum 25(OH)D concentration above 12 ng/mL [5].

COVID-19 infection causes a systemic hyperimmune response triggered by the hypoxic environment due to respiratory failure and the ensuing cytokine storm [6]. During this acute phase, metabolic changes in macronutrients, as well as micronutrients, could be expected; vitamin D metabolites are no exception to this [7]. Serum 25(OH)D concentration reflects vitamin D body stores, and, under normal circumstances, it is the best indicator of vitamin D status [8]. However, the role of 25(OH)D concentration as a biomarker for acute inflammatory illness is controversial. The literature suggests that significant variations in 25(OH)D levels may occur within hours in acutely ill patients. Many pathophysiological mechanisms have been proposed, e.g., a direct effect of inflammation, hemodilution, decreased synthesis of binding proteins, or renal wasting of 25(OH)D [9]. As such, a single-point assessment of 25(OH)D may be inaccurate in estimating vitamin D status during acute phase response [7]. Consistent with this, in one study, the prevalence of vitamin D inadequacy increased from 38% before total hip arthroplasty even to 68% the day after surgery [10].

To the best of our knowledge, there is no study prospectively evaluating serum 25(OH)D levels during acute COVID-19 infection. The objective of our study was to examine whether serum 25(OH)D levels change during the evolution of acute COVID-19 pneumonia and to explore the possibility of determining an optimal time window for 25(OH)D concentration assessment during acute illness.

2. Materials and Methods

This study was undertaken as a prospective cohort study. Patients with acute COVID-19 pneumonia hospitalized in the internal medicine department between 1 November 2021 and 31 December 2021 were recruited. Exclusion criteria were as follows:

1. Patients with no need for supplemental oxygen;
2. Patients not meeting the criteria for severe disease;
3. COVID-19 pneumonia was not the primary diagnosis upon admission;
4. Patients with another acute infection, (e.g., urinary tract infection) during the monitoring period.

Demographic characteristics, comorbidities, hematological and biochemical laboratory results on admission, information regarding the intensity of care during hospitalization, and pharmacological treatment before (including vitamin D supplementation) and during hospitalization were collected from electronic medical records and discharge summaries by two physicians using a standardized approach. All patients received pharmacological and supportive measures according to interim COVID-19 guidance of treatment approved by the University Hospital Bratislava. These guidelines were based on current Centers for Disease Control and Prevention (USA) recommendations for treatment. (https://www.cdc.gov/coronavirus/2019-ncov/hcp/clinical-guidance-management-patients.html, accessed on 1 October 2021). Patients were not supplemented with vitamin D preparations during the monitoring period. Severe infection was defined as clinical signs of pneumonia and one of the following: respiratory rate >30 breaths/minute; severe respiratory distress; or oxygen saturation <90% on room air.

A first venous blood sample was taken on admission (day 0) and then every 24 h for the subsequent four days: the second sample at the 24th hour (day 1), the third sample at the 48th hour (day 2), the fourth sample at the 72nd hour (day 3), and the fifth sample at the 96th hour (day 4), respectively.

Serum 25(OH)D concentrations (in ng/mL) were obtained using an automated electrochemiluminescence system (Eclesys Vitamin D Total II, 2019, Roche Diagnostics GmBH, Mannheim, Germany). The detection limit of serum 25(OH)D was 3 ng/mL. The complete blood count testing was performed on an automated analyzer (Mindray BC-6800 Plus Auto Hematology Analyzer). CRP was analyzed by immunoturbidimetric quantitative assay (Roche, CRP4 Cobas, module c 501). Other tests such as liver tests, kidney functions, and serum minerals were performed on each patient and assessed with commercial standardized tests. The presence of the SARS-COV-2 virus was assessed by a real-time reverse transcriptase-polymerase chain reaction (RT-PCR) test on nasopharyngeal swab. Blood oxygen saturation was assessed using arterial blood sampling ± 2 h from venous blood sampling.

Statistical analysis was performed using the statistics software Analyse-it (Leeds, UK) v 5.40.2 or R (v 3.6.0). Continuous data were expressed as the mean $\pm$ standard error of the mean (SEM) if normally distributed or as median and interquartile range if not normally distributed. The Shapiro–Francia test was used to test the normality of the distributions of studied parameters. Changes between time points were assessed with Student's paired *t*-test and Wilcoxon test in normally and not normally distributed variables, respectively. Relationships between investigated parameters were calculated using the Pearson correlation coefficient or binominal logistic regression. *p*-values less than 0.05 were considered to be statistically significant.

The study was conducted in accordance with the Declaration of Helsinki and approved by the Institutional Ethics Committee of University Hospital Bratislava (ethical approval code: EK/011/2021) on 1 January 2021.

3. Results

The baseline characteristics of the patients are shown in Table 1. Twenty-two patients were included (six females, and sixteen males; the median age of 60.6 years). Twelve patients (55%) have a history of arterial hypertension, seven patients (32%) have a history of diabetes mellitus, two patients (9%) have a history of coronary artery disease, and five patients (23%) have a history of chronic kidney disease. The mean value of body mass index (BMI) was 29.74 kg/m^2; ten patients (45%) were overweight (BMI > 25 kg/m^2) and eight patients were obese (BMI > 30 kg/m^2). All patients required supplemental oxygen, of whom twelve patients (55%) needed high-flow oxygen via nasal cannula. Two patients required invasive mechanical ventilation. Four patients eventually died after completing the protocol.

Table 1. Baseline characteristics of participants.

Variable	n = 22	Females (n = 6)	Males (n = 16)
Females/males (n, %)	6 (27%)/16 (73%)	N/A	N/A
Survivors/non-survivors (n, %)	18 (82%)/4 (18%)	5 (83%)/1 (17%)	13 (81%)/3 (19%)
Age (years) (mean $\pm$ SD)	60.5 $\pm$ 14.4	66.8 $\pm$ 12	58.3 $\pm$ 15
Body mass index (kg/m^2), median (IQR)	29.74 (7.3)	26.5 (8.8)	28.9 (7.34)
Arterial hypertension (n, %)	12 (55%)	3 (50%)	9 (56%)
Diabetes mellitus (n, %)	7 (32%)	2 (33%)	5 (31%)
Coronary artery disease (n, %)	2 (9%)	1 (17%)	1 (6%)
Chronic kidney disease (n, %)	5 (23%)	2 (33%)	3 (19%)
Number of symptomatic days before hospitalization, mean $\pm$ SD	7.45 $\pm$ 3.1	7.6 $\pm$ 3.4	7 $\pm$ 2.6
Days of dyspnea before hospitalization, median (IQR)	2.45 (1.1)	2 (1.0)	2 (3.2)
Need for supplemental oxygen (n, %)	22 (100%)	6 (100%)	16 (100%)
Need for high flow oxygen (n, %)	12 (55%)	3 (33%)	9 (56%)
Number of days of high flow oxygen, n (IQR)	7 (6)	4 (6)	2.5 (6.1)
Invasive mechanical ventilation (n, %)	2 (9%)	0 (0%)	2 (13%)
Vitamin D supplementation before hospitalization (n, %)	10 (45%)	2 (33%)	8 (50%)

The patients were recruited between 1 November and 31 December. Symptoms of COVID-19 (fever, chills, cough, nausea, malaise, myalgias, cefalea) were present 7.45 days before hospitalization (mean value). Dyspnea had been present for more than 48 h in 36% and less than 48 h in 64% of the patients. The baseline 25(OH)D concentration was 30.71 ng/mL; 59% of patients were 25(OH)D sufficient (>30 ng/mL), 18% were 25(OH)D insufficient (30–20 ng/mL) and 5% were 25(OH)D deficient (<20 ng/mL). A total of 45%

of patients used vitamin D supplements before hospitalization. There was no association between 25(OH)D status upon admission and mortality, the need for high flow nasal oxygen, or duration of symptoms. The differences in baseline 25(OH)D concentration between selected subgroups of patients upon admission are displayed in Table 2.

Table 2. The difference in 25-hydroxyvitamin D (25[OH]D) concentration between selected subgroups of patients upon admission.

Variable	25(OH)D (ng/mL) upon Admission	*p*-Value
Sex		0.43
Males	32.3 ± 16.2	
Females	26.3 ± 12.9	
Body mass index		0.47
>30 kg/m^2	27.5 ± 16.2	
<30 kg/m^2	32.5 ± 15.1	
End of hospitalization		0.69
Death	33.6 ± 13.2	
Discharge	30.1 ± 16.1	
Arterial hypertension		0.57
Yes	28.9 ± 10.4	
No	32.7 ± 20.2	
Diabetes mellitus		0.73
Yes	29.03 ± 12.5	
No	31.5 ± 16.9	
Coronary artery disease		0.61
Yes	25.2 ± 5.4	
No	31.2 ± 16.1	
Chronic kidney disease		0.18
Yes	22.5 ± 10.2	
No	33.1 ± 16.1	
Chronic pulmonary disease		0.26
Yes	18.9 ± 19.4	
No	31.8 ± 15.01	
Need for high flow oxygen		0.36
Yes	33.5 ± 16.6	
No	27.4 ± 13.8	
Vitamin D supplementation before hospitalization		0.0005
Yes	41.85 ± 11.45	
No	21.4 ± 11.7	

A significant fall in mean 25(OH)D concentration from admission to day 2 (first 48 h) was observed (30.7 ng/mL vs. 26.4 ng/mL; $p < 0.0001$). No subsequent significant fall in 25(OH)D concentration was observed between day 2 and 3 (26.4 ng/mL vs. 25.9 ng/mL; $p = 0.2300$) and day 3 and day 4 (25.8 ng/mL vs. 25.9 ng/mL; $p = 0.7026$). The evolution of serum 25(OH)D levels change is displayed in Figure 1.

Upon admission, serum 25(OH)D levels were higher in males compared to females, although this difference was not statistically significant ($p = 0.43$). A decline in serum 25(OH)D levels was observed in males, as well as females. In males, the decline in 25(OH)D concentration in the first 48 h was more significant compared to females ($p < 0.005$ vs. $p < 0.05$, respectively). The decline between days 2 and 3 was observed in males, but no further decline was seen in females. The 25(OH)D kinetics in males and females is displayed in Figure 2, respectively.

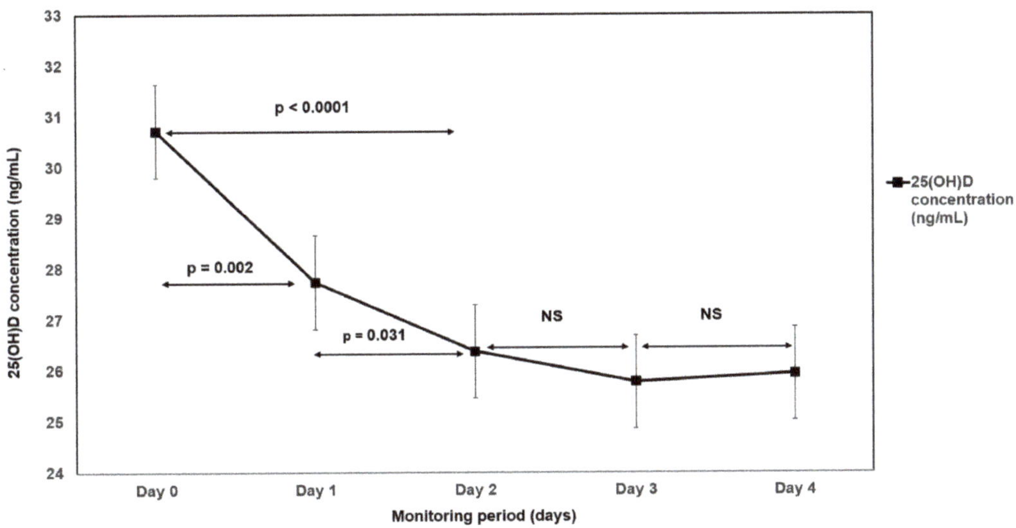

Figure 1. The evolution of serum 25-hydroxyvitamin D (25[OH]D) levels change during the study period.

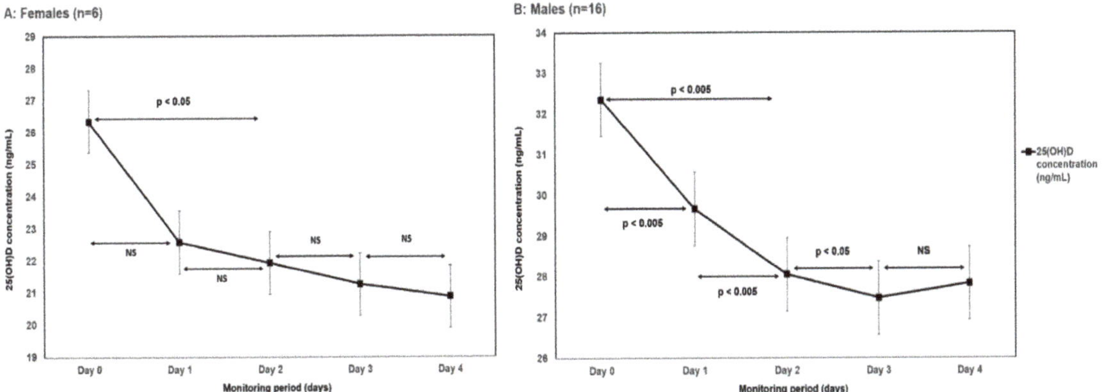

Figure 2. Serum 25-hydroxyvitamin D (25[OH]D) levels kinetics during the study period in females (**A**) and males (**B**), respectively.

In the majority of patients (n = 17), serum 25(OH)D levels decreased during the monitoring period, see Figure 3.

The absolute 25(OH)D change between hospital admission and day 4 was 4.8 ng/mL ($p < 0.0001$) and was not associated with mortality ($p = 0.211$), the need for high flow oxygen ($p = 0.647$), or duration of symptoms ($p = 0.14$). On day 4, the number of patients with 25(OH)D inadequacy (<30 ng/mL) increased by 18% ($p = 0.018$). The kinetics of the 25(OH)D concentration during hospitalization was compared between survivors and non-survivors (Figure 4), between patients treated with high flow oxygen and conventional oxygen (Figure 5), and between shorter and longer duration of symptoms than the median (Figure 6). In survivors, the decline in 25(OH)D levels during the first 48 h was highly statistically significant ($p < 0.0001$) in comparison to non-survivors, where no significant change in 25(OH)D concentration was observed ($p = 0.2$). Irrespective of the mode of oxygen therapy, a statistically significant decline in 25(OH)D levels during the first 48 h of hospitalization was observed (high-flow oxygen: $p = 0.018$; conventional oxygen: $p = 0.04$). Patients with a shorter duration of symptoms before hospitalization have a slightly higher decline in 25(OH)D concentration during the first 48 h compared to patients with a longer duration of symptoms (<7 days of symptoms: $p = 0.001$ vs. >7 days of symptoms: $p = 0.01$).

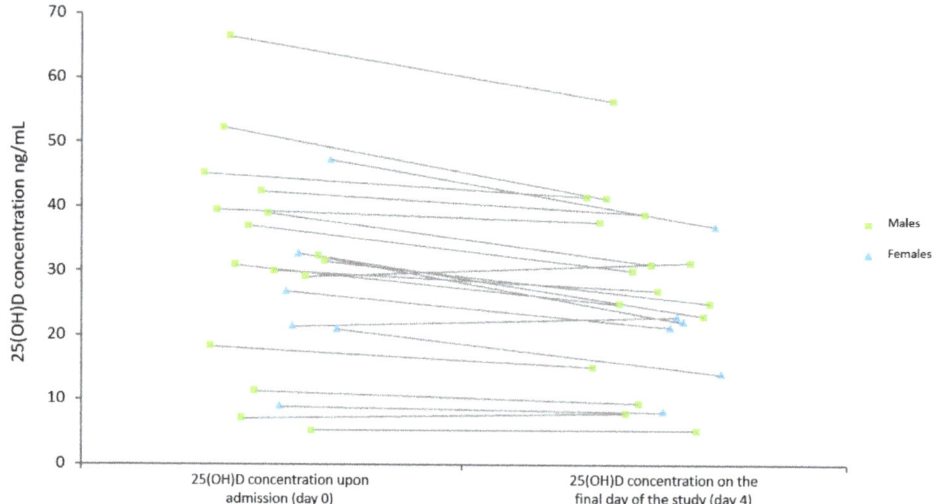

Figure 3. The comparison of 25-hydroxyvitamin D (25[OH]D) status in all patients between day of admission and day 4 of the study.

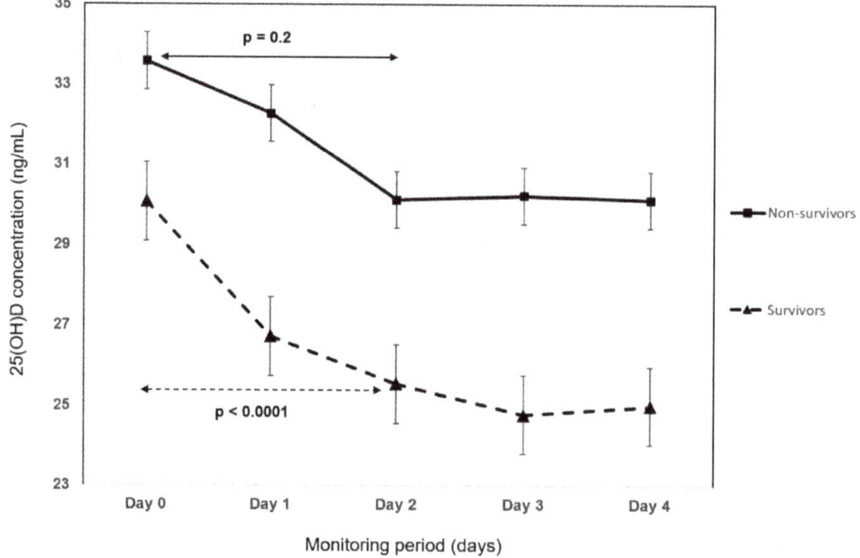

Figure 4. The kinetics of 25-hydroxyvitamin D (25[OH]D) concentration during hospitalization between survivors and non-survivors. *p*-values of the changes during the first 48 h are displayed.

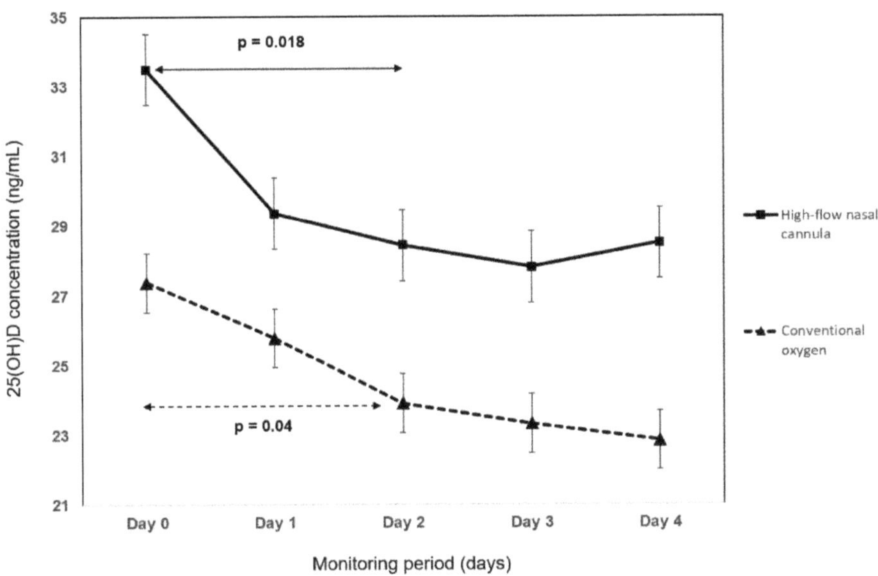

Figure 5. The kinetics of 25-hydroxyvitamin D (25[OH]D) concentration between patients treated with high-flow oxygen and conventional oxygen. *p*-values of the changes during the first 48 h are displayed.

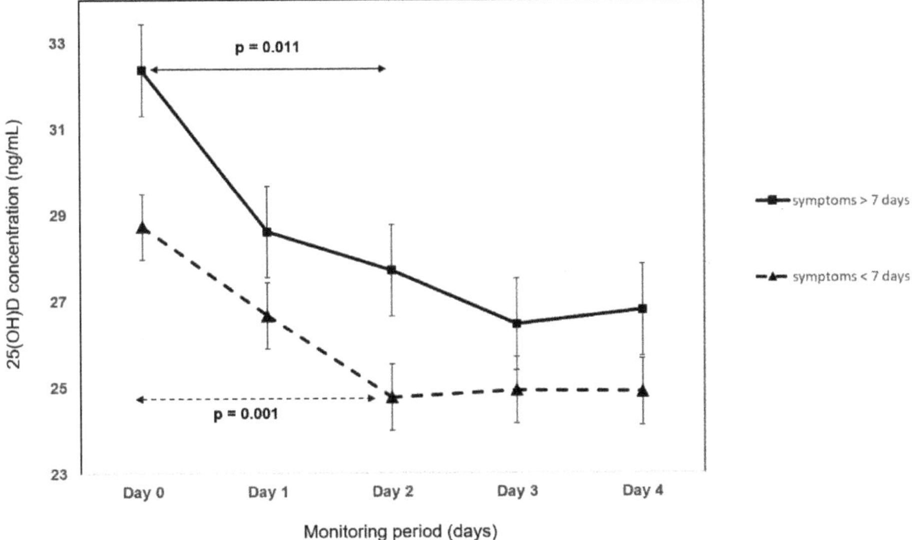

Figure 6. The kinetics of 25-hydroxyvitamin D (25[OH]D) concentration during hospitalization between shorter and longer duration of symptoms than median (median = 7 days).

A significant fall in creatinine, albumin, hemoglobin, and hematocrit was observed (day 1 vs. day 5, all $p < 0.05$). Both albumin and hemoglobin significantly decreased during the first 48 h after the hospital admission ($p < 0.05$ and $p < 0.005$, respectively). For albumin and hemoglobin concentration kinetics during the study period, see Supplementary Figure S1.

Regarding the markers of inflammation, C-reactive protein (CRP) and interleukin-6 (IL-6) significantly decreased during the monitoring period (day 0 vs. day 4, both $p < 0.0005$). We observed a highly significant rapid fall in IL-6 concentration in the first 24 h, followed by an increase in concentration and a slight subsequent decrease. For CRP and IL-6 concentration kinetics during the study period, see Supplementary Figure S2. There was a significant increase in neutrophils between admission and day 1 and a subsequent decrease in the number of neutrophils from day 1 to day 3 (all $p < 0.0001$). Lymphocytes did not change significantly during the first 48 h; after the first 48 h, a statistically significant increase in lymphocyte concentration was observed (day 0 vs. day 4; $p < 0.05$). Monocytes increased significantly during the monitoring period ($p < 0.005$). The kinetics of the concentration of neutrophils, lymphocytes, and monocytes are displayed in Supplementary Figure S3. The changes in concentration of all parameters during the monitoring period are displayed in Table 3. For all laboratory parameters of each patient during the monitoring period see Supplementary Tables.

Table 3. The changes in concentration of all parameters during the monitoring period compared to the baseline value on day 0 (admission).

Variable (Mean)	Day 0	Day 1	Day 2	Day 3	Day 4
25(OH)D (ng/mL) ± SD	30.7 ± 15.4	27.7 ± 13.5 **	26.4 ± 12.7 **	25.8 ± 12.5 **	25.9 ± 13.05 **
Neutrophils (10 × 9/L) (IQR)	6.0 (6.7)	8.2 (8.1) **	9.4 (6.3) *	8.7 (5.2)	7.9 (4.6)
Lymphocytes (10 × 9/L)	0.90 ± 0.42	0.96 ± 0.4	0.99 ± 0.53	1.19 ± 0.62 *	1.29 ± 0.68 *
Monocytes (10 × 9/L) (IQR)	0.32 (0.23)	0.54 (0.3) **	0.5 (0.3) **	0.58 (0.27) **	0.53 (0.27) **
C-reactive protein (mg/L) (IQR)	162.8 (102.3)	94.5 (100.5) **	66.2 (72.2) **	51.2 (53.3) **	28.9 (43.1) **
Interleukin-6 (ng/L) (IQR)	69.8 (148)	30.3 (43.4) **	35.2 (69.7) *	31.6 (61.7)*	11.1 (44.8) **
Procalcitonin (ng/mL) (IQR)	0.28 (0.8)	0.19 (0.34) *	0.13 (0.23)	0.12 (0.26)*	0.095 (0.21) **
Lactate dehydrogenase (IU) ± SD	8.85 ± 4.4	8.83 ± 4.5	8.28 ± 4.1	8.28 ± 4.15	7.45 ± 3.82 *
Neutrophil to lymphocyte ratio (IQR)	7(10.2)	10.5(8.1) *	10(11)	7.5(6.5)	6.5 (10)
Calcium (mmol/L) (IQR)	2.12 (0.26)	2.16 (0.18)	2.02 (0.19)	2.17 (0.15)	2.16 (0.17)
Ionized calcium (mmol/L) ± SD	1.137 ± 0.1	1.16 ± 0.07 *	1.17 ± 0.1 *	1.18 ± 0.08 *	1.18 ± 0.1 *
Blood urea nitrogen (mmol/L) (IQR)	6.6 (6.5)	7.5 (6.8)	8.5 (8.2)	8.2 (3.8)	7.6 (3.1)
Creatinine (umol/L) (IQR)	94.3 (43.3)	73.1 (31.1)	81.2 (19.6)	78.2 (11.6)	79.3 (15.5) *
Albumin (g/L) ± SD	34.7 ± 4.6	32.9 ± 3.8 *	32.1 ± 3.8 *	31.9 ± 3.9 *	31.8 ± 4.1 *
Hemoglobin (g/L) ± SD	138 ± 15	136 ± 15	132 ± 16.8 **	133 ± 15.9 *	133 ± 17 *
Hematocrit ± SD	0.42 ± 0.05	0.41 ± 0.05	0.403 ± 0.05 **	0.409 ± 0.05 *	0.409 ± 0.05 *
Alanine aminotransferase (ukat/L)	1.43 ± 2.03	1.5 ± 1.7	1.65 ± 2.04	1.91 ± 1.83	2.1 ± 2.17
Alkaline phosphatase (ukat/L) (IQR)	1.5 (1.38)	1.44 (1.02)	1.69 (0.92)	1.71 (0.96)	1.61 (0.91)

** $p < 0.005$. * $p < 0.05$.

4. Discussion

In this study, we found a significant reduction in circulating 25(OH)D concentration of ~16% within two days of hospitalization in patients with severe COVID-19 pneumonia. As a result, the proportion of patients with vitamin D inadequacy increased from 41% to 59%. In this pilot study, the change in 25(OH)D was unrelated to changes in CRP and IL-6. Serum 25(OH)D levels tend to be significantly higher in males than in females across all BMI groups [11]. In our cohort of patients, we also found a difference in 25(OH)D concentration between females and males upon admission. Females had lower 25(OH)D levels, although this difference was not statistically significant. A significant decline in 25(OH)D concentration during the first 48 h after the hospital admission was observed in both sexes. In males, this decline was more significant and observed even after the first 48 h.

The existing literature associates inflammation of various causes with low 25(OH)D. Most studies to date considered healthcare-associated intervention, i.e., elective surgery, as an inflammatory stimulus. Except for one study, which retrospectively assessed 25(OH)D concentration during malarial infection [12], there is no study investigating 25(OH)D changes in the context of acute inflammatory reaction caused by an infectious disease.

Serum 25(OH)D concentration decreases hours (6–48) after the surgical procedure, and the maximum change in 25(OH)D concentration could be as high as 40% compared to the baseline status [13,14]. CRP was most commonly used as an inflammatory marker. Due to the correlation of increased CRP concentration with reduced 25(OH)D levels in these studies, many have suggested that serum 25(OH)D is simply a negative acute phase reactant [14,15]. However, the observed changes could result from surgery and anesthetic management rather than inflammation itself [10]. Nevertheless, a decrease in 25(OH)D levels, accompanied by an increase in CRP, was also observed during the first days after acute pancreatitis [16] and after the intravenous infusion of bisphosphonates [17].

In contrast, neither change in 25(OH)D serum levels nor significant correlation between markers of inflammation were observed in patients after acute myocardial infarction [18] and severe malarial infection [12], respectively. However, in both studies, serum 25(OH)D levels were measured several days after symptom onset, which might blunt the ability to see a decline.

Interestingly, even though our patients were symptomatic for 7.45 days and experienced dyspnea for 2.45 days before hospitalization, we still observed a significant decline in 25(OH)D. The CRP was at its peak at admission, and its concentration decreased during the study period, presumably because of treatment with immunomodulatory drugs. Notably, severe illness in people with COVID-19 typically occurs approximately 8–12 days after symptoms onset [19]. The most common symptom is dyspnea, accompanied by hypoxemia. Importantly, in the severe phase of COVID-19, the pulmonary lesions generally peaked 6–11 days after the symptom onset [20]. It is thus possible that excessive local pulmonary inflammation was close to its peak at the time of hospital admission, and functional vitamin D deficiency was observed.

In this regard, it is plausible that vitamin D deficiency would be related to tissue requirement. In such a case, the circulating 25(OH)D pool represents a substrate reservoir for conversion to active metabolites (1,25[OH]2D) at the pulmonary tissue level during times of acute stress [21]. Several previous studies have suggested that the vitamin D metabolic pathways may influence the development of acute respiratory distress syndrome (ARDS) by various mechanisms [22,23]. Interestingly, Abrishami et al. have observed that higher levels of 25(OH)D were associated with significantly less total lung involvement on chest computed tomography in hospitalized COVID-19 patients [24].

Factors other than inflammation were also identified as a potential cause of the 25(OH)D decline during acute illness. For example, intravenous fluid administration has been associated with reduced 25(OH)D levels. Krishnan et al. showed that acute fluid loading rather than inflammation might have a more profound effect on 25(OH)D concentration during the early phase of acute illness [25]. However, Reid et al. observed that after the administration of 3 liters of intravenous fluid over 24 h after knee arthroplasty, the concentration of 25(OH)D dropped by 40% and was accompanied only by a 15% reduction in albumin concentration and 10% in vitamin D binding protein concentration (VDBP) [14]. As such, acute fluid loading does not seem to play a significant role in our cohort of patients with considerably less intravenous fluid given during the first 48 h after hospital admission.

Hypoalbuminemia is a common condition in patients with serious illnesses [26]. Albumin binds approximately 15% of serum 25(OH)D; thus, hypoalbuminemia of critically ill patients could be responsible for approximately 15% of the variation 25(OH)D [27].

Binding proteins are also essential for 25(OH)D reabsorption at the renal tubules; therefore, loss of these proteins may lead to subsequent renal wasting of 25(OH)D during acute illness [7]. Indeed, the decrease in serum VDBP during systemic inflammatory response is significantly associated with increased urinary loss of VDBP [13].

It has been suggested that glucocorticoid administration may decrease 25(OH)D levels [28]. All of our patients received six milligrams of intravenous dexamethasone daily during the study duration. Dexamethasone increases renal expression of vitamin D-24-hydroxylase, which degrades vitamin D metabolites such as 25(OH)D [29]. After 24 h of therapy with dexamethasone, a significant abundance of 24-hydroxylase was observed [30]. Additionally, glucocorticoids enhance direct 24-hydroxylase transcription via cooperation between glucocorticoid receptor C/EBPbeta and vitamin D receptor (VDR) [31]. Thus, the possibility of lowering 25(OH)D levels because of the infusion of glucocorticoids cannot be ruled out in patients with COVID-19 and hypoxemia. Consistent with this, an observed increase in the number of neutrophils could also be connected with systemic glucocorticoid administration. The white blood count rises after glucocorticoid administration mainly because of neutrophilic migration from the endothelial lining of the blood vessels, which manifests within 5–24 h following administration and can persist during therapy [32]. However, specifically in critically ill COVID-19 patients, treatment with dexamethasone was actually associated with lower neutrophil proportions in bronchoalveolar lavage compared to patients without dexamethasone, thus highlighting the critical role of neutrophils in the pathophysiology of ARDS in COVID-19 [33].

Our study has several limitations. Above all, we evaluated a small group of patients. Our primary aim was to observe changes in serum 25(OH)D levels; thus, this present study was underpowered to see whether an association between mortality and magnitude of change in 25(OH)D serum levels exists. Secondly, we did not have a control group of patients. Additionally, treatment with glucocorticoids is currently a therapeutic option supported by the highest level of evidence; therefore, we also did not have a cohort of patients without i.v. glucocorticoid treatment. Thus, we cannot rule out the possible effect of hospitalization or i.v. glucocorticoid administration on the evaluated laboratory parameters. Thirdly, we did not know the serum 25(OH)D concentration of patients prior to hospitalization. We also did not evaluate VDBP; thus, we cannot exclude some variations of 25(OH)D connected with a possible decrease in VDBP. Our study also has several strengths. We have prospectively evaluated a homogeneous, clearly defined group of patients with acute severe COVID-19 pneumonia. The 25(OH)D concentration was evaluated with the same method and technique in all patients at precisely defined periods. To the best of our knowledge, this is the first study in which a change of 25(OH)D concentration was evaluated for a longer period of time in hospitalized patients with COVID-19. Additionally, this is also the first study with a prospective design regarding changes of 25(OH)D in the context of an acute infectious disease.

In conclusion, this study showed that serum 25(OH)D concentration decreases significantly during the first 48 h after hospital admission in acutely ill COVID-19 patients. The number of patients with 25(OH)D inadequacy increased by 18% during the monitoring period. After the first 48 h, no significant change in 25(OH)D concentration was observed, which could have practical implications for vitamin D status assessment during acute COVID-19. Whether low 25(OH)D in COVID-19 reflects functional vitamin D deficiency and has a causal link to worse prognosis in COVID-19 or represents only a laboratory phenomenon remains to be elucidated in further prospective randomized trials of vitamin D supplementation.

Supplementary Materials: The following supporting information can be downloaded at: https://www.mdpi.com/article/10.3390/nu14122362/s1, Figure S1: The change of mean values of albumin and hemoglobin during the monitoring period. *p*-values of the change during the first 48 h are displayed. Figure S2: The changes in mean values of C-reactive protein and Interleukin-6 during the monitoring period. *p*- values of the change during the first 48 h are displayed. Figure S3: The changes in mean values of the neutrophils, lymphocytes, and monocytes during the monitoring period. *p*-values of the change during the first 48 h are displayed. Table S1: The mean values of 25-hydroxyvitamin D (25[OH]D) of each patient during the monitoring period (day 0–4). Table S2: The mean values of C-reactive protein, interleukin-6, procalcitonin, and lactate dehydrogenase of each patient during the monitoring period (day 0–4). Table S3: The mean values of neutrophils, lymphocytes, monocytes, and neutrophil to lymphocyte ratio of each patient during the monitoring period (day 0–4). Table S4: The mean values of albumin, calcium, ionized calcium, hemoglobin, and hematocrit of each patient during the monitoring period (day 0–4). Table S5: The mean values of selected kidney and liver markers of each patient during the monitoring period (day 0–4).

Author Contributions: Conceptualization, J.S. and J.P.; methodology, J.S.; software, F.M.; validation, E.T., P.J. and J.P.; formal analysis, M.K.; investigation, J.S. and S.N.; resources, E.T.; data curation, J.P.; writing—original draft preparation, J.S.; writing—review and editing, J.P. and N.B.; visualization, J.S.; supervision, P.J.; project administration, J.P. and S.N. All authors have read and agreed to the published version of the manuscript.

Funding: This research received no external funding.

Institutional Review Board Statement: The study was conducted in accordance with the Declaration of Helsinki, and approved by the Institutional Ethics Committee of University Hospital Bratislava (ethical approval code: EK/011/2021) on 1 January 2021.

Informed Consent Statement: Informed consent was obtained from all subjects involved in the study.

Data Availability Statement: The data presented in this study are available on request from the corresponding author.

Acknowledgments: We would like to thank all our patients, who agreed to participate in the study.

Conflicts of Interest: The authors declare no conflict of interest.

References

1. Grant, W.B.; Lahore, H.; McDonnell, S.L.; Baggerly, C.A.; French, C.B.; Aliano, J.L.; Bhattoa, H.P. Evidence that Vitamin D Supplementation Could Reduce Risk of Influenza and COVID-19 Infections and Deaths. *Nutrients* **2020**, *12*, 988. [CrossRef]
2. D'Avolio, A.; Avataneo, V.; Manca, A.; Cusato, J.; De Nicolò, A.; Lucchini, R.; Keller, F.; Cantù, M. 25-Hydroxyvitamin D Concentrations Are Lower in Patients with Positive PCR for SARS-CoV-2. *Nutrients* **2020**, *12*, 1359. [CrossRef] [PubMed]
3. Campi, I.; Gennari, L.; Merlotti, D.; Mingiano, C.; Frosali, A.; Giovanelli, L.; Torlasco, C.; Pengo, M.F.; Heilbron, F.; Soranna, D.; et al. Vitamin D and COVID-19 severity and related mortality: A prospective study in Italy. *BMC Infect. Dis.* **2021**, *21*, 1–13. [CrossRef]
4. Angelidi, A.M.; Belanger, M.J.; Lorinsky, M.K.; Karamanis, D.; Chamorro-Pareja, N.; Ognibene, J.; Palaiodimos, L.; Mantzoros, C.S. Vitamin D Status Is Associated With In-Hospital Mortality and Mechanical Ventilation: A Cohort of COVID-19 Hospitalized Patients. *Mayo Clin. Proc.* **2021**, *96*, 875–886. [CrossRef] [PubMed]
5. Smaha, J.; Kužma, M.; Brázdilová, K.; Nachtmann, S.; Jankovský, M.; Pastírová, K.; Gažová, A.; Jackuliak, P.; Killinger, Z.; Kyselovič, J.; et al. Patients with COVID-19 pneumonia with 25(OH)D levels lower than 12 ng/ml are at increased risk of death. *Int. J. Infect. Dis.* **2022**, *116*, 313–318. [CrossRef] [PubMed]
6. Batabyal, R.; Freishtat, N.; Hill, E.; Rehman, M.; Freishtat, R.; Koutroulis, I. Metabolic dysfunction and immunometabolism in COVID-19 pathophysiology and therapeutics. *Int. J. Obes.* **2021**, *45*, 1163–1169. [CrossRef]
7. Quraishi, S.A.; Camargo, C.A., Jr. Vitamin D in acute stress and critical illness. *Curr. Opin. Clin. Nutr. Metab. Care* **2012**, *15*, 625–634. [CrossRef]
8. Pludowski, P.; Takacs, I.; Boyanov, M.; Belaya, Z.; Diaconu, C.C.; Mokhort, T.; Zherdova, N.; Rasa, I.; Payer, J.; Pilz, S. Clinical Practice in the Prevention, Diagnosis and Treatment of Vitamin D Deficiency: A Central and Eastern European Expert Consensus Statement. *Nutrients* **2022**, *14*, 1483. [CrossRef]
9. Silva, M.C.; Furlanetto, T.W. Does serum 25-hydroxyvitamin D decrease during acute-phase response? A systematic review. *Nutr. Res.* **2015**, *35*, 91–96. [CrossRef] [PubMed]
10. Binkley, N.; Coursin, D.; Krueger, D.; Iglar, P.; Heiner, J.; Illgen, R.; Squire, M.; Lappe, J.; Watson, P.; Hogan, K. Surgery alters parameters of vitamin D status and other laboratory results. *Osteoporos. Int.* **2016**, *28*, 1013–1020. [CrossRef]
11. Muscogiuri, G.; Barrea, L.; Di Somma, C.; Laudisio, D.; Salzano, C.; Pugliese, G.; de Alteriis, G.; Colao, A.; Savastano, S. Sex Differences of Vitamin D Status across BMI Classes: An Observational Prospective Cohort Study. *Nutrients* **2019**, *11*, 3034. [CrossRef] [PubMed]
12. Newens, K.; Filteau, S.; Tomkins, A. Plasma 25-hydroxyvitamin D does not vary over the course of a malarial infection. *Trans. R. Soc. Trop. Med. Hyg.* **2006**, *100*, 41–44. [CrossRef]
13. Louw, J.A.; Werbeck, A.; Louw, M.E.J.; Kotze, T.J.V.W.; Cooper, R.; Labadarios, D. Blood vitamin concentrations during the acute-phase response. *Crit. Care Med.* **1992**, *20*, 934–941. [CrossRef] [PubMed]

14. Reid, D.; Toole, B.J.; Knox, S.; Talwar, D.; Harten, J.; O'Reilly, D.S.J.; Blackwell, S.; Kinsella, J.; McMillan, D.C.; Wallace, A.M.; et al. The relation between acute changes in the systemic inflammatory response and plasma 25-hydroxyvitamin D concentrations after elective knee arthroplasty. *Am. J. Clin. Nutr.* **2011**, *93*, 1006–1011. [CrossRef]
15. Waldron, J.L.; Ashby, H.L.; Cornes, M.P.; Bechervaise, J.; Razavi, C.; Thomas, O.L.; Chugh, S.; Deshpande, S.; Ford, C.; Gama, R. Vitamin D: A negative acute phase reactant. *J. Clin. Pathol.* **2013**, *66*, 620–622. [CrossRef] [PubMed]
16. Bang, U.C.; Novovic, S.; Andersen, A.M.; Fenger, M.; Hansen, M.B.; Jensen, J.-E.B. Variations in Serum 25-Hydroxyvitamin D during Acute Pancreatitis: An Exploratory Longitudinal Study. *Endocr. Res.* **2011**, *36*, 135–141. [CrossRef]
17. Bertoldo, F.; Pancheri, S.; Zenari, S.; Boldini, S.; Giovanazzi, B.; Zanatta, M.; Valenti, M.T.; Carbonare, L.D.; Cascio, V.L. Serum 25-hydroxyvitamin D levels modulate the acute-phase response associated with the first nitrogen-containing bisphosphonate infusion. *J. Bone Miner. Res.* **2010**, *25*, 447–454. [CrossRef] [PubMed]
18. Barth, J.H.; Field, H.P.; Mather, A.N.; Plein, S. Serum 25 hydroxy-vitamin D does not exhibit an acute phase reaction after acute myocardial infarction. *Ann. Clin. Biochem. Int. J. Lab. Med.* **2012**, *49*, 399–401. [CrossRef] [PubMed]
19. Li, X.; Ma, X. Acute respiratory failure in COVID-19: Is it "typical" ARDS? *Crit. Care* **2020**, *24*, 198. [CrossRef]
20. Wang, Y.; Dong, C.; Hu, Y.; Li, C.; Ren, Q.; Zhang, X.; Shi, H.; Zhou, M. Temporal Changes of CT Findings in 90 Patients with COVID-19 Pneumonia: A Longitudinal Study. *Radiology* **2020**, *296*, E55–E64. [CrossRef] [PubMed]
21. Lee, P. Vitamin D metabolism and deficiency in critical illness. *Best Pr. Res. Clin. Endocrinol. Metab.* **2011**, *25*, 769–781. [CrossRef]
22. Zheng, S.; Yang, J.; Hu, X.; Li, M.; Wang, Q.; Dancer, R.C.; Parekh, D.; Gao-Smith, F.; Thickett, D.R.; Jin, S. Vitamin D attenuates lung injury via stimulating epithelial repair, reducing epithelial cell apoptosis and inhibits TGF-β induced epithelial to mesenchymal transition. *Biochem. Pharmacol.* **2020**, *177*, 113955. [CrossRef]
23. Kloc, M.; Ghobrial, R.M.; Lipińska-Opałka, A.; Wawrzyniak, A.; Zdanowski, R.; Kalicki, B.; Kubiak, J.Z. Effects of vitamin D on macrophages and myeloid-derived suppressor cells (MDSCs) hyperinflammatory response in the lungs of COVID-19 patients. *Cell. Immunol.* **2020**, *360*, 104259. [CrossRef] [PubMed]
24. Abrishami, A.; Dalili, N.; Torbati, P.M.; Asgari, R.; Arab-Ahmadi, M.; Behnam, B.; Sanei-Taheri, M. Possible association of vitamin D status with lung involvement and outcome in patients with COVID-19: A retrospective study. *Eur. J. Nutr.* **2020**, *60*, 2249–2257. [CrossRef] [PubMed]
25. Krishnan, A.; Ochola, J.; Mundy, J.; Jones, M.; Kruger, P.; Duncan, E.; Venkatesh, B. Acute fluid shifts influence the assessment of serum vitamin D status in critically ill patients. *Crit. Care* **2010**, *14*, R216. [CrossRef]
26. Erstad, B.L. Albumin disposition in critically Ill patients. *J. Clin. Pharm. Ther.* **2018**, *43*, 746–751. [CrossRef] [PubMed]
27. Bikle, D.D.; Schwartz, J. Vitamin D Binding Protein, Total and Free Vitamin D Levels in Different Physiological and Pathophysiological Conditions. *Front. Endocrinol.* **2019**, *10*, 317. [CrossRef] [PubMed]
28. Skversky, A.L.; Kumar, J.; Abramowitz, M.K.; Kaskel, F.J.; Melamed, M.L. Association of Glucocorticoid Use and Low 25-Hydroxyvitamin D Levels: Results from the National Health and Nutrition Examination Survey (NHANES): 2001–2006. *J. Clin. Endocrinol. Metab.* **2011**, *96*, 3838–3845. [CrossRef]
29. Akeno, N.; Matsunuma, A.; Maeda, T.; Kawane, T.; Horiuchi, N. Regulation of vitamin D-1alpha-hydroxylase and -24-hydroxylase expression by dexamethasone in mouse kidney. *J. Endocrinol.* **2000**, *164*, 339–348. [CrossRef] [PubMed]
30. Kurahashi, I.; Matsunuma, A.; Kawane, T.; Abe, M.; Horiuchi, N. Dexamethasone Enhances Vitamin D-24-Hydroxylase Expression in Osteoblastic (UMR-106) and Renal (LLC-PK1) Cells Treated with 1α,25-Dihydroxyvitamin D3. *Endocrine* **2002**, *17*, 109–118. [CrossRef]
31. Dhawan, P.; Christakos, S. Novel regulation of 25-hydroxyvitamin D3 24-hydroxylase (24(OH)ase) transcription by glucocorticoids: Cooperative effects of the glucocorticoid receptor, C/EBPβ, and the Vitamin D receptor in 24(OH)ase transcription. *J. Cell. Biochem.* **2010**, *110*, 1314–1323. [CrossRef] [PubMed]
32. Ronchetti, S.; Ricci, E.; Migliorati, G.; Gentili, M.; Riccardi, C. How Glucocorticoids Affect the Neutrophil Life. *Int. J. Mol. Sci.* **2018**, *19*, 4090. [CrossRef] [PubMed]
33. Voicu, S.; Malissin, I.; Pepin-Lehaleur, A.; Sutterlin, L.; Naim, G.; M'Rad, A.; Guerin, E.; Ekherian, J.; Deye, N.; Adle-Biassette, H.; et al. Cytological patterns of bronchoalveolar lavage fluid in mechanically ventilated COVID-19 patients on extracorporeal membrane oxygenation. *Clin. Respir. J.* **2022**, *16*, 329–334. [CrossRef] [PubMed]

Review

Hypercalcemia in Pregnancy Due to CYP24A1 Mutations: Case Report and Review of the Literature

Stefan Pilz [1,*], Verena Theiler-Schwetz [1], Pawel Pludowski [2], Sieglinde Zelzer [3], Andreas Meinitzer [3], Spyridon N. Karras [4], Waldemar Misiorowski [5], Armin Zittermann [6], Winfried März [3,7,8] and Christian Trummer [1]

[1] Department of Internal Medicine, Division of Endocrinology and Diabetology, Medical University of Graz, Auenbruggerplatz 15, 8036 Graz, Austria; verena.schwetz@medunigraz.at (V.T.-S.); christian.trummer@medunigraz.at (C.T.)

[2] Department of Biochemistry, Radioimmunology and Experimental Medicine, The Children's Memorial Health Institute, 04-730 Warsaw, Poland; p.pludowski@ipczd.pl

[3] Clinical Institute of Medical and Chemical Laboratory Diagnostics, Medical University of Graz, 8036 Graz, Austria; sieglinde.zelzer@medunigraz.at (S.Z.); andreas.meinitzer@medunigraz.at (A.M.); winfried.maerz@synlab.de (W.M.)

[4] National Scholarship Foundation, 55535 Thessaloniki, Greece; karraspiros@yahoo.gr

[5] Department of Endocrinology, Centre of Postgraduate Medical Education, Bielanski Hospital, 01-809 Warsaw, Poland; wmisiorowski@cmkp.edu.pl

[6] Clinic for Thoracic and Cardiovascular Surgery, Herz- und Diabeteszentrum Nordrhein-Westfalen (NRW), Ruhr University Bochum, 32545 Bad Oeynhausen, Germany; azittermann@hdz-nrw.de

[7] SYNLAB Academy, Synlab Holding Deutschland GmbH, 68159 Mannheim, Germany

[8] V th Department of Medicine (Nephrology, Hypertensiology, Rheumatology, Endocrinology, Diabetology, Lipidology), Medical Faculty Mannheim, University of Heidelberg, 68167 Mannheim, Germany

[*] Correspondence: stefan.pilz@medunigraz.at; Tel.: +43-316-3858-1143

Citation: Pilz, S.; Theiler-Schwetz, V.; Pludowski, P.; Zelzer, S.; Meinitzer, A.; Karras, S.N.; Misiorowski, W.; Zittermann, A.; März, W.; Trummer, C. Hypercalcemia in Pregnancy Due to CYP24A1 Mutations: Case Report and Review of the Literature. *Nutrients* **2022**, *14*, 2518. https://doi.org/10.3390/nu14122518

Academic Editor: Suzuki Atsushi

Received: 8 May 2022
Accepted: 14 June 2022
Published: 17 June 2022

Publisher's Note: MDPI stays neutral with regard to jurisdictional claims in published maps and institutional affiliations.

Abstract: Pathogenic mutations of CYP24A1 lead to an impaired catabolism of vitamin D metabolites and should be considered in the differential diagnosis of hypercalcemia with low parathyroid hormone concentrations. Diagnosis is based on a reduced 24,25-dihydroxyvitamin D to 25-hydroxyvitamin D ratio and confirmed by genetic analyses. Pregnancy is associated with an upregulation of the active vitamin D hormone calcitriol and may thus particularly trigger hypercalcemia in affected patients. We present a case report and a narrative review of pregnant women with CYP24A1 mutations (13 women with 29 pregnancies) outlining the laboratory and clinical characteristics during pregnancy and postpartum and the applied treatment approaches. In general, pregnancy triggered hypercalcemia in the affected women and obstetric complications were frequently reported. Conclusions on drugs to treat hypercalcemia during pregnancy are extremely limited and do not show clear evidence of efficacy. Strictly avoiding vitamin D supplementation seems to be effective in preventing or reducing the degree of hypercalcemia. Our case of a 24-year-old woman who presented with hypercalcemia in the 24th gestational week delivered a healthy baby and hypercalcemia resolved while breastfeeding. Pathogenic mutations of CYP24A1 mutations are rare but should be considered in the context of vitamin D supplementation during pregnancy.

Keywords: vitamin D; CYP24A1; pregnancy; hypercalcemia; fertility; lactation; intoxication; supplementation; idiopathic infantile hypercalcemia

1. Introduction

Accurate diagnosis and treatment of hypercalcemia during pregnancy is crucial because it is associated with an increased risk of adverse health outcomes for the mother and the fetus [1–3]. Hypercalcemia during pregnancy is relatively rare and its further differential diagnosis is dependent on the prevailing parathyroid hormone (PTH) level. Primary hyperparathyroidism (PHPT) that is characterized by elevated or inappropriately high parathyroid hormone (PTH) concentrations seems to be the main cause of hypercalcemia during pregnancy [1–3]. Pregnant women with hypercalcemia and reduced PTH concentrations may suffer from a variety of diseases including, e.g., disorders of vitamin D metabolism, vitamin D intoxication per se, malignancy, granulomatous diseases (sarcoidosis and tuberculosis), pseudohyperparathyroidism (due to elevated PTH related peptide (PTHrP) levels), milk-alkali syndrome (due to excess intake of calcium and antacid drugs), etc. posing a diagnostic and therapeutic challenge [2–6].

In 2011, Schlingmann et al. used a candidate gene approach to evaluate the cause of idiopathic infantile hypercalcemia, a condition that had a particular high incidence in Great Britain in the 1950s during a time of high vitamin D supplementation and food fortification [7]. They identified loss-of-function mutations in cytochrome-P450 family 24 subfamily A member 1 (CYP24A1) with an indication for autosomal recessive inheritance in affected children who were particularly prone to develop hypercalcemia after high dose vitamin D supplementation [7]. This CYP24A1 gene encodes the mitochondrial inner membrane P450 24-hydroxylase enzyme responsible for the catabolism of vitamin D metabolites [7–9]. It catalyzes the conversion of 25-hydroxyvitamin D (25(OH)D), the main circulating vitamin D metabolite that is used for the classification of vitamin D status, and of 1α,25-dihydroxyvitamin D ($1,25(OH)_2D$), also called calcitriol or the active vitamin D hormone, to inactive (or less active) metabolites [8,9]. In detail, 25(OH)D is converted to 24,25-dihydroxyvitamin D ($24,25(OH)_2D$) and $1,25(OH)_2D$ is converted to 1,24,25-trihydroxyvitamin D as an initial step of their catabolism [8,9].

Most patients with idiopathic infantile hypercalcemia probably remain unrecognized as there are only a few hundred cases described in the literature (i.e., 221 cases in a recent systematic review) with a Polish study suggesting a population disease frequency of 1 in 32,465 births [4,10]. Of note, the term idiopathic infantile hypercalcemia is meanwhile considered a misnomer for patients with pathogenic mutations of CYP24A1 as these patients can develop hypercalcemia across their whole life-span [4]. Due to the increasing use of vitamin D supplements and food fortification, knowledge on pathogenic CYP24A1 mutations in terms of their clinical relevance are of public health interest. Importantly, pregnancy may be a trigger factor for the development of hypercalcemia in affected patients with pathogenic CYP24A1 mutations, but data on this topic are extremely rare and there exists a knowledge gap regarding guidance for the management of affected patients in this setting [4].

In this work, we present the case of a pregnant woman with hypercalcemia due to pathogenic mutations in CYP24A1. In addition, we perform a literature review to summarize the current knowledge on this issue by outlining the laboratory and clinical characteristics during pregnancy and postpartum and the applied treatment approaches. Based on the totality of available evidence we aim to provide some guidance for the management of affected women during pregnancy, postpartum and lactation.

2. Case Report

In summer 2021, a 24-year-old pregnant woman at 24 weeks of gestation was referred to our outpatient clinic of the Department of Endocrinology & Diabetology of the Medical University of Graz, Austria, for further diagnostics and treatment of hypercalcemia. A total serum calcium of 3.5 mmol/L had been incidentally discovered at a routine laboratory measurement about 1 week before the initial visit in our outpatient clinic. At that time, she had complained of polyuria, increased thirst and fatigue, and hydronephrosis grade 2 to 3 on the left side had been diagnosed at an external hospital, followed by an insertion of a double J catheter. Kidney stones or nephrolithiasis had, however, not been detected at any time in her life. She was already advised to stop the intake of a multimicronutrient supplement containing 800 international units (IU) of vitamin D3 (20 µg) that she had taken daily until then.

At our outpatient clinic, she reported no symptoms and no major previous diseases, except, an abortion in the 6th week of gestation two years before. She took a daily iron and magnesium supplement, and levothyroxine (LT4) 75 μg daily had been prescribed about 1 week before due to hypothyroxinemia with subsequently normal thyroid function tests throughout pregnancy with this treatment. Her laboratory report yielded hypercalcemia with reduced PTH concentrations, and we recommended the avoidance of vitamin D supplements and sun exposure, low calcium intake and high oral fluid intake, and arranged another appointment for a more extensive laboratory work-up (see Table 1 for selected laboratory results at our department with laboratory methods as described in previous publications [11–14]).

Table 1. Selected characteristics of the case patient with pathogenic CYP24A1 mutations at the endocrine outpatient clinic.

Parameter (Unit)	Reference Range	1st Visit	2nd Visit	3rd Visit	4th Visit	5th Visit
Gestational week		24	27	31	33	Two months after giving birth
Albumin adjusted serum calcium (mmol/L)	2.20 to 2.65	3.08	3.00	3.07	2.95	2.35
Ionized serum calcium (mmol/L)	1.15 to 1.35	1.57	1.51	1.57	1.50	1.29
Total serum calcium (mmol/L)	2.20 to 2.65	2.97	2.93	2.99	2.84	2.58
Serum phosphate (mmol/L)	0.84 to 1.45	0.74	0.72	0.89	0.95	1.19
Serum magnesium (mmol/L)	0.70 to 1.10	0.55	0.55	0.55	0.66	0.74
Serum creatinine (mg/dL)	up to 1.00	0.78	0.82	0.88	0.75	1.02
eGFR (CKD-EPI) (ml/min/1.73 m^2)	90 to 120	106	100	92	111	76
Spot urine calcium/creatinine ratio (mmol/mmol)	up to 0.60	1.02	0.82	0.25	0.54	0.29
Parathyroid hormone (pg/mL)	15.0 to 65.0	8.0	7.1	7.6		8.4
25-hydroxyvitamin D (nmol/L) *	75 to 150	87	75	75		45
1.25-dihydroxyvitamin D (pmol/L) *	52 to 267	279	325	295		73
Bone-specific alkaline phosphatase (μg/L)	4.7 to 27.0	6.9				19.0
Osteocalcin (ng/mL)	1.0 to 35.0	19.3	24.3	29.6		42.6
Procollagen type 1 N-terminal propetide (ng/mL)	15 to 49	50.0	81.3	87.9		94.4
C-terminal telopeptide of type 1 collagen (ng/mL)	0.03 to 0.37	0.29	0.35	0.67		0.55
Fibroblast-growth-factor-23 (pg/mL)	14.0 to 48.0		176.1	156.6		56.0
Parathyroid hormone-related peptide (pmol/L)	0.0 to 1.3	1.2	1.2	0.5		
25-hydroxyvitamin D$_3$ (nmol/L) **	NA			90.2		71.7
25-hydroxyvitamin D$_2$ (nmol/L) **	NA			1.8		2.7
25-hydroxyvitamin D$_2$ + D$_3$ (nmol/L) **	75 to 150			92.0		74.4
24,25-dihydroxyvitamin D$_3$ (nmol/L) **	NA			0.66		0.13
24,25-hydroxyvitamin D$_3$ to 25-hydoxyvitamin D$_3$ ratio (%) **	>3			0.73		0.18

eGFR (CKD-EPI: estimated glomerular filtration rate (Chronic Kidney Disease Epidemiology Collaboration); * measured by immunoassays; ** measured by a liquid chromatography tandem mass spectrometry (LC-MS/MS) method.

Even after the follow-up appointment, we could not identify a causal disease for her PTH-independent hypercalcemia. We did not have an indication for tuberculosis (normal gamma-interferon test), for sarcoidosis (normal soluble interleukin-2 receptor and angiotensin-converting-enzyme (ACE)), for pseudohyperparathyroidism or malignancy (normal PTHrP), for familial hypocalciuric hypercalcemia (FHH) (hypercalciuria and low PTH) and no clear clinical and laboratory indication for hereditary hypophosphatemic rickets with hypercalciuria (HHRH) (tubular resorption of phosphate: 80% to 90%), or other diseases (a magnetic resonance imaging of the chest and abdomen was recommended by us but not performed). We advised to start with a potassium supplement (i.e., Reducto-spezial® 3 times 1 tablet daily) due to its potential to reduce intestinal calcium absorption, increased her magnesium supplement dose and made another (3rd) appointment to test for vitamin D metabolites in order to evaluate for the presence of 24-hydroxylase deficiency caused by pathogenic mutations of CYP24A1. As she had a reduced 24,25-dihydroxyvitamin D to 25(OH)D ratio, we made a final (4th) appointment at our outpatient clinic for genetic analyses and confirmed that she was compound heterozygous for two previously described pathogenic variants of CYP24A1, i.e., c.443T > C (p.Leu148Pro) and c.1226T > C (p.Leu409Ser). At 40 weeks of gestation, labor was induced in an external hospital due to suspected fetal macrosomia and the patient delivered a healthy daughter (4160 g and 54 cm) with a normal total serum calcium of 2.33 mmol/L and 2.25 mmol/L on day 2 and 4 of her life, respectively, and an uncomplicated normal development thereafter. Maternal total serum calcium concentrations were 2.67, 2.60, and 2.77 mmol/L on the day of giving birth, one and three days thereafter, respectively. Breastfeeding was started and continued by the mother until the last appointment at our department, more than two months after giving birth. At that appointment, she presented symptom free with normal serum calcium concentrations. The decision to not refrain from breastfeeding as most previous cases was partially based on the clinical experience of one of our co-authors from Poland (W.M.) that women with pathogenic CYP24A1 mutations showed a gradual decline in serum calcium concentrations while breastfeeding (unpublished observation). Usual vitamin D supplementation with 400 IU per day was recommended for the newborn daughter and she has been developing well without any complications. Follow-up laboratory tests in a few months were recommended for the mother and the newborn to check for serum calcium and creatinine levels.

The patient gave written informed consent for this publication and was included in the Graz Endocrinology Registry Study that was approved by the ethics committee at the Medical University of Graz, Austria.

3. Literature Review

A literature review in PubMed using the search terms "CYP24A1" and "pregnancy" was performed on 13 April 2022 to identify articles presenting data on pregnant women with pathogenic CYP24A1 mutations causing hypercalcemia. Out of 97 articles, we identified 9 eligible manuscripts by this search strategy and by screening their reference lists, we retrieved two additional publications so that 11 articles were included in our work [15–25]. Overall, we retrieved 13 women with pathogenic CYP24A1 mutations who had reports on 29 pregnancies. Selected characteristics of the pregnancies of the affected women are presented in Table 2. Obstetric complications were frequently reported but the vast majority of women delivered a healthy baby with usually either no or just transient complications, mainly due to disturbed calcium metabolism. We could not identify a clear pattern regarding changes of the magnitude of hypercalcemia as a function of pregnancy time. Due to the low number of cases and pregnancies we refrained from performing any statistical analyses to estimate relative risks for obstetric complications in women with pathogenic CYP24A1 mutations as opposed to the general pregnancy population.

Regarding treatment of hypercalcemia during pregnancy and postpartum there are no randomized controlled trials (RCTs) available so that any treatment approach was only based on a risk benefit estimation and mainly on expert opinions. The general treatment recommendation for most patients was, of course, to strictly avoid vitamin D supplementation and also to minimize sunlight (ultraviolet-B) induced vitamin D synthesis in the skin. Avoidance of vitamin D supplements during pregnancy in affected women seems to have a great effect on serum calcium. In this context, one woman used vitamin D supplements during her first pregnancy and had serum calcium concentrations up to 3.3 mmol/L, but remained normocalcemic throughout her second pregnancy during which she avoided vitamin D supplements [24]. Nevertheless, several pregnant women developed hypercalcemia during pregnancy despite avoidance of vitamin D supplement intake [17]. In addition, a high oral fluid intake and dietary calcium restriction were usually advised during pregnancy. Given that total serum calcium levels decrease during pregnancy as a consequence of reduced serum albumin levels due to volume expansion, it is generally recommended to measure albumin adjusted and/or ionized (free) calcium to assess calcium levels in pregnancy. Specific drug treatments of these patients during pregnancy are shown in Table 3.

Table 2. Characteristics of pregnancies of women with pathogenic CYP24A1 mutations.

Case Number of the Mother	Reference Number	Age (Years)	Number of Fetuses	Peak Serum Calcium in Pregnancy (mmol/L) *	Major Maternal Pregnancy Complications	Type of Delivery **	Major Maternal Postpartum Complications	Live Birth	Breast-Feeding	Major Newborn Complications
1	[15]	27	1	Not reported	Pre-eclampsia, polyhydramnios	Caesarian section	Acute kidney injury	Yes	Yes	Symptomatic hypercalcemia at 5 days
	[16]	32	1	Ionized serum calcium > 1.5	Hypertension, worsening renal function	Caesarian section	Worsening renal function	Yes	No	Symptomatic hypocalcemia at 3 months
2	[17]	23	1	NA	None	NA	Pre-eclampsia, hypercalcemic crisis	Yes	NA	Convulsions, hypoglycemia, necrotizing enterocolitis
		NA	1	2.89	Hypertension	Vaginal	Not reported	Yes	NA	Hypoglycemia, hypercalcemia
		NA	1	3.44	Symptomatic hypercalcemia	Vaginal	None	Yes	NA	Hypercalcemia
		NA	1	2.88	None	Vaginal	Hypertension, hypercalcemic crisis	Yes	NA	Hypoglycemia
		NA	1	2.92	Hypertension	Vaginal	Hypercalcemic crisis	Yes	NA	None
3	[17]	21	1	2.87	None	Vaginal	Hypercalcemia	Yes	NA	Hypercalcemia
		NA	1	2.83	Hypertension	NA	Hypercalcemia	Yes	NA	Hypercalcemia
4	[18]	47	2	3.11	Hypertension, diabetes	Caesarian section	Hypercalcemia	Yes for both	No	None
5	[18]	36	2	NA	NA	NA	Hypercalcemic crisis	Yes for both	NA	None
		NA	1	NA	None	NA	None	Yes	NA	None
		NA	1	NA	None	NA	None	Yes	NA	None
6	[19]	32	1	3.27	Pre-eclampsia	Caesarian section	Hypertension, acute kidney injury	Yes	NA	Mild hypercalcemia
7	[19]	32	NA	NA	Nephrolithiasis	NA	NA	NA	NA	NA
8	[20]	20	2	3.07	Hypertension	Vaginal	Hypercalcemic crisis, acute pancreatitis	No	No	Not alive (intrauterine demise at 26 weeks)
		20	1	2.87	Acute pancreatitis	Vaginal	None	Yes	No	None

Table 2. *Cont.*

Case Number of the Mother	Reference Number	Age (Years)	Number of Fetuses	Peak Serum Calcium in Pregnancy (mmol/L) *	Major Maternal Pregnancy Complications	Type of Delivery **	Major Maternal Postpartum Complications	Live Birth	Breast-Feeding	Major Newborn Complications
9	[21]	33	2	3.4	Hypertension	Caesarian section	NA	Yes for one	NA	Development disorder, anorectal malformation, asymptomatic hypercalcemia
10	[25]	20	2	3.82	Acute pancreatitis	NA	Acute pancreatitis	No	No	Not alive (intrauterine demise at 26 weeks)
		20	1	2.99	Pre-eclampsia, acute pancreatitis	Vaginal	None	Yes	No	None
11	[22]	24	NA	3.07	NA	NA	NA	NA	NA	NA
		26	NA	3.07	NA	NA	NA	NA	NA	NA
		27	NA	2.92	NA	NA	NA	NA	NA	NA
		28	1	3.04	Pre-eclampsia	NA	NA	Yes	Yes	Slight hypocalcemia
12	[23]	NA	1	2.92	Intrauterine growth retardation	NA	NA	Yes	NA	None
		NA	1	2.92	Intrauterine growth retardation	NA	NA	Yes	NA	None
		35	2	2.99	Rupture of membranes	Vaginal and caesarian section	Hypercalcemic crisis	Yes	NA	None
13	[24]	Mid 20	1	3.3	Idiopathic cholestasis	NA	NA	Yes	NA	None
		End 20	1	2.6	Idiopathic cholestasis	NA	NA	Yes	NA	None

NA, not available; * Total or albumin adjusted serum/plasma calcium is indicated; ** Most cases with NA may have had vaginal delivery but this was often not clearly indicated; Hypercalcemic crisis was usually defined as serum calcium of at least 3.5 mmol/L plus symptoms.

Table 3. Specific drug treatments of women with pathogenic CYP24A1 mutations during pregnancy.

Case Number of the Mother	Reference Number	Intravenous Hydration	Loop Diuretics	Glucocorticoids	Phosphate Supplements	Calcitonin
1	[15]	No *	No	No	No	No
	[16]	**Yes**	No	**Yes**	**Yes**	**Yes**
2	[17]	No	No	No	No	No
		No	No	No	No	No
		No	**Yes**	No	No	No
		No	No	No	No	No
		No	No	No	No	No
3	[17]	No	No	No	No	No
		No	No	No	No	No
4	[18]	**Yes**	**Yes**	**Yes**	No	No
5	[18]	No	No	No	No	No
		No	No	No	No	No
		No	No	No	No	No
6	[19]	**Yes**	No	**Yes**	No	No
7	[19]	No	No	No	No	No
8	[20]	No	No	No	No	No
		Yes	No	No	**Yes **	**Yes**
9	[21]	No	**Yes**	No	No	No
10	[25]	No	No	No	No	No
		No	No	No	No	No
11	[22]	No	No	No	No	No
		No	No	No	No	No
		No	No	No	No	No
		No	No	No	No	No
12	[23]	No	No	No	No	No
		No	No	No	No	No
		No	No	No	No	No
13	[24]	No	No	**Yes**	No	No
		No	No	No	No	No

* If not specifically indicated in the manuscript we assumed no treatment if it was not described; ** Omeprazole was also prescribed to reduce intestinal calcium absorption.

We assumed that there was no specific treatment when it was not clearly indicated in the respective publication. During 29 pregnancies identified in the literature, the applied treatment approaches were intravenous hydration in four, loop diuretics in three, glucocorticoids in four, phosphate supplements in two (in one case in combination with omeprazole) and calcitonin in two pregnancies, respectively. Conclusions on efficacies of these treatments cannot be made due to the observational nature of these limited data and the challenge to disentangle potential effects of specific drugs (a) from each other when used in combination, (b) from general treatment approaches such as stopping vitamin D supplement intake, and (c) from the natural course of the disease. In general, serum calcium levels were not significantly and consistently reduced or normalized after initiation of the above-described specific drug treatments except for some cases in which vitamin D supplement intake was stopped in parallel (e.g., case numbers 6 and 9 in Table 3) [19,21]. Of note, some authors concluded on the missing effect of their treatment such as in terms of glucocorticoids (case number 13 in Table 3) [24]. Glucocorticoids are known to modulate vitamin D metabolism resulting in lower calcitriol levels by effects including induction of 24-hydroxylation and reduce intestinal calcium absorption, but there are still several knowledge gaps regarding this issue. Importantly, glucocorticoids may not be effective in treating hypercalcemia in patients with CYP24A1 mutations as reported previously for nonpregnant patients [26]. It should also be noted that there was no use of bisphosphonates or denosumab during pregnancy, probably because bisphosphonates cross the placenta and because there is hardly any experience with denosumab during human pregnancy. Data from case reports on the use of bisphosphonates before conception or during pregnancy do not report major teratogenic effects, but some cases of transient neonatal hypocalcemia and low birth weight occurred [27–30]. Calcitonin does not cross the placenta but its calcium lowering effects diminish after a few days due to tachyphylaxis (i.e., downregulation of its receptors on osteoclasts). Specific drug treatments after delivery (i.e., postpartum) are outlined in Table 4.

Table 4. Specific drug treatments of women with pathogenic CYP24A1 mutations postpartum.

Case Number of the Mother	Reference Number	Intravenous Hydration	Loop Diuretics	Glucocorticoids	Potassium Supplements	Calcitonin	Denosumab	Bisphosphonates
1	[15]	No *	No	No	No	No	No	No
	[16]	No	No	No	No	Yes	No	No
2	[17]	Yes	Yes	No	No	Yes	No	Yes
		No	No	No	No	No	No	No
		No	Yes	No	No	No	No	Yes
		Yes	No	No	No	No	No	No
		Yes	No	No	No	No	No	No
3	[17]	Yes	No	No	No	No	No	No
		Yes	No	No	No	No	No	No
4	[18]	Yes	Yes	Yes	No	No	Yes	No
5	[18]	Yes	No	Yes	No	No	Yes	Yes
		No	No	No	No	No	No	No
		No	No	No	No	No	No	No
6	[19]	Yes	No	Yes	No	No	No	No
7	[19]	No	No	No	No	No	No	No
8	[20]	Yes	No	Yes	No	Yes	No	No
		No	No	No	No	Yes	No	No
9	[21]	No	No	No	No	No	No	No
10	[25]	Yes	No	No	No	Yes	No	No
		No	No	No	No	No	No	No
11	[22]	No	No	No	No	No	No	No
		No	No	No	No	No	No	No
		No	No	No	No	No	No	No
		No	No	No	No	No	No	No
12	[23]	No	No	No	No	No	No	No
		No	No	No	No	No	No	No
		Yes	No	Yes	No	No	No	Yes
13	[24]	No	No	No	No	No	No	No
		No	No	No	No	No	No	No

* If not specifically indicated in the manuscript we assumed no treatment if it was not described.

Regarding the specific treatments of the women after delivery there were 11 cases treated with intravenous hydration, three cases treated with loop diuretics, four cases treated with glucocorticoids, five cases treated with calcitonin, two cases treated with denosumab and four cases treated with bisphosphonates. Conclusions on the efficacy of these treatments are limited based on the observational data with its inherent limitations as noted above, and in particular due to the natural course of serum calcium fluctuations postpartum. In this context, some but not all, reports documented an increase in serum calcium within the first days to few weeks after delivery. Following this immediate postpartum period, a gradual improvement, i.e., reduction of serum calcium concentrations, was reported over the next several weeks to a few months in virtually all cases with available data. Interestingly, in two cases, denosumab treatment in the postpartum period was followed by a significant decrease in serum calcium concentrations after only a few days, while preceding treatments with, e.g., bisphosphonates, furosemide and glucocorticoids were insufficient to control hypercalcemia (case numbers 4 and 5 in Table 4) [18].

The vast majority of newborns of mothers with hypercalcemia due to CYP24A1 mutations in pregnancy is not affected by idiopathic infantile hypercalcemia, as this disease typically follows an autosomal recessive inheritance pattern. In the newborns, hypercalcemia was detected in some cases that usually resolved after several days to a few weeks as well as transient hypocalcemia, and some cases reported on transient hypoglycemia (see Table 4). One newborn was compound heterozygous for pathogenic CYP24A1 mutations and developed symptomatic hypercalcemia after receiving 50,000 IU of vitamin D2 on day 1 of his life [15]. It should also be noted that the clinical significance of heterozygote carriers is not clear at present [8]. Some publications report on a mild, yet clinically significant, phenotype of some heterozygote carriers with slight hypercalcemia and hypercalciuria [8]. Therefore, common vitamin D doses for rickets prevention, e.g., 400 IU per day, should not be exceeded in these newborns and screening for hypercalcemia may be prudent as recommended for our case.

4. Discussion

We have presented a case report and results of a systematic literature review on pregnant women with pathogenic CYP24A1 mutations. It is evident that pregnancy with its associated changes in vitamin D metabolism triggers or enhances hypercalcemia in affected women. Obstetric complications (e.g., arterial hypertension, pre-eclampsia and hypercalcemic crisis) were frequently reported, but most pregnancies resulted in live births (see Table 2). Although there were also some, usually minor, complications in the newborns (e.g., hypercalcemia, hypocalcemia or hypoglycemia) immediately after birth (see Table 2), we can conclude that the vast majority of newborns will be healthy and have a normal development in their further life as reported by most of the cases.

In general, pathogenic CYP24A1 mutations are relatively rare but in view of the high prevalence of vitamin D supplementation during preconception and pregnancy, affected women are at particularly high risk of severe hypercalcemia and associated obstetrics complications. From a pathophysiologic point of view, pregnancy is associated with about a doubling to tripling of serum $1,25(OH)_2D$ (calcitriol) concentrations that seems to be important to increase intestinal calcium absorption in order to meet the mineral (calcium) demands of the growing fetus [1,31]. Therefore, it appears logical that in this setting of increased "vitamin D activation", i.e., hydroxylation of $25(OH)D$ to $1,25(OH)_2D$, a disease with impaired vitamin D catabolism, may be aggravated in pregnancy. In this context, we support the suggestion of a European expert consensus to measure serum calcium concentrations as part of otherwise indicated routine screening programs or visits during preconception and early pregnancy [1]. Such an approach targets to detect alterations in calcium metabolism that may, beyond CYP24A1 mutations, of course, be important to diagnose parathyroid disorders and related diseases. In the case of pathogenic CYP24A1 mutations, it appears logical to assume that the degree of hypercalcemia may be, as in the case of PHPT, associated with the risk of obstetric complications although we cannot definitely claim this due to relatively few reported cases. In this context, it is well known from patients with PHPT and other related diseases that hypercalcemia per se can cause glomerular hyperfiltration with hypercalciuria and increased risk of kidney stones (urolithiasis/nephrolithiasis), worsening of kidney function or pancreatitis. These latter complications have also been reported by some of the cases (see Table 2). During pregnancy, calcium is transported by the placenta to the fetus with the consequence that hypercalcemia of the mother is also causing hypercalcemia of the fetus. This may in turn suppress PTH in the fetus and may explain why newborns of mothers with CYP24A1 mutations are at risk of both, hypercalcemia due to transfer of calcium from the mother during pregnancy and of hypocalcemia due to suppression of PTH. Therefore, we suggest, as it is recommended for newborns of mothers with PHPT, to measure serum or ionized calcium concentrations in the newborns every second day starting on day two for about 1 to 2 weeks [1]. Considering the uncertainty regarding the clinical significance of heterozygote carriers for CYP24A1 mutations it may also be justified to re-check the serum calcium concentrations in the children again several weeks to a few months after birth, in order to capture hypercalcemic episodes that may be triggered by usual vitamin D supplementation. It may also appear reasonable to measure blood glucose levels in the newborns in the first days after birth as some cases of hypoglycemia have been reported [17]. Serum calcium concentrations should also be measured in the affected mothers in the first week after giving birth (with re-measurements of serum calcium depending on the initial value) as the calcium transfer via the placenta is immediately stopped after delivery and may thus further aggravate hypercalcemia in the mother.

Regarding treatment approaches during pregnancy, it is logical and also seems to be highly effective that women with pathogenic CYP24A1 mutations are advised to stop any vitamin D supplementation and aim to minimize other sources of vitamin D supply. In particular sunlight (ultraviolet-B (UV-B)) induced vitamin D synthesis in the skin should be avoided. Notably, despite our advice to minimize vitamin D supply in our case patient there were no major changes in serum calcium concentrations, but we noticed a significant decrease in urinary calcium excretion. Data interpretation of this single case is, however, challenging as we, of course, have no data on the natural course of the disease without any intervention. In addition, a sufficient oral fluid intake and avoidance of overwhelming calcium supply can be recommended. Intravenous hydration has been used as a treatment approach for several women with pathogenic CYP24A1 mutations during pregnancy and postpartum, but although this treatment is generally safe and efficient for treatment of hypercalcemia, risk of volume overload (edema) should be kept in mind [32]. Apart from this, other specific drug treatments should only be used on an individual basis during pregnancy by considering and balancing the potential risks and benefits. Overall, the existing literature does not clearly support the efficacy of any of these treatments so that we would rather be cautious when using them during pregnancy. In our case woman, we would have initiated a specific drug treatment for hypercalcemia in case of albumin adjusted serum calcium concentrations above 3.5 mmol/L. We personally consider albumin adjusted serum calcium concentrations from 3.0 to 3.5 mmol/L as a range in which specific drug treatments to lower calcium levels should be seriously considered if other measures are not effective and if symptoms or adverse consequences of hypercalcemia emerge. Breastfeeding was not established in most women after delivery as lactation might potentially aggravate hypercalcemia. In our case report we did, however, observe a gradual decline in serum calcium concentration while breastfeeding. Therefore, it appears reasonable that breastfeeding should not be a priori banned in women with pathogenic CYP24A1 mutations. Regarding treatment options for severe hypercalcemia in the postpartum period, denosumab appeared to be effective in reducing serum calcium concentrations.

Beyond pregnancy and the postpartum period, data on the long-term perspective of patients with pathogenic CYP24A1 mutations suggest that affected patients may be at increased risk of chronic kidney disease, nephrocalcinosis and nephrolithiasis (kidney stones) despite avoidance of sun exposure, vitamin D and calcium supplementation [33]. It may thus be prudent to suggest regular screening for kidney function, nephrocalcinosis and nephrolithiasis in affected patients.

Azole agents (e.g., fluconazole or ketoconazole) and rifampin have not been used to treat hypercalcemia during pregnancy and postpartum in patients with pathogenic CYP24A1 mutations, but have been suggested as potential candidates for long-term treatment in this setting [9]. Azole agents that are usually used to treat fungal infections are inhibitors of cytochrome P450-enzymes that also inhibit 1-alpha-hydroxylase (cytochrome-P450 family 27 subfamily B member 1; CYP27B1) and thus the conversion (activation) of 25(OH)D to 1,25(OH)$_2$D. Case reports describe their successful use in patients with pathogenic CYP24A1 mutations [9,34]. Fluconazole may be preferred over ketoconazole due to generally fewer side effects, in particular less hepatotoxicity. The tuberculosis drug rifampin (also termed rifampicin) has also been successfully used to treat patients with pathogenic CYP24A1 mutations [35]. This drug induces the enzyme cytochrome-P450 family 3 subfamily A member 4 (CYP3A4) that inactivates vitamin D metabolites by an alternative degradative pathway to CYP24A1. Consequently, it has been recommended to avoid use of medications and foods (e.g., starfruit, pomegranate, and white grapefruit) that can inhibit CYP3A4 in patients with pathogenic CYP24A1 mutations. Rifampin has an excellent safety profile but may also be hepatotoxic in some cases [35]. Fluconazole and rifampin might be considered for treatment of hypercalcemia during pregnancy as there is much experience with these drugs in treating infections in pregnant women, but this requires careful consideration of the risks and benefits for the individual patient [36,37].

Given that the prevalence of pathogenic CYP24A1 mutations is very low and considering the beneficial effects of preventing and treating vitamin D deficiency, we do support the current recommendations for vitamin D intakes and supplementation [31,38–42]. The existing risk of severe hypercalcemia with high vitamin D bolus doses as it has been applied for rickets prevention in the UK in the 1950s or in Poland and former East Germany in the 1980s, in individuals with pathogenic CYP24A1 mutations, points towards caution with high dose vitamin D supplementation [33]. Awareness must be increased among clinicians that in the event of hypercalcemia with low PTH concentrations and in patients with nephrolithiasis and/or nephrocalcinosis, pathogenic CYP24A1 mutations should be considered as a potential underlying disease. If pathogenic CYP24A1 mutations are suspected, the measurement of vitamin D metabolites, i.e., the $24,25(OH)_2D_3$ to $25(OH)D_3$ ratio (or vice versa) is recommended and if it is pathologic in terms of relatively low $24,25(OH)2D_3$ concentrations for the prevailing $25(OH)D_3$ status, genetic analyses should be performed to establish the diagnosis. In general, individuals without pathogenic CYP24A1 mutations do have a $25(OH)D_3$ to $24,25(OH)_2D_3$ ratio >25 and those with pathogenic mutations do have a respective ratio of > 80 [43,44]. Of note, there are different approaches regarding calculations and cut-offs for this ratio published, pointing to the need for harmonization of this issue in the future [45–47].

It is a limitation of our work that it was not based on an a priori registered systematic review, but we consider it as a main strength that our paper is the first to specifically address the issue of pathogenic CYP24A1 mutations in pregnancy. We are aware that our conclusions are based on observational data with all their inherent limitations, but we do hope that our work may provide some guidance for clinicians regarding the management of pregnant women with pathogenic CYP24A1 mutations.

5. Conclusions

Pregnant women with pathogenic CYP24A1 mutations are at particular high risk of hypercalcemia and therewith associated complications. Minimizing vitamin D supply in these women seems to be highly effective as a therapeutic approach, whereas the existing data are insufficient and limited to draw firm conclusions on specific drug treatments in affected women. It seems reasonable to screen for serum calcium concentrations in pregnant women in order to accurately diagnose and treat cases with pathogenic CYP24A1 mutations and other disorders of calcium metabolism. Affected women can be informed that the most likely outcome of their pregnancy is that they will have a healthy infant, but ionized calcium should be measured in the newborns. Long-term screening with reference to kidney function and nephrolithiasis seems to be reasonable in patients with pathogenic CYP24A1 mutations, but whether any chronic drug treatment should be advised requires further investigations.

Author Contributions: Writing—original draft preparation, S.P.; all authors S.P., V.T.-S., P.P., S.Z., A.M., S.N.K., W.M. (Waldemar Misiorowski), A.Z., W.M. (Winfried März) and C.T. contributed to writing—review and editing. All authors have read and agreed to the published version of the manuscript.

Funding: This research received no external funding.

Institutional Review Board Statement: Not applicable.

Informed Consent Statement: Not applicable.

Data Availability Statement: All available data are presented in the manuscript.

Conflicts of Interest: The authors declare no conflict of interest.

References

1. Bollerslev, J.; Rejnmark, L.; Zahn, A.; Heck, A.; Appelman-Dijkstra, N.M.; Cardoso, L.; Hannan, F.M.; Cetani, F.; Sikjaer, T.; Formenti, A.M.; et al. European Expert Consensus on Practical Management of Specific Aspects of Parathyroid Disorders in Adults and in Pregnancy: Recommendations of the ESE Educational Program of Parathyroid Disorders. *Eur. J. Endocrinol.* **2021**, *186*, R33–R63. [CrossRef] [PubMed]
2. Kovacs, C.S. Maternal Mineral and Bone Metabolism during Pregnancy, Lactation, and Post-Weaning Recovery. *Physiol. Rev.* **2016**, *96*, 449–547. [CrossRef] [PubMed]
3. Dandurand, K.; Ali, D.S.; Khan, A.A. Hypercalcemia in Pregnancy. *Endocrinol. Metab. Clin. N. Am.* **2021**, *50*, 753–768. [CrossRef] [PubMed]
4. Cappellani, D.; Brancatella, A.; Morganti, R.; Borsari, S.; Baldinotti, F.; Caligo, M.A.; Kaufmann, M.; Jones, G.; Marcocci, C.; Cetani, F. Hypercalcemia due to CYP24A1 mutations: A systematic descriptive review. *Eur. J. Endocrinol.* **2021**, *186*, 137–149. [CrossRef] [PubMed]
5. Molin, A.; Lemoine, S.; Kaufmann, M.; Breton, P.; Nowoczyn, M.; Ballandonne, C.; Coudray, N.; Mittre, H.; Richard, N.; Ryckwaert, A.; et al. Overlapping Phenotypes Associated with CYP24A1, SLC34A1, and SLC34A3 Mutations: A Cohort Study of Patients with Hypersensitivity to Vitamin D. *Front. Endocrinol.* **2021**, *12*, 736240. [CrossRef] [PubMed]
6. Tsourdi, E.; Anastasilakis, A.D. Parathyroid Disease in Pregnancy and Lactation: A Narrative Review of the Literature. *Biomedicines* **2021**, *9*, 475. [CrossRef]
7. Schlingmann, K.P.; Kaufmann, M.; Weber, S.; Irwin, A.; Goos, C.; John, U.; Misselwitz, J.; Klaus, G.; Kuwertz-Broking, E.; Fehrenbach, H.; et al. Mutations in CYP24A1 and idiopathic infantile hypercalcemia. *N. Engl. J. Med.* **2011**, *365*, 410–421. [CrossRef]
8. Carpenter, T.O. CYP24A1 loss of function: Clinical phenotype of monoallelic and biallelic mutations. *J. Steroid Biochem. Mol. Biol.* **2017**, *173*, 337–340. [CrossRef]
9. De Paolis, E.; Scaglione, G.L.; De Bonis, M.; Minucci, A.; Capoluongo, E. CYP24A1 and SLC34A1 genetic defects associated with idiopathic infantile hypercalcemia: From genotype to phenotype. *Clin. Chem. Lab. Med.* **2019**, *57*, 1650–1667. [CrossRef]
10. Pronicka, E.; Ciara, E.; Halat, P.; Janiec, A.; Wojcik, M.; Rowinska, E.; Rokicki, D.; Pludowski, P.; Wojciechowska, E.; Wierzbicka, A.; et al. Biallelic mutations in CYP24A1 or SLC34A1 as a cause of infantile idiopathic hypercalcemia (IIH) with vitamin D hypersensitivity: Molecular study of 11 historical IIH cases. *J. Appl. Genet.* **2017**, *58*, 349–353. [CrossRef]
11. Zelzer, S.; Meinitzer, A.; Enko, D.; Simstich, S.; Le Goff, C.; Cavalier, E.; Herrmann, M.; Goessler, W. Simultaneous determination of 24,25- and 25,26-dihydroxyvitamin D3 in serum samples with liquid-chromatography mass spectrometry—A useful tool for the assessment of vitamin D metabolism. *J. Chromatogr. B Analyt. Technol. Biomed. Life Sci.* **2020**, *1158*, 122394. [CrossRef] [PubMed]
12. Zelzer, S.; Pruller, F.; Curcic, P.; Sloup, Z.; Holter, M.; Herrmann, M.; Mangge, H. Vitamin D Metabolites and Clinical Outcome in Hospitalized COVID-19 Patients. *Nutrients* **2021**, *13*, 2129. [CrossRef] [PubMed]
13. Trummer, C.; Schwetz, V.; Pandis, M.; Grubler, M.R.; Verheyen, N.; Gaksch, M.; Zittermann, A.; Marz, W.; Aberer, F.; Lang, A.; et al. Effects of Vitamin D Supplementation on IGF-1 and Calcitriol: A Randomized-Controlled Trial. *Nutrients* **2017**, *9*, 623. [CrossRef] [PubMed]
14. Pilz, S.; Gaksch, M.; Kienreich, K.; Grubler, M.; Verheyen, N.; Fahrleitner-Pammer, A.; Treiber, G.; Drechsler, C.; ó Hartaigh, B.; Obermayer-Pietsch, B.; et al. Effects of vitamin D on blood pressure and cardiovascular risk factors: A randomized controlled trial. *Hypertension* **2015**, *65*, 1195–1201. [CrossRef]
15. McBride, L.; Crosthwaite, A.; Houlihan, C.; Stark, Z.; Rodda, C. Rare cause of maternal and neonatal hypercalcaemia. *J. Paediatr. Child Health* **2019**, *55*, 232–235. [CrossRef]
16. McBride, L.; Houlihan, C.; Quinlan, C.; Messazos, B.; Stark, Z.; Crosthwaite, A. Outcomes Following Treatment of Maternal Hypercalcemia Due to CYP24A1 Pathogenic Variants. *Kidney Int. Rep.* **2019**, *4*, 888–892. [CrossRef]
17. Hedberg, F.; Pilo, C.; Wikner, J.; Torring, O.; Calissendorff, J. Three Sisters with Heterozygous Gene Variants of CYP24A1: Maternal Hypercalcemia, New-Onset Hypertension, and Neonatal Hypoglycemia. *J. Endocr. Soc.* **2019**, *3*, 387–396. [CrossRef]
18. Arnold, N.; O'Toole, V.; Huynh, T.; Smith, H.C.; Luxford, C.; Clifton-Bligh, R.; Eastman, C.J. Intractable hypercalcaemia during pregnancy and the postpartum secondary to pathogenic variants in CYP24A1. *Endocrinol. Diabetes Metab. Case Rep.* **2019**, *2019*. [CrossRef]
19. Griffin, T.P.; Joyce, C.M.; Alkanderi, S.; Blake, L.M.; O'Keeffe, D.T.; Bogdanet, D.; Islam, M.N.; Dennedy, M.C.; Gillan, J.E.; Morrison, J.J.; et al. Biallelic CYP24A1 variants presenting during pregnancy: Clinical and biochemical phenotypes. *Endocr. Connect.* **2020**, *9*, 530–541. [CrossRef]
20. Woods, G.N.; Saitman, A.; Gao, H.; Clarke, N.J.; Fitzgerald, R.L.; Chi, N.W. A Young Woman with Recurrent Gestational Hypercalcemia and Acute Pancreatitis Caused by CYP24A1 Deficiency. *J. Bone Miner. Res.* **2016**, *31*, 1841–1844. [CrossRef]
21. Romasovs, A.; Jaunozola, L.; Berga-Svitina, E.; Daneberga, Z.; Miklasevics, E.; Pirags, V. Hypercalcemia and CYP24A1 Gene Mutation Diagnosed in the 2nd Trimester of a Twin Pregnancy: A Case Report. *Am. J. Case Rep.* **2021**, *22*, e931116. [CrossRef] [PubMed]
22. Shah, A.D.; Hsiao, E.C.; O'Donnell, B.; Salmeen, K.; Nussbaum, R.; Krebs, M.; Baumgartner-Parzer, S.; Kaufmann, M.; Jones, G.; Bikle, D.D.; et al. Maternal Hypercalcemia Due to Failure of 1,25-Dihydroxyvitamin-D3 Catabolism in a Patient with CYP24A1 Mutations. *J. Clin. Endocrinol. Metab.* **2015**, *100*, 2832–2836. [CrossRef] [PubMed]

23. Dinour, D.; Davidovits, M.; Aviner, S.; Ganon, L.; Michael, L.; Modan-Moses, D.; Vered, I.; Bibi, H.; Frishberg, Y.; Holtzman, E.J. Maternal and infantile hypercalcemia caused by vitamin-D-hydroxylase mutations and vitamin D intake. *Pediatr. Nephrol.* **2015**, *30*, 145–152. [CrossRef] [PubMed]

24. Macdonald, C.; Upton, T.; Hunt, P.; Phillips, I.; Kaufmann, M.; Florkowski, C.; Soule, S.; Jones, G. Vitamin D supplementation in pregnancy: A word of caution. Familial hypercalcaemia due to disordered vitamin D metabolism. *Ann. Clin. Biochem.* **2020**, *57*, 186–191. [CrossRef] [PubMed]

25. Kwong, W.T.; Fehmi, S.M. Hypercalcemic Pancreatitis Triggered by Pregnancy with a CYP24A1 Mutation. *Pancreas* **2016**, *45*, e31–e32. [CrossRef]

26. Colussi, G.; Ganon, L.; Penco, S.; De Ferrari, M.E.; Ravera, F.; Querques, M.; Primignani, P.; Holtzman, E.J.; Dinour, D. Chronic hypercalcaemia from inactivating mutations of vitamin D 24-hydroxylase (CYP24A1): Implications for mineral metabolism changes in chronic renal failure. *Nephrol. Dial. Transplant.* **2014**, *29*, 636–643. [CrossRef]

27. Stathopoulos, I.P.; Liakou, C.G.; Katsalira, A.; Trovas, G.; Lyritis, G.G.; Papaioannou, N.A.; Tournis, S. The use of bisphosphonates in women prior to or during pregnancy and lactation. *Hormones* **2011**, *10*, 280–291. [CrossRef]

28. Sokal, A.; Elefant, E.; Leturcq, T.; Beghin, D.; Mariette, X.; Seror, R. Pregnancy and newborn outcomes after exposure to bisphosphonates: A case-control study. *Osteoporos. Int.* **2019**, *30*, 221–229. [CrossRef]

29. Machairiotis, N.; Ntali, G.; Kouroutou, P.; Michala, L. Clinical evidence of the effect of bisphosphonates on pregnancy and the infant. *Horm. Mol. Biol. Clin. Investig.* **2019**, *40*, 20190021. [CrossRef]

30. Green, S.B.; Pappas, A.L. Effects of maternal bisphosphonate use on fetal and neonatal outcomes. *Am. J. Health Syst. Pharm.* **2014**, *71*, 2029–2036. [CrossRef]

31. Pilz, S.; Zittermann, A.; Obeid, R.; Hahn, A.; Pludowski, P.; Trummer, C.; Lerchbaum, E.; Perez-Lopez, F.R.; Karras, S.N.; Marz, W. The Role of Vitamin D in Fertility and during Pregnancy and Lactation: A Review of Clinical Data. *Int. J. Environ. Res. Public Health* **2018**, *15*, 2241. [CrossRef] [PubMed]

32. Appelman-Dijkstra, N.M.; Ertl, D.A.; Zillikens, M.C.; Rjenmark, L.; Winter, E.M. Hypercalcemia during pregnancy: Management and outcomes for mother and child. *Endocrine* **2021**, *71*, 604–610. [CrossRef] [PubMed]

33. Janiec, A.; Halat-Wolska, P.; Obrycki, L.; Ciara, E.; Wojcik, M.; Pludowski, P.; Wierzbicka, A.; Kowalska, E.; Ksiazyk, J.B.; Kulaga, Z.; et al. Long-term outcome of the survivors of infantile hypercalcaemia with CYP24A1 and SLC34A1 mutations. *Nephrol. Dial. Transplant.* **2021**, *36*, 1484–1492. [CrossRef] [PubMed]

34. Sayers, J.; Hynes, A.M.; Srivastava, S.; Dowen, F.; Quinton, R.; Datta, H.K.; Sayer, J.A. Successful treatment of hypercalcaemia associated with a CYP24A1 mutation with fluconazole. *Clin. Kidney J.* **2015**, *8*, 453–455. [CrossRef] [PubMed]

35. Hawkes, C.P.; Li, D.; Hakonarson, H.; Meyers, K.E.; Thummel, K.E.; Levine, M.A. CYP3A4 Induction by Rifampin: An Alternative Pathway for Vitamin D Inactivation in Patients with CYP24A1 Mutations. *J. Clin. Endocrinol. Metab.* **2017**, *102*, 1440–1446. [CrossRef]

36. Liu, D.; Zhang, C.; Wu, L.; Zhang, L.; Zhang, L. Fetal outcomes after maternal exposure to oral antifungal agents during pregnancy: A systematic review and meta-analysis. *Int. J. Gynaecol. Obstet.* **2020**, *148*, 6–13. [CrossRef]

37. Orazulike, N.; Sharma, J.B.; Sharma, S.; Umeora, O.U.J. Tuberculosis (TB) in pregnancy—A review. *Eur. J. Obstet. Gynecol. Reprod. Biol.* **2021**, *259*, 167–177. [CrossRef]

38. Maretzke, F.; Bechthold, A.; Egert, S.; Ernst, J.B.; Melo van Lent, D.; Pilz, S.; Reichrath, J.; Stangl, G.I.; Stehle, P.; Volkert, D.; et al. Role of Vitamin D in Preventing and Treating Selected Extraskeletal Diseases-An Umbrella Review. *Nutrients* **2020**, *12*, 969. [CrossRef]

39. Pludowski, P.; Takacs, I.; Boyanov, M.; Belaya, Z.; Diaconu, C.C.; Mokhort, T.; Zherdova, N.; Rasa, I.; Payer, J.; Pilz, S. Clinical Practice in the Prevention, Diagnosis and Treatment of Vitamin D Deficiency: A Central and Eastern European Expert Consensus Statement. *Nutrients* **2022**, *14*, 1483. [CrossRef]

40. Pilz, S.; Trummer, C.; Theiler-Schwetz, V.; Grubler, M.R.; Verheyen, N.D.; Odler, B.; Karras, S.N.; Zittermann, A.; Marz, W. Critical Appraisal of Large Vitamin D Randomized Controlled Trials. *Nutrients* **2022**, *14*, 303. [CrossRef]

41. Buttriss, J.L.; Lanham-New, S.A.; Steenson, S.; Levy, L.; Swan, G.E.; Darling, A.L.; Cashman, K.D.; Allen, R.E.; Durrant, L.R.; Smith, C.P.; et al. Implementation strategies for improving vitamin D status and increasing vitamin D intake in the UK: Current controversies and future perspectives: Proceedings of the 2nd Rank Prize Funds Forum on vitamin D. *Br. J. Nutr.* **2021**, *127*, 1567–1587. [CrossRef] [PubMed]

42. Perez-Lopez, F.R.; Pilz, S.; Chedraui, P. Vitamin D supplementation during pregnancy: An overview. *Curr. Opin. Obstet. Gynecol.* **2020**, *32*, 316–321. [CrossRef] [PubMed]

43. Molin, A.; Baudoin, R.; Kaufmann, M.; Souberbielle, J.C.; Ryckewaert, A.; Vantyghem, M.C.; Eckart, P.; Bacchetta, J.; Deschenes, G.; Kesler-Roussey, G.; et al. CYP24A1 Mutations in a Cohort of Hypercalcemic Patients: Evidence for a Recessive Trait. *J. Clin. Endocrinol. Metab.* **2015**, *100*, E1343–E1352. [CrossRef]

44. Kaufmann, M.; Gallagher, J.C.; Peacock, M.; Schlingmann, K.P.; Konrad, M.; DeLuca, H.F.; Sigueiro, R.; Lopez, B.; Mourino, A.; Maestro, M.; et al. Clinical utility of simultaneous quantitation of 25-hydroxyvitamin D and 24,25-dihydroxyvitamin D by LC-MS/MS involving derivatization with DMEQ-TAD. *J. Clin. Endocrinol. Metab.* **2014**, *99*, 2567–2574. [CrossRef] [PubMed]

45. Fabregat-Cabello, N.; Farre-Segura, J.; Huyghebaert, L.; Peeters, S.; Le Goff, C.; Souberbielle, J.C.; Cavalier, E. A fast and simple method for simultaneous measurements of 25(OH)D, 24,25(OH)2D and the Vitamin D Metabolite Ratio (VMR) in serum samples by LC-MS/MS. *Clin. Chim. Acta* **2017**, *473*, 116–123. [CrossRef] [PubMed]

46. Francic, V.; Ursem, S.R.; Dirks, N.F.; Keppel, M.H.; Theiler-Schwetz, V.; Trummer, C.; Pandis, M.; Borzan, V.; Grubler, M.R.; Verheyen, N.D.; et al. The Effect of Vitamin D Supplementation on its Metabolism and the Vitamin D Metabolite Ratio. *Nutrients* **2019**, *11*, 2539. [CrossRef]
47. Tang, J.C.Y.; Nicholls, H.; Piec, I.; Washbourne, C.J.; Dutton, J.J.; Jackson, S.; Greeves, J.; Fraser, W.D. Reference intervals for serum 24,25-dihydroxyvitamin D and the ratio with 25-hydroxyvitamin D established using a newly developed LC-MS/MS method. *J. Nutr. Biochem.* **2017**, *46*, 21–29. [CrossRef]

Article

Effect of Cholecalciferol Supplementation on the Clinical Features and Inflammatory Markers in Hospitalized COVID-19 Patients: A Randomized, Open-Label, Single-Center Study

Tatiana L. Karonova [1,*], Ksenia A. Golovatyuk [1], Igor V. Kudryavtsev [1,2], Alena T. Chernikova [1], Arina A. Mikhaylova [1], Arthur D. Aquino [1], Daria I. Lagutina [1], Ekaterina K. Zaikova [1], Olga V. Kalinina [1], Alexey S. Golovkin [1], William B. Grant [3] and Evgeny V. Shlyakhto [1]

[1] Almazov National Medical Research Centre, 197341 Saint Petersburg, Russia; ksgolovatiuk@gmail.com (K.A.G.); igorek1981@yandex.ru (I.V.K.); arabicaa@gmail.com (A.T.C.); armikhaylova@yandex.ru (A.A.M.); akino97@bk.ru (A.D.A.); daria.lagutina.i@yandex.ru (D.I.L.); catherine3452@yandex.ru (E.K.Z.); olgakalinina@mail.ru (O.V.K.); golovkin_a@mail.ru (A.S.G.); shlyakhto_ev@almazovcentre.ru (E.V.S.)

[2] Institute of Experimental Medicine, 197376 Saint Petersburg, Russia

[3] Sunlight, Nutrition, and Health Research Center, P.O. Box 641603, San Francisco, CA 94164-1603, USA; williamgrant08@comcast.net

[*] Correspondence: karonova@mail.ru

Citation: Karonova, T.L.; Golovatyuk, K.A.; Kudryavtsev, I.V.; Chernikova, A.T.; Mikhaylova, A.A.; Aquino, A.D.; Lagutina, D.I.; Zaikova, E.K.; Kalinina, O.V.; Golovkin, A.S.; et al. Effect of Cholecalciferol Supplementation on the Clinical Features and Inflammatory Markers in Hospitalized COVID-19 Patients: A Randomized, Open-Label, Single-Center Study. *Nutrients* **2022**, *14*, 2602. https://doi.org/10.3390/nu14132602

Academic Editor: Sara Baldassano

Received: 31 May 2022
Accepted: 21 June 2022
Published: 23 June 2022

Publisher's Note: MDPI stays neutral with regard to jurisdictional claims in published maps and institutional affiliations.

Abstract: Recent studies showed that a low 25-hydroxyvitamin D (25(OH)D) level was associated with a higher risk of morbidity and severe course of COVID-19. Our study aimed to evaluate the effects of cholecalciferol supplementation on the clinical features and inflammatory markers in patients with COVID-19. A serum 25(OH)D level was determined in 311 COVID-19 patients. Among them, 129 patients were then randomized into two groups with similar concomitant medication. Group I (n = 56) received a bolus of cholecalciferol at a dose of 50,000 IU on the first and the eighth days of hospitalization. Patients from Group II (n = 54) did not receive the supplementation. We found significant differences between groups with the preferential increase in serum 25(OH)D level and Δ 25(OH)D in Group I on the ninth day of hospitalization ($p < 0.001$). The serum 25(OH)D level on the ninth day was negatively associated with the number of bed days (r = −0.23, $p = 0.006$); we did not observe other clinical benefits in patients receiving an oral bolus of cholecalciferol. Moreover, in Group I, neutrophil and lymphocyte counts were significantly higher ($p = 0.04$; $p = 0.02$), while the C-reactive protein level was significantly lower on the ninth day of hospitalization ($p = 0.02$). Patients with supplementation of 100,000 IU of cholecalciferol, compared to those without supplementation, showed a decrease in the frequencies of CD38++CD27 transitional and CD27−CD38+ mature naive B cells ($p = 0.006$ and $p = 0.02$) and an increase in the level of CD27−CD38− DN B cells ($p = 0.02$). Thus, the rise in serum 25(OH)D level caused by vitamin D supplementation in vitamin D insufficient and deficient patients may positively affect immune status and hence the course of COVID-19.

Keywords: COVID-19; SARS-CoV-2; vitamin D; 25(OH)D; B cell subsets; inflammatory markers

1. Introduction

Since the beginning of the COVID-19 pandemic, a large amount of data has been accumulated describing the vitamin D status in patients with COVID-19 and its impact on the course and prognosis of a new coronavirus infection. A meta-analysis by Kazemi et al. has illustrated that, in 10 of 12 studies, patients with confirmed COVID-19 had a lower 25(OH)D serum concentration compared to the control group [1].

It was shown that a low 25(OH)D level was associated with a higher risk of morbidity and a severe course of acute respiratory viral infection [2]. It was suggested that a vitamin D deficiency might be one of the modifiable risk factors for new coronavirus infection, worsening the course and prognosis of the disease. Studies over the past two years have found that a low serum 25(OH)D level was associated with the risk of SARS-CoV-2 infection [3] and with the severe course of COVID-19 [4]. Thus, in a study by Mercola et al., patients with a serum 25(OH)D level > 30 ng/mL had mild disease, while patients with a serum 25(OH)D level below 30 ng/mL had a higher severity of COVID-19 and higher mortality rates [4].

Many of the immune cells express vitamin D receptors. Therefore, by its binding, vitamin D can modulate both innate and acquired immune responses [5]. The protective role of vitamin D is linked to several mechanisms: a decrease in neutrophil activity; suppression of the exaggerated activity of type 1 T-helper cells, thus preventing cytokine storm development; and a direct positive effect of the active form of vitamin D on the expression of ACE2, the host receptor for SARS-CoV-2 [6–8].

Previous studies have shown that most hospitalized patients with COVID-19 had vitamin D deficiency, which was associated with a 3.79-fold increase in the risk of a severe course of COVID-19 and a 4.07-fold increase in the risk of a fatal outcome. The threshold of a serum 25(OH)D level of 11.4 ng/mL was associated with mortality, as well as with stimulation of type 2 T-helper cells and downregulation of T-helper 17 cell polarization [9–11].

Cholecalciferol supplementation, especially in patients with vitamin D insufficiency and deficiency, demonstrated effectiveness in preventing acute respiratory viral infections and COVID-19 [8]. A meta-analysis that included 72 observational and four randomized interventional trials confirmed a correlation between 25(OH)D level and disease severity and mortality. Moreover, clinical benefits, such as a reduction in inflammatory markers, were associated with the addition of vitamin D supplementation to standard COVID-19 therapy [12].

Therefore, we can hypothesize that vitamin D supplementation may be beneficial to reduce morbidity and mortality in COVID-19 patients. Currently, the development of methods for the prevention and treatment of a new coronavirus infection continues. Further studies are needed to evaluate the additive therapeutic efficacy of cholecalciferol in combination with the first-line therapy for COVID-19.

So, this study aimed to evaluate the effects of cholecalciferol supplementation on the clinical features and inflammatory markers in patients with COVID-19.

2. Materials and Methods

2.1. Patients

We analyzed the vitamin D status of 311 patients hospitalized with COVID-19 (161 men and 150 women). One hundred and twenty-nine patients from 311 patients were randomly included in the interventional study. All patients signed informed consent for participation. The randomized single-center open-label study was performed from 30 November 2020 to 20 March 2021, when the Almazov National Medical Research Centre (St. Petersburg, Russia) was transformed into an infectious hospital for COVID-19 patients. The local Ethics Committee of the Almazov National Medical Research Centre approved this study (protocol No. 1011-20-02C, 30 November 2020), which complied with the Declaration of Helsinki. This study was registered on clinicaltrials.gov (NCT number: NCT05166005).

The inclusion criteria included: age from 18 to 75 years, confirmed diagnosis of COVID-19 (polymerase chain reaction (PCR)-test and/or chest computed tomography (CT) scan), and signed informed consent. Subjects with daily vitamin D intake of 1000 IU and higher or who had contraindications to vitamin D supplementation were not included. Additional exclusion criteria were clinically significant kidney pathology with an eGFR of less than 45 mL/min/1.73 m^2; gastrointestinal and liver diseases; granulomatous diseases; oncology diseases (less than 5 years); immunodeficiency disorders; and addiction to drugs and alcohol. We did not include pregnant or breastfeeding women. Potential subjects with other circumstances considered inappropriate by the investigator were not allowed to participate in the study. All participants were unvaccinated, since general vaccination was not yet available.

One hundred and twenty-nine patients were randomized by random number table into two groups depending on vitamin D supplementation (water-soluble cholecalciferol): Group I received a bolus of cholecalciferol at a dose of 50,000 IU on the 1st and the 8th day of hospitalization, with the total dose being 100,000 IU; Group II received no supplementation.

2.2. Clinical Data

We analyzed the following clinical data: height, weight, body mass index (BMI), and co-morbidities. We assessed the severity of the disease by oxygen supplementation, SpO2, the time between symptom onset and hospitalization, intensive care unit admission rates, and bed days. The disease severity was classified according to the following criteria: mild illness—temperature < 38 °C, absence of shortness of breath, dyspnea or normal chest computed tomography (CT); moderate illness—temperature > 38 °C, SpO2 < 95%, C-reactive protein (CRP) > 10 mg/L, CT—1 or 2; and severe illness—hemodynamic instability, SpO2 < 93%, CT—3 or 4 [13].

2.3. Laboratory Tests

Laboratory parameters of serum 25(OH)D level, complete blood count, and the acute phase proteins, including CRP, lactate dehydrogenase (LDH), and ferritin, were measured at baseline. On the 9th day of vitamin D supplementation, we assessed serum 25(OH)D level, complete blood count, and CRP level.

We measured serum 25(OH)D levels using a chemiluminescence immunoassay on microparticles (Abbott Architect i8000, Chicago, IL, USA); the reference interval was 3.4–155.9 ng/mL, intra-assay coefficient of variation ranged from 1.60% to 5.92%, and inter-assay coefficient of variation ranged from 2.15% to 2.63%. Blood samples for 25(OH)D measurements were taken in the morning from the cubital vein, centrifuged, aliquoted, and stored in a freezer at a temperature of −70 °C before the laboratory testing. According to a current vitamin D supplementation guideline, vitamin D status was considered normal when the 25(OH)D level was ≥30 ng/mL (≥75 nmol/L); for insufficiency, the 25(OH)D level was ≥20 and <30 ng/mL (≥50 and <75 nmol/L); for deficiency, the 25(OH)D level was <20 ng/mL (<50 nmol/L) [14].

We used a Cobas Integra 400analyzer (Roche Diagnostics GmbH, Mannheim, Germany) and corresponding diagnostic kits to determine the CRP level (reference range 0–5 mg/L) and LDH level (reference range 133–225 units/L). Ferritin level was measured on an Abbott Architect c8000 analyzer (Chicago, IL, USA; reference range, 64–111 nmol/L).

2.4. Instrumental Data

To detect pneumonia, we used CT scans without intravenous contrast enhancement. The volume of lung tissue lesions was described as CT-1, lesion volume < 25%; CT-2, lesion volume 25–50%; CT-3, lesion volume 50–75%; and CT-4, lesion volume > 75% [13].

2.5. Concomitant Medication

Concomitant medication for COVID-19 was analyzed for each patient group. Treatment was carried out according to the local guidelines [13]. The analysis included evaluation of anti-IL-6 receptor monoclonal antibodies and glucocorticoids (GC) administration, as well as calculation of total GC dose from the 1st until the 9th day.

2.6. Immunological Data

The frequencies of peripheral blood B cell subsets in patients with COVID-19 were analyzed by flow cytometry. Staining protocol, reagents, and gating strategy were described in detail earlier [15]. We classified B cell subsets using CD27 and CD38 co-expression, as was suggested by P. Hanley et al. [16]. As a result, we identified six main B cell subsets, including CD27−CD38++ transitional B cells, CD27++CD38++ circulating plasma cell precursors, mature naïve and mature activated B cells (CD27−CD38+ and CD27+CD38+, respectively), CD27−CD38− double-negative or DN B cells and, finally, CD27+CD38− resting memory B cells.

2.7. Study Objective

The primary outcomes were changes in serum 25(OH)D level, complete blood count, CRP level in peripheral blood, and B cell subsets on the 9th day of hospitalization compared to the first day. The secondary endpoint was to evaluate the effects of cholecalciferol supplementation on the severity of the disease, oxygen supplementation, intensive care unit admission rates, and clinical outcomes. The additional secondary endpoint was the duration of hospitalization.

2.8. Statistical Analysis

For sample size calculation, we used Power and Sample Size software (Sealedenvelope, 2022; London, UK). At the 95% confidence level, 80% power, and 10% estimated dropout rate from the study protocol after randomization, the optimal sample size was determined as 118 participants (59 per group).

Statistical processing was conducted using Jamovi Software, version 2.3.2 (Jamovi project, 2022; Sydney, Australia). Results are presented as the median (Me) and interquartile range [25%; 75%]. A Mann–Whitney U-test was carried out to compare the means of the two groups. The statistically significant differences between the Me [25%; 75%] on the 1st day and on the 9th day of hospitalization were assessed using the Wilcoxon W test. Spearman's correlation coefficient was used for associations between quantitative parameters. A p-value of <0.05 was the criterion for the statistical reliability of the obtained results.

3. Results

Vitamin D status was determined in 311 patients with confirmed COVID-19. Sixty-nine patients (22.2%) had a normal vitamin D status, 57 (18.3%) had an insufficiency, and 185 (59.5%) had a deficiency. The serum 25(OH)D levels were measured simultaneously in all stored samples after the interventional study.

One hundred and twenty-nine COVID-19 patients were randomly chosen from 311 and included in the interventional study: Group I (n = 65) and Group II (n = 64) (Figure 1). The study design is illustrated in Figure 1.

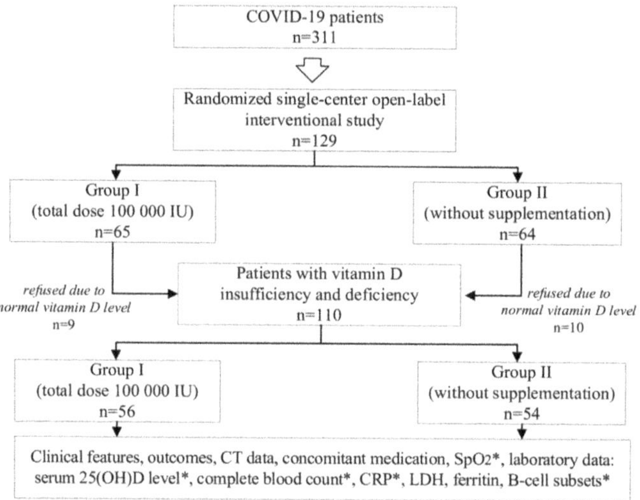

Figure 1. Study design. CT, computed tomography; SpO2, oxygen saturation; 25(OH)D, 25-hydroxyvitamin D; CRP, C-reactive protein; LDH, lactate dehydrogenase; *, both on the 1st and at the 9th day.

Patients' baseline characteristics are presented in Table 1.

Table 1. Patients' baseline characteristics ($n = 129$).

Parameters	Group I $n = 65$	Group II $n = 64$	p
Age, years, Me and IQR [25; 75]	57 [51; 66]	64 [55; 70]	0.03
Gender, female, n (%)	31 (47.7)	32 (50.0)	0.86
Days from symptoms onset to hospitalization, days, Me and IQR [25; 75]	8 [6;10]	8 [6;10]	0.37
Severe clinical course, n (%)	13 (20)	13 (20)	0.36
CT lung involvement, %, Me and IQR [25; 75]	39 [30; 50]	30 [20; 45]	0.06
CT grading, n (%) *0* *1* *2* *3* *4*	*4 (6)* *10 (15)* *37 (57)* *12 (17)* *2 (3)*	*2 (3)* *20 (30)* *33 (52)* *6 (10)* *3 (5)*	*0.29*
SpO2, %, Me and IQR [25; 75]	95 [92; 97]	95 [92; 97]	0.51
Supplemental Oxygenation, n (%)	38 (58.4)	32 (50)	0.35
BMI, kg/m^2, Me and IQR [25; 75]	29.5 [25.5; 32.9]	28.9 [25.5; 31.4]	0.41
Obesity, n (%)	28 (43.1)	22 (34.9)	0.42
DM type 2, n (%)	17 (26.2)	24 (38.1)	0.84
AH, n (%)	46 (70.8)	49 (76.6)	0.31
IHD, n (%)	16 (24.6)	14 (21.9)	0.12
Neutrophils, $\times 10^9$/L, Me and IQR [25; 75]	4.5 [2.4; 7.1]	4.2 [2.9; 5.9]	0.80
Lymphocytes, $\times 10^9$/L, Me and IQR [25; 75]	1.3 [0.8; 1.5]	1.04 [0.7; 1.4]	0.25
NLR, Me and IQR [25; 75]	3.7 [2.5; 7.6]	4.3 [2.7; 8]	0.15
CRP, mg/L, Me and IQR [25; 75]	48 [21; 134]	49 [18; 107]	0.73
Ferritin, ng/mL, Me and IQR [25; 75]	610 [243; 610]	446.1 [237; 825]	0.12
LDH, μ/L, Me and IQR [25; 75]	351 [261; 483]	327.5 [265; 495]	0.80
25(OH)D, ng/mL, Me and IQR [25; 75]	17.8 [11.7; 25.4]	15.4 [11.0; 22.9]	0.47
Vitamin D status, n (%) *Normal* *Insufficiency* *Deficiency*	*9 (13.8)* *20 (30.8)* *36 (55.4)*	*10 (15.6)* *11 (17.2)* *43 (67.2)*	*0.07*

CT, computed tomography; BMI, body mass index; DM, diabetes mellitus; IHD, ischemic heart disease; AH, arterial hypertension; NLR, neutrophil/lymphocyte ratio; CRP, C-reactive protein; LDH, lactate dehydrogenase; Me, median; IQR, interquartile range.

The groups were comparable and had no significant differences in baseline parameters, including serum 25(OH)D level, and clinical course of the disease, including CT data and oxygenation parameters ($p > 0.05$). Patients in Group I were significantly younger than patients in Group II ($p = 0.03$).

Participants in Group I had a normal serum 25(OH)D level in nine cases (13.8%), 20 patients (30.8%) had an insufficient level, and 36 (55.4%) had a deficiency. In Group II, 10 patients (15.6%) had a normal level, 11 (17.2%) patients had an insufficiency, and 43 (67.2%) had a deficiency. Thus, 19 (14.7%) of the 129 participants had a normal serum 25(OH)D level on the 1st day of hospital admission. Therefore, their data were not considered for further analysis of the cholecalciferol supplementation effect on the clinical and laboratory parameters (Table 2).

Table 2. Patients' baseline characteristics with vitamin D insufficiency and deficiency ($n = 110$).

Parameters	Group I $n = 56$	Group II $n = 54$	p
Age, years, Me and IQR [25; 75]	58 [50; 65]	64 [55; 70]	0.03
25(OH)D, ng/mL, Me and IQR [25; 75]	16.4 [11.0; 21.8]	13.9 [9.7; 17.4]	0.08
Vitamin D status, *n* (%) *Insufficiency* *Deficiency*	20 (36) 36 (64)	11 (20) 43 (80)	0.07
CT lung involvement, %, Me and IQR [25; 75]	42 [30; 48.5]	32.5 [20.5; 45]	0.21
CT grading, *n* (%) 1 2 3 4	11 (19.6) 33 (58.9) 11 (19.6) 1 (1.9)	19 (35.2) 26 (48.1) 6 (11.1) 3 (5.6)	0.77
SpO2, %, Me and IQR [25; 75]	95 [92; 97]	95 [92; 97]	0.50
Supplemental Oxygenation, *n* (%)	38 (68)	32 (59)	0.35
Neutrophils, $\times 10^9$/L, Me and IQR [25; 75]	4.3 [2.9; 6.0]	4.3 [2.9; 5.8]	0.53
Lymphocytes, $\times 10^9$/L, Me and IQR [25; 75]	1.3 [0.9; 1.5]	1.0 [0.7; 1.3]	0.16
NLR, Me + IQR [25; 75]	3.5 [2.2; 5.3]	4.7 [2.6; 7.3]	0.09
CRP, mg/L, Me and IQR [25; 75]	48.2 [22.7; 135.3]	47.5 [17.5; 99.0]	0.97
Ferritin, ng/mL, Me and IQR [25; 75]	559 [217; 925]	365 [229; 765]	0.21
LDH, μ/L, Me and IQR [25; 75]	351 [261; 516]	327 [261; 496]	0.84
Concomitant medication *GC* *Dexamethasone, n (%)* *Dexamethasone, mg* *Prednisolone, n (%)* *Prednisolone, mg*	*47 (84)* *136 [72; 214]* *15 (26.7)* *1295 [846; 1658]*	*43 (80)* *149 [112; 234]* *11 (20.3)* *1140 [375; 1703]*	*0.67* *0.99* *0.53* *0.60*
Anti-IL-6 receptor monoclonal antibodies, *n* (%) *Olokizumab, n (%)* *Levilimab, n (%)* *Tocilizumab, n (%)*	16 (28.5) *13 (23.2)* *2 (3.5)* *4 (7.14)*	18 (33.3) *15 (27.7)* *2 (3.7)* *3 (5.55)*	0.59 *0.39* *0.38* *0.74*
Anticoagulant therapy, *n* (%)	56 (100)	54 (100)	-
Antibiotics therapy, *n* (%)	8 (12.5)	11 (20.4)	0.26

25(OH)D, 25-hydroxyvitamin D; SpO2, oxygen saturation; NLR, neutrophil/lymphocyte ratio; CRP, C-reactive protein; LDH, lactate dehydrogenase; Me, median; IQR, interquartile range.

After the exclusion of subjects with a normal baseline 25(OH)D level, the groups remained comparable and had no significant differences in baseline parameters. There were no significant differences in serum 25(OH)D levels between the groups ($p = 0.08$). The analysis of the concomitant medication showed the absence of significant differences in the groups ($p > 0.05$). However, patients in Group I were still younger than patients in Group II ($p = 0.03$).

After the initiation of vitamin D supplementation, we performed a comparative analysis between the groups to assess the parameters on the 9th day of hospitalization (Table 3).

In Group I ($n = 56$), on the 9th day after 100,000 IU cholecalciferol supplementation, the median serum 25(OH)D level was 22.8 ng/mL [17.7;27.7]. The absolute and the relative Δ 25(OH)D were 6.2 ng/mL [2.4;11] and 40.7% [14.0;78.4], respectively. At the same time, in Group II ($n = 54$), the median level of 25(OH)D on the 9th day was 10.6 ng/mL [8.4;14.9]. The absolute and the relative Δ 25(OH)D were 2.6 ng/mL [−4.3;0] and −18.2% [−28.8;0], respectively.

Table 3. Patients' characteristics on the 9th day of hospitalization (*n* = 110).

Parameters	Group I *n* = 56	Group II *n* = 54	*p*
Vitamin D status, *n* (%)			
Normal	13 (23)	1 (2)	
Insufficiency	20 (36)	3 (6)	
Deficiency	23 (41)	50 (92)	<0.001
25(OH)D, ng/mL, Me and IQR [25; 75]	22.8 [17.7; 27.7]	10.6 [8.4; 14.9]	<0.001
Bed days, Me and IQR [25; 75]	18 [14; 22]	17 [14; 23]	0.87
Discharged, *n* (%)	56 (100)	54 (100)	0.93
ICU admission rates, *n* (%)	0	3 (6)	-
SpO2, %, Me and IQR [25; 75]	97 [96; 98]	97 [96; 98]	0.56
Supplemental Oxygenation, *n* (%)	27 (48)	28 (52)	0.70
Neutrophils, $\times 10^9$/L, Me and IQR [25; 75]	8.6 [5.1; 10.6]	6.4 [5.2; 8.6]	0.04
Lymphocytes, $\times 10^9$/L, Me and IQR [25; 75]	1.8 [1.3; 2.6]	1.58 [1.0; 2.0]	0.02
NLR, Me and IQR [25; 75]	4.5 [2.6; 6.9]	4.4 [2.7; 7.0]	0.71
CRP, mg/L, Me and IQR [25; 75]	2 [0.8; 4.7]	3 [1; 9]	0.02

25(OH)D, 25-hydroxyvitamin D; ICU, intensive care unit; SpO2, oxygen saturation; NLR, neutrophil/lymphocyte ratio; CRP, C-reactive protein; LDH, lactate dehydrogenase; Me, median; IQR, interquartile range.

Thus, we found significant differences on the 9th day of hospitalization between the groups in vitamin D status, serum 25(OH)D level, and Δ 25(OH)D (*p* < 0.001) (Figure 2). Furthermore, the serum 25(OH)D level on the 9th day was negatively associated with the number of bed days (r = −0.23, *p* = 0.006).

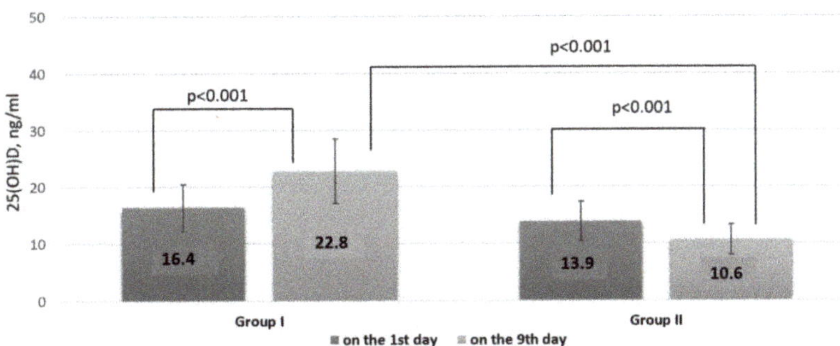

Figure 2. Serum 25(OH)D level before and after supplementation with 100,000 IU of cholecalciferol.

When comparing the results of the complete blood count, neutrophil and lymphocyte counts were significantly higher in Group I (*p* = 0.04; *p* = 0.02) (Figure 3).

Additionally, the CRP level on the 9th day of hospitalization was significantly lower among patients in Group I (*p* = 0.02). There was also a negative association between CRP and serum 25(OH)D level (r = −0.28, *p* = 0.02).

We next addressed whether the circulating B cell subsets were stable in groups I and II or changed between the 1st and 9th days of hospitalization. We also observed that the subpopulation of CD38++CD27 transitional B cells was decreased, while mature activated B cells (CD27+CD38+) and resting memory B cells (CD27+CD38−) frequencies increased in most subjects from both groups of COVID-19 patients (Figure 4). Furthermore, the relative numbers of mature naive CD27−CD38+ B cells, circulating plasmablast precursors CD27++CD38++, and double-negative CD27−CD38− B cells did not change significantly over time.

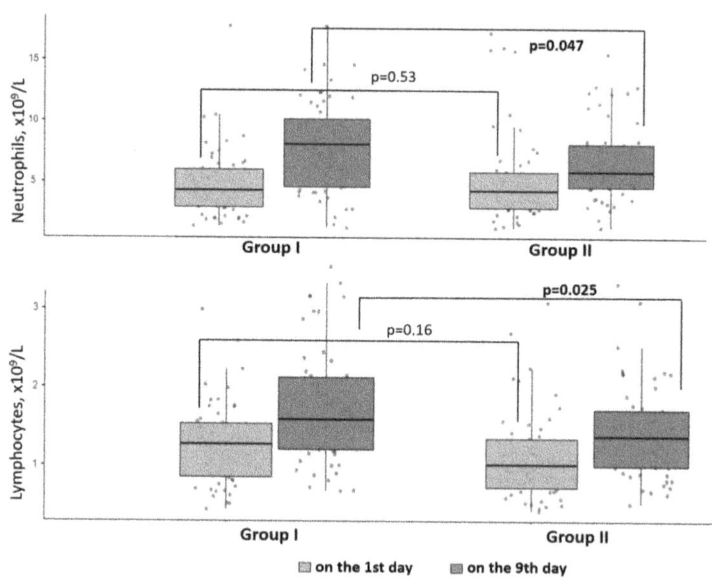

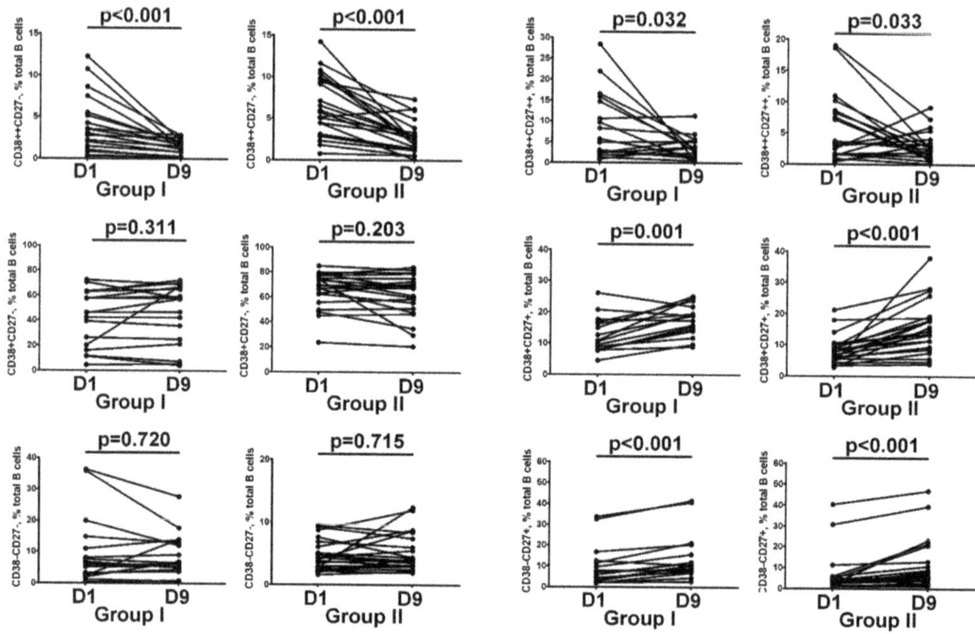

Figure 3. Neutrophil and lymphocyte counts before and after 100,000 IU cholecalciferol supplementation.

Figure 4. Main B cell subsets frequencies before and after 100,000 IU cholecalciferol supplementation. Numbers represent the percentages of the indicated B cell subset among the total B cell population. Each pair of connected points represents an individual subject. Day 1 vs. day 9 post-hospitalization intra-individual patient samples were compared by Wilcoxon matched-pairs signed rank test with two-tailed *p* value.

Finally, we compared the relative numbers of B cell subsets in Groups I and II on day 9 post-hospitalization. We found that after 100,000 IU cholecalciferol supplementation, patients with COVID-19 had decreased frequencies of CD38++CD27 transitional and mature naive CD27−CD38+ B cells if compared to Group II (1.43% (0.79; 2.08) vs. 2.74% (1.43; 3.91), p = 0.006 and 57.57% (25.15; 66.82) vs. 67.03% (51.16; 74.71), p = 0.02, respectively). We also noticed that the level of CD27−CD38− DN B cells was increased in patients from Group I when compared to Group II patients (6.21% (4.96; 12.91) vs. 4.19% (3.04; 7.33), p = 0.02) (Figure 5).

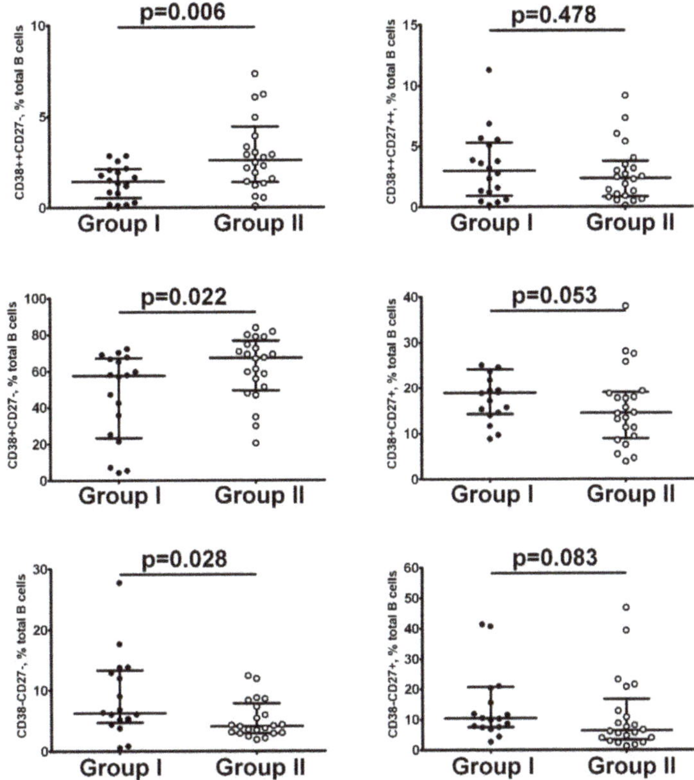

Figure 5. The impact of 100,000 IU cholecalciferol supplementation on B cell subset frequencies on the 9th day of hospitalization. Black circles—patients from Group I (n = 18); open circles—patients from Group II (n = 22). Numbers represent the percentages of the indicated B cell subset among the total B cell population. Each dot represents individual subject. Horizontal bars depict the group medians and interquartile ranges (Med (Q25; Q75)). Statistical analysis was performed with the Mann–Whitney U test.

4. Discussion

Recently published systematic reviews and meta-analyses have demonstrated that vitamin D insufficiency and deficiency are highly prevalent in patients with moderate and severe COVID-19 [1,17]. At present, there are sufficient data to demonstrate that a low serum 25(OH)D concentration increases the disease severity and risk of death in patients with COVID-19 [10,18]. A study conducted in the United States showed that patients with a positive SARS-CoV-2 test and a 25(OH)D concentration of 15 ng/mL compared to 40 ng/mL had a 20% greater risk of hospitalization (p = 0.009) and an increased risk of mortality by 53% (p = 0.001) [19]. The same results were obtained in a study of 311 hospitalized COVID-

19 patients: low serum 25(OH)D concentrations were found in patients with poorer clinical outcomes compared to those with a moderate and mild clinical course ($p = 0.001$) [10].

There is no definitive position regarding the additive therapeutic efficacy of vitamin D combined with standard treatment. Many studies showed a positive effect of vitamin D supplementation on the course and prognosis of COVID-19 [20–24]. In the observational study of Ling et al., cholecalciferol treatment using high-dose booster therapy (approximately $\geq$ 280,000 IU over a period of up to 7 weeks) was associated with a reduced risk of COVID-19 mortality in the cohort of 444 patients [25]. Torres et al. demonstrated that a daily dose of 10,000 IU of cholecalciferol increased serum 25(OH)D levels to 29 ng/mL on average vs. 19 ng/mL in the group receiving 2000 IU/day, after 7 and 14 days of treatment ($p < 0.0001$). The beneficial effect of supplementation with 10,000 IU/day was observed in participants with COVID-19 and acute respiratory distress syndrome who stayed in the hospital for 8 days. In contrast, those who received 2000 IU/day stayed for 29 days ($p = 0.03$) [26]. Another study using cholecalciferol at doses ranging from 224,000 to 500,000 IU over 3–14 days, in 132 COVID-19 patients with baseline serum 25(OH)D level of <30 ng/mL, showed a significant decrease in 14-day mortality (OR for survival: 2.14, 95% CI: 1.06 to 4.33, $p = 0.03$), when 25(OH)D levels of 31 ± 12 ng/mL on the 7th day and 35 ± 11 ng/mL on the 14th day were achieved [27].

A study from Spain included 527 patients with COVID-19; among them, 79 patients received calcifediol treatment (532 µg on entry and then 266 µg on days 3, 7, 14, 21, and 28). The calcifediol treatment was associated with significantly lower in-hospital mortality during the first 30 days [28]. The results of a large population-based study that compared patients receiving cholecalciferol or calcifediol (>250 µg of cholecalciferol or calcifediol as a bolus dose) showed that achieving serum 25(OH)D levels of $\geq$30 ng/mL improved the clinical outcomes of COVID-19 [29].

Calcifediol had better effects on COVID-19 outcome due to its ability to rapidly increase serum 25(OH)D levels compared to cholecalciferol. For example, the Gonen trial study found very little effect with high-dose vitamin D3 [27], while the Spanish studies generally demonstrate high beneficial effects for calcifediol [28]. The rapid response to treatment is of great importance since the main therapeutic effects of vitamin D are linked to the reduction in the viability and replication of SARS-CoV-2 and to the down-regulation of the production of the pro-inflammatory cytokines, thus decreasing the risk of the cytokine storm that causes organ damage [28,29].

We initiated this study assuming that vitamin D supplementation in hospitalized subjects might improve clinical outcomes and decrease inflammatory markers. At the end of the study, we showed that the use of 100,000 IU cholecalciferol supplementation in addition to standard COVID-19 therapy led to an increase in serum 25(OH)D levels by 40.7%, thus preventing vitamin D deficiency in the acute period of COVID-19. On the other hand, we have found no differences in mortality, ICU admission rates, and the average time of hospital stay between the supplemented and control groups. However, these results may be related to the fact that we did not achieve the recommended 25(OH)D level of 40 to 60 ng/mL. Similar results were obtained in other studies; there were no statistically significant differences between the groups receiving only standard therapy and the groups supplemented with vitamin D [24,30]. For example, in a study of 240 hospitalized patients with moderate and severe COVID-19, a single dose of vitamin D supplementation (200,000 IU) did not significantly reduce bed days, ICU admission rates, or mortality. These conflicting findings can probably be explained by the different doses and frequencies of administration, size samples, and principle of the sampling of patients included in the studies.

Even before the pandemic, there was evidences that GC therapy was associated with a decrease in serum 25(OH)D levels [31]. This is also a common phenomenon in patients with moderate and severe COVID-19. The results of the present study partially support these findings, confirming a negative dynamic of serum 25(OH)D levels in patients who received GC therapy without vitamin D supplementation.

The severity of COVID-19 is determined by the hyperactive immune response and cytokine storm, accompanied by dysregulated innate and adaptive immune responses [15,32]. Previous studies showed that COVID-19 patients deficient in vitamin D demonstrated higher levels of acute phase reactants compared to those with a normal serum 25(OH)D concentration [10,21]. Rastogi et al. showed a significant decrease in fibrinogen levels after cholecalciferol supplementation at a dose of 60,000 IU daily, with a therapeutic target of 25(OH)D > 50 ng/mL [22].

High serum CRP levels are key markers of disease progression and a risk factor for mortality in patients with severe COVID-19, and indicate the development of a cytokine storm. In the present study, patients receiving cholecalciferol therapy had a significant decrease in CRP level compared to patients without supplementation on the 9th day of hospitalization ($p = 0.02$). In addition, a negative correlation was found between the serum 25(OH)D level and CRP level on the 9th day of hospitalization (r = -0.28, $p = 0.02$).

Viral infections lead to dynamic changes in the peripheral blood leukocyte count and its subsets. A sustained decrease in the peripheral blood lymphocyte count is considered as an early indicator of severe/critically ill patients with COVID-19 [33]. Our previous study showed that vitamin D deficiency was associated with a greater decrease in absolute lymphocyte count [11]. When we compared complete blood counts on the 9th day of hospitalization, the group receiving a vitamin D bolus dose had higher levels of neutrophils and leukocytes ($p = 0.04$; $p = 0.02$). Similar results were previously described in the randomized, placebo-controlled clinical trial. Treatment with oral vitamin D resulted in a significant increase in the lymphocyte percentage and a decrease in the neutrophil-to-lymphocyte ratio in the patients [34].

Different types of innate and adaptive immune cells, including antigen-presenting dendritic cells, tissue-resident macrophages, and peripheral blood circulating monocytes, as well as T- and B-lymphocytes, express the vitamin D receptor and are able to modify their functional activity in response to vitamin D stimulation [35,36]. Currently, within adaptive immune responses, the effects of vitamin D have been studied well for T cells and their separate subsets, including CD4+ T cells, CD8+ T cells, and regulatory T cells [37,38]. However, the data on the influence of vitamin D on B cell functions are scarce.

Previously it was found that vitamin D receptor signaling could reduce or even prevent activation of B cells due to modulation of NF-κB mediated activation of naïve B cells [39] Moreover, vitamin D inhibited the proliferation of activated B cells, induced their apoptosis, and down-regulated generation of plasma cells and post-switch memory B cells [40]. Finally, vitamin D receptor activation induced in vitro production of anti-inflammatory IL-10 by B cells [41]. Thus, vitamin D deficiency could dramatically influence B cell functions and negatively affect humoral immune responses [42]. Interestingly, low levels of vitamin D have been closely linked with increased morbidity in several infectious diseases [43]. Oppositely, increased levels of vitamin D showed an inverse association with the development of several autoimmune diseases, such as systemic lupus erythematosus, thyrotoxicosis, multiply sclerosis, iridocyclitis, Crohn's disease, ulcerative colitis, psoriasis vulgaris etc., pointing to its anti-inflammatory role during immune responses [44]. Furthermore, during acute SARS-CoV-2 infection, vitamin D, in addition to standard therapies, significantly reduced several inflammation markers including the concentrations of C-reactive protein, IL-6, ferritin, and the neutrophil-to-lymphocyte ratio [45], as well as faster improvement in some clinical symptoms [46] and the SaO2/FiO2 ratio [47].

Previously, it was noted that almost all innate and adaptive immune cell subsets were altered during acute COVID-19 [48–50]. Furthermore, it was noted that key alterations in peripheral blood circulating B cell populations during the acute phase of SARS-CoV-2 infection were linked to the decreased levels of the relative and absolute number of B cells, increased frequencies of plasma cell precursors, atypical CD21-negative and activated B cells etc. [49,51,52]. Currently, we have shown that vitamin D supplementation could effectively decrease the relative numbers of activated CD38++CD27 transitional B cells and 'naïve' B cells in patients with acute COVID-19. Similarly, in systemic lupus erythematosus, vitamin D down-regulated B cell activity in vivo and in vitro [40,53,54]. Thus, altered B cell activation in circulation and vitamin D deficiency could be associated with an increased risk of autoimmune disorders and poor outcomes of infectious diseases. Finally, an increased level of activated B cells and altered B cell subset composition was detected in the peripheral blood of patients recovered after COVID-19 [55,56]; thus, future studies will be needed to investigate whether supplementation with regular vitamin D can prevent or reduce the risk of developing severe pathologies associated with 'post-COVID' syndrome.

To sum up, most hospitalized patients with moderate and severe COVID-19 had vitamin D deficiency or insufficiency. An increase in the serum 25(OH)D level may positively affect the course of COVID-19 and patient laboratory parameters. We demonstrated that bolus doses of cholecalciferol resulted in an increased serum 25(OH)D level, neutrophil and lymphocyte counts, and decreased CRP levels in the acute period of COVID-19. In addition, intake of 100,000 IU cholecalciferol was associated with a decline in CD38++CD27 transitional and CD27−CD38+ mature naive B cells and a rise in CD27−CD38− DN B cells. The present study corroborates the therapeutic efficacy of cholecalciferol as a supplement to standard therapy for COVID-19.

5. Limitations

Possible study limitations include the small sample size and application of small doses of vitamin D, thus the inability to reach the target 25(OH)D level. Moreover, in the present study, the treatment group was slightly younger than the control group; although, at the same time both groups did not differ in other assessed parameters. In addition, the study treatment design was open-label.

Author Contributions: T.L.K.: conceptualization, project administration, writing and editing; K.A.G.: data collection and writing; I.V.K.: data analysis and interpretation and writing; A.T.C.: data analysis and interpretation, and writing (review and editing); A.A.M.: data analysis and writing; A.D.A.: writing and editing; D.I.L.: data analysis and writing; E.K.Z.: writing and editing; O.V.K.: conceptualization, project administration and data analysis; A.S.G.: conceptualization, writing and editing; W.B.G.: conceptualization and project administration; E.V.S.: conceptualization and project administration. All authors have read and agreed to the published version of the manuscript.

Funding: This work was financially supported by the Ministry of Science and Higher Education of the Russian Federation (Agreement No. 075-15-2022-301).

Institutional Review Board Statement: The study was conducted according to the guidelines of the Declaration of Helsinki and approved by the Almazov National Medical Research Centre Local Ethics Committee (1011-20-02C). The study has been registered on clinicaltrials.gov as "Severity of COVID-19 and Vitamin D Supplementation"; ID: NCT05166005.

Informed Consent Statement: Informed consent was obtained from all the subjects involved in the study.

Data Availability Statement: The data generated and analyzed during this study are included in this published article. Additional information is available from the corresponding author on reasonable request.

Acknowledgments: We acknowledge the healthcare professionals from the Almazov National Medical Research Centre for their assistance with collecting the data.

Conflicts of Interest: The Sunlight; Nutrition; and Health Research Center (W.B.G.) receives funding from Bio-Tech Pharmacal Inc. (Fayetteville; AR; USA). The other authors declare no conflict of interest.

References

1. Kazemi, A.; Mohammadi, V.; Aghababaee, S.K.; Golzarand, M.; Clark, C.C.T.; Babajafari, S. Association of Vitamin D status with SARS-CoV-2 infection or COVID-19 severity: A systematic review and meta-analysis. *Adv. Nutr.* **2021**, *12*, 1636–1658. [CrossRef] [PubMed]
2. Berry, D.J.; Hesketh, K.; Power, C.; Hypponen, E. Vitamin D status has a linear association with seasonal infections and lung function in British adults. *Br. J. Nutr.* **2011**, *106*, 1433–1440. [CrossRef] [PubMed]

3. Kaufman, H.W.; Niles, J.K.; Kroll, M.H.; Bi, C.; Holick, M.F. SARS-CoV-2 positivity rates associated with circulating 25-hydroxyvitamin D levels. *PLoS ONE* **2020**, *15*, e0239252. [CrossRef] [PubMed]

4. Mercola, J.; Grant, W.B.; Wagner, C.L. Evidence Regarding Vitamin D and Risk of COVID-19 and Its Severity. *Nutrients* **2020**, *12*, 3361. [CrossRef]

5. Khammissa, R.; Fourie, J.; Motswaledi, M.H.; Ballyram, R.; Lemmer, J.; Feller, L. The Biological Activities of Vitamin D and Its Receptor in Relation to Calcium and Bone Homeostasis, Cancer, Immune and Cardiovascular Systems, Skin Biology, and Oral Health. *BioMed Res. Int.* **2018**, *2018*, 9276380. [CrossRef]

6. Agraz-Cibrian, J.M.; Giraldo, D.M.; Urcuqui-Inchima, S. 25-Dihydroxyvitamin D3 induces formation of neutrophil extracellular trap-like structures and modulates the transcription of genes whose products are neutrophil extracellular trap-associated proteins: A pilot study. *Steroids* **2019**, *141*, 14–22. [CrossRef]

7. Bouillon, R.; Marcocci, C.; Carmeliet, G.; Bikle, D.; White, J.H.; Dawson-Hughes, B.; Lips, P.; Munns, C.F.; Lazaretti-Castro, M.; Giustina, A.; et al. Skeletal and extraskeletal actions of Vitamin D: Current evidence and outstanding questions. *Endocr. Rev.* **2019**, *40*, 1109–1151. [CrossRef]

8. Xu, J.; Yang, J.; Chen, J.; Luo, Q.; Zhang, Q.; Zhang, H. Vitamin D alleviates lipopolysaccharide-induced acute lung injury via regulation of the renin-angiotensin system. *Mol. Med. Rep.* **2017**, *16*, 7432–7438. [CrossRef]

9. Karonova, T.L.; Andreeva, A.T.; Golovatuk, K.A.; Bykova, E.S.; Simanenkova, A.V.; Vashukova, M.A.; Grant, W.B.; Shlyakhto, E.V. Low 25(OH)D Level Is Associated with Severe Course and Poor Prognosis in COVID-19. *Nutrients* **2021**, *13*, 3021. [CrossRef]

10. Karonova, T.L.; Kudryavtsev, I.V.; Golovatyuk, K.A.; Aquino, A.D.; Kalinina, O.V.; Chernikova, A.T.; Zaikova, E.K.; Lebedev, D.A.; Bykova, E.S.; Golovkin, A.S.; et al. Vitamin D Status and Immune Response in Hospitalized Patients with Moderate and Severe COVID-19. *Pharmaceuticals* **2022**, *15*, 305. [CrossRef]

11. Carpagnano, G.E.; Di Lecce, V.; Quaranta, V.N.; Zito, A.; Buonamico, E.; Capozza, E.; Palumbo, A.; Di Gioia, G.; Valerio, V.N.; Resta, O. Vitamin D deficiency as a predictor of poor prognosis in patients with acute respiratory failure due to COVID-19. *J. Endocrinol. Investig.* **2021**, *44*, 765–771. [CrossRef] [PubMed]

12. Dissanayake, H.A.; de Silva, N.L.; Sumanatilleke, M.; de Silva, S.D.N.; Gamage, K.K.K.; Dematapitiya, C.; Kuruppu, D.C.; Ranasinghe, P.; Pathmanathan, S.; Katulanda, P. Prognostic and therapeutic role of vitamin D in COVID-19: Systematic review and meta-analysis. *J. Clin. Endocrinol. Metab.* **2021**, *107*, 1484–1502. [CrossRef] [PubMed]

13. Prevention, diagnosis and treatment of new coronavirus infection (COVID-19). In *Temporary Guidelines*; Version 9; Moscow, Russia, 2020. (In Russian)

14. Pludowski, P.; Holick, M.F.; Grant, W.B.; Konstantynowicz, J.; Mascarenhas, M.R.; Haq, A.; Povoroznyuk, V.; Balatska, N.; Barbosa, A.P.; Karonova, T.; et al. Vitamin D supplementation guidelines. *J. Steroid Biochem. Mol. Biol.* **2018**, *175*, 125–135. [CrossRef] [PubMed]

15. Golovkin, A.; Kalinina, O.; Bezrukikh, V.; Aquino, A.; Zaikova, E.; Karonova, T.; Melnik, O.; Vasilieva, E.; Kudryavtsev, I. Imbalanced Immune Response of T-Cell and B-Cell Subsets in Patients with Moderate and Severe COVID-19. *Viruses* **2021**, *13*, 1966. [CrossRef]

16. Hanley, P.; Sutter, J.A.; Goodman, N.G.; Du, Y.; Sekiguchi, D.R.; Meng, W.; Rickels, M.R.; Naji, A.; Luning Prak, E.T. Circulating B cells in type 1 diabetics exhibit fewer maturation-associated phenotypes. *Clin. Immunol.* **2017**, *183*, 336–343. [CrossRef]

17. Kaya, M.O.; Pamukçu, E.; Yakar, B. The role of vitamin D deficiency on COVID-19: A systematic review and meta-analysis of observational studies. *Epidemiol. Health* **2021**, *43*, e2021074. [CrossRef]

18. Grant, W.B.; Anouti, F.A.; Boucher, B.J.; Dursun, E.; Gezen-Ak, D.; Jude, E.B.; Karonova, T.; Pludowski, P. A Narrative Review of the Evidence for Variations in Serum 25-Hydroxyvitamin D Concentration Thresholds for Optimal Health. *Nutrients* **2022**, *14*, 639. [CrossRef]

19. Seal, K.H.; Bertenthal, D.; Carey, E.; Grunfeld, C.; Bikle, D.D.; Lu, C.M. Association of Vitamin D Status and COVID-19-Related Hospitalization and Mortality. *J. Gen. Intern. Med.* **2022**, *37*, 853–861. [CrossRef]

20. Annweiler, G.; Corvaisier, M.; Gautier, J.; Dubée, V.; Legrand, E.; Sacco, G.; Annweiler, C. Vitamin D Supplementation Associated to Better Survival in Hospitalized Frail Elderly COVID-19 Patients: The GERIA-COVID Quasi-Experimental Study. *Nutrients* **2020**, *12*, 3377. [CrossRef]

21. Efird, J.T.; Anderson, E.J.; Jindal, C.; Redding, T.S.; Thompson, A.D.; Press, A.M.; Upchurch, J.; Williams, C.D.; Choi, Y.M.; Suzuki, A. The Interaction of Vitamin D and Corticosteroids: A Mortality Analysis of 26,508 Veterans Who Tested Positive for SARS-CoV-2. *Int. J. Environ. Res. Public Health* **2022**, *19*, 447. [CrossRef]

22. Rastogi, A.; Bhansali, A.; Khare, N.; Suri, V.; Yaddanapudi, N.; Sachdeva, N.; Puri, G.D.; Malhotra, P. Short term, high-dose vitamin D supplementation for COVID-19 disease: A randomised, placebo-controlled, study (SHADE study). *Postgrad. Med. J.* **2022**, *98*, 87–90. [CrossRef]

23. Shah, K.; Varna, V.P.; Sharma, U.; Mavalankar, D. Does vitamin D supplementation reduce COVID-19 severity?: A systematic review. *QJM Int. J. Med.* **2022**, *15*, hcac040. [CrossRef] [PubMed]

24. Murai, I.H.; Fernandes, A.L.; Sales, L.P.; Pinto, A.J.; Goessler, K.F.; Duran, C.S.C.; Silva, C.B.R.; Franco, A.S.; Macedo, M.B.; Dalmolin, H.H.H.; et al. Effect of a Single High Dose of Vitamin D3 on Hospital Length of Stay in Patients With Moderate to Severe COVID-19: A Randomized Clinical Trial. *JAMA* **2021**, *325*, 1053–1060. [CrossRef] [PubMed]

25. Ling, S.F.; Broad, E.; Murphy, R.; Pappachan, J.M.; Pardesi-Newton, S.; Kong, M.F.; Jude, E.B. High-Dose Cholecalciferol Booster Therapy is Associated with a Reduced Risk of Mortality in Patients with COVID-19: A Cross-Sectional Multi-Centre Observational Study. *Nutrients* **2020**, *12*, 3799. [CrossRef] [PubMed]

26. Torres, M.; Casado, G.; Vigón, L.; Rodríguez-Mora, S.; Mateos, E.; Ramos-Martín, F.; López-Wolf, D.; Sanz-Moreno, J.; Ryan-Murua, P.; Taboada-Martínez, M.L.; et al. Changes in the immune response against SARS-CoV-2 in individuals with severe COVID-19 treated with high dose of vitamin D. *Biomed. Pharmacother.* **2022**, *150*, 112965. [CrossRef] [PubMed]

27. Gonen, M.S.; Alaylioglu, M.; Durcan, E.; Ozdemir, Y.; Sahin, S.; Konukoglu, D.; Nohut, O.K.; Urkmez, S.; Kucukece, B.; Balkan, I.I.; et al. Rapid and Effective Vitamin D Supplementation May Present Better Clinical Outcomes in COVID-19 (SARS-CoV-2) Patients by Altering Serum INOS1, IL1B, IFNg, Cathelicidin-LL37, and ICAM1. *Nutrients* **2021**, *13*, 4047. [CrossRef]

28. Alcala-Diaz, J.F.; Limia-Perez, L.; Gomez-Huelgas, R.; Martin-Escalante, M.D.; Cortes-Rodriguez, B.; Zambrana-Garcia, J.L.; Entrenas-Castillo, M.; Perez-Caballero, A.I.; López-Carmona, M.D.; Garcia-Alegria, J.; et al. Calcifediol Treatment and Hospital Mortality Due to COVID-19: A Cohort Study. *Nutrients* **2021**, *13*, 1760. [CrossRef]

29. Oristrell, J.; Oliva, J.C.; Casado, E.; Subirana, I.; Domínguez, D.; Toloba, A.; Balado, A.; Grau, M. Vitamin D supplementation and COVID-19 risk: A population-based, cohort study. *J. Endocrinol. Investig.* **2022**, *45*, 167–179. [CrossRef]

30. Soliman, A.R.; Abdelaziz, T.S.; Fathy, A. Impact of Vitamin D Therapy on the Progress COVID-19: Six Weeks Follow-Up Study of Vitamin D Deficient Elderly Diabetes Patients. *Proc. Singap. Healthc.* **2021**, *31*, 1–5. [CrossRef]

31. Mazziotti, G.; Formenti, A.M.; Frara, S.; Doga, M.; Giustina, A. Vitamin D and Glucocorticoid-Induced Osteoporosis. *Front. Horm. Res.* **2018**, *50*, 149–160. [CrossRef]

32. Kudryavtsev, I.; Kalinina, O.; Bezrukikh, V.; Melnik, O.; Golovkin, A. The significance of phenotyping and quantification of plasma extracellular vesicles levels using high-sensitivity flow cytometry during covid-19 treatment. *Viruses* **2021**, *13*, 767. [CrossRef]

33. Gao, Y.D.; Ding, M.; Dong, X.; Zhang, J.J.; Kursat Azkur, A.; Azkur, D.; Gan, H.; Sun, Y.L.; Fu, W.; Li, W.; et al. Risk factors for severe and critically ill COVID-19 patients: A review. *Allergy* **2021**, *76*, 428–455. [CrossRef] [PubMed]

34. Maghbooli, Z.; Sahraian, M.A.; Jamalimoghadamsiahkali, S.; Asadi, A.; Zarei, A.; Zendehdel, A.; Varzandi, T.; Mohammadnabi, S.; Alijani, N.; Karimi, M.; et al. Treatment With 25-Hydroxyvitamin D3 (Calcifediol) Is Associated With a Reduction in the Blood Neutrophil-to-Lymphocyte Ratio Marker of Disease Severity in Hospitalized Patients With COVID-19: A Pilot Multicenter, Randomized, Placebo-Controlled, Double-Blinded Clinical Trial. *Endocr. Pract. Off. J. Am. Coll. Endocrinol. Am. Assoc. Clin. Endocrinol.* **2021**, *27*, 1242–1251. [CrossRef]

35. Deluca, H.F.; Cantorna, M.T. Vitamin D: Its role and uses in immunology. *FASEB J.* **2001**, *15*, 2579–2585. [CrossRef] [PubMed]

36. Ao, T.; Kikuta, J.; Ishii, M. The Effects of Vitamin D on Immune System and Inflammatory Diseases. *Biomolecules* **2021**, *11*, 1624. [CrossRef]

37. Hayes, C.E.; Hubler, S.L.; Moore, J.R.; Barta, L.E.; Praska, C.E.; Nashold, F.E. Vitamin D Actions on CD4(+) T Cells in Autoimmune Disease. *Front. Immunol.* **2015**, *6*, 100. [CrossRef]

38. Todosenko, N.; Vulf, M.; Yurova, K.; Khaziakhmatova, O.; Mikhailova, L.; Litvinova, L. Causal Links between Hypovitaminosis D and Dysregulation of the T Cell Connection of Immunity Associated with Obesity and Concomitant Pathologies. *Biomedicines* **2021**, *9*, 1750. [CrossRef]

39. Geldmeyer-Hilt, K.; Heine, G.; Hartmann, B.; Baumgrass, R.; Radbruch, A.; Worm, M. 1,25-dihydroxyvitamin D3 impairs NF-κB activation in human naïve B cells. *Biochem. Biophys. Res. Commun.* **2011**, *407*, 699–702. [CrossRef]

40. Chen, S.; Sims, G.P.; Chen, X.X.; Gu, Y.Y.; Chen, S.; Lipsky, P.E. Modulatory effects of 1,25-dihydroxyvitamin D3 on human B cell differentiation. *J. Immunol.* **2007**, *179*, 1634–1647. [CrossRef]

41. Heine, G.; Niesner, U.; Chang, H.D.; Steinmeyer, A.; Zügel, U.; Zuberbier, T.; Radbruch, A.; Worm, M. 1,25-dihydroxyvitamin D(3) promotes IL-10 production in human B cells. *Eur. J. Immunol.* **2008**, *38*, 2210–2218. [CrossRef]

42. Rolf, L.; Muris, A.H.; Hupperts, R.; Damoiseaux, J. Vitamin D effects on B cell function in autoimmunity. *Ann. N. Y. Acad. Sci.* **2014**, *1317*, 84–91. [CrossRef] [PubMed]

43. Amrein, K.; Scherkl, M.; Hoffmann, M.; Neuwersch-Sommeregger, S.; Köstenberger, M.; Tmava Berisha, A.; Martucci, G.; Pilz, S.; Malle, O. Vitamin D deficiency 2.0: An update on the current status worldwide. *Eur. J. Clin. Nutr.* **2020**, *74*, 1498–1513. [CrossRef] [PubMed]

44. Murdaca, G.; Tonacci, A.; Negrini, S.; Greco, M.; Borro, M.; Puppo, F.; Gangemi, S. Emerging role of vitamin D in autoimmune diseases: An update on evidence and therapeutic implications. *Autoimmun. Rev.* **2019**, *18*, 102350. [CrossRef]

45. Lakkireddy, M.; Gadiga, S.G.; Malathi, R.D.; Karra, M.L.; Raju, I.; Ragini; Chinapaka, S.; Baba, K.; Kandakatla, M. Author Correction: Impact of daily high dose oral vitamin D therapy on the inflammatory markers in patients with COVID-19 disease. *Sci. Rep.* **2021**, *11*, 17652. [CrossRef] [PubMed]

46. Sabico, S.; Enani, M.A.; Sheshah, E.; Aljohani, N.J.; Aldisi, D.A.; Alotaibi, N.H.; Alshingetti, N.; Alomar, S.Y.; Alnaami, A.M.; Amer, O.E.; et al. Effects of a 2-Week 5000 IU versus 1000 IU Vitamin D3 Supplementation on Recovery of Symptoms in Patients with Mild to Moderate Covid-19: A Randomized Clinical Trial. *Nutrients* **2021**, *13*, 2170. [CrossRef]

47. Elamir, Y.M.; Amir, H.; Lim, S.; Rana, Y.P.; Lopez, C.G.; Feliciano, N.V.; Omar, A.; Grist, W.P.; Via, M.A. A randomized pilot study using calcitriol in hospitalized COVID-19 patients. *Bone* **2022**, *154*, 116175. [CrossRef]

48. Taefehshokr, N.; Taefehshokr, S.; Heit, B. Mechanisms of Dysregulated Humoral and Cellular Immunity by SARS-CoV-2. *Pathogens* **2020**, *9*, 1027. [CrossRef]

49. Mangge, H.; Kneihsl, M.; Schnedl, W.; Sendlhofer, G.; Curcio, F.; Domenis, R. Immune Responses against SARS-CoV-2—Questions and Experiences. *Biomedicines* **2021**, *9*, 1342. [CrossRef]
50. Gusev, E.; Sarapultsev, A.; Solomatina, L.; Chereshnev, V. SARS-CoV-2-Specific Immune Response and the Pathogenesis of COVID-19. *Int. J. Mol. Sci.* **2022**, *23*, 1716. [CrossRef]
51. Shafqat, A.; Shafqat, S.; Salameh, S.A.; Kashir, J.; Alkattan, K.; Yaqinuddin, A. Mechanistic Insights Into the Immune Pathophysiology of COVID-19; An In-Depth Review. *Front. Immunol.* **2022**, *13*, 835104. [CrossRef]
52. Kudryavtsev, I.; Rubinstein, A.; Golovkin, A.; Kalinina, O.; Vasilyev, K.; Rudenko, L.; Isakova-Sivak, I. Dysregulated Immune Responses in SARS-CoV-2-Infected Patients: A Comprehensive Overview. *Viruses* **2022**, *14*, 1082. [CrossRef] [PubMed]
53. Terrier, B.; Derian, N.; Schoindre, Y.; Chaara, W.; Geri, G.; Zahr, N.; Mariampillai, K.; Rosenzwajg, M.; Carpentier, W.; Musset, L.; et al. Restoration of regulatory and effector T cell balance and B cell homeostasis in systemic lupus erythematosus patients through vitamin D supplementation. *Arthritis Res. Ther.* **2012**, *17*, R221. [CrossRef]
54. Dankers, W.; Colin, E.M.; van Hamburg, J.P.; Lubberts, E. Vitamin D in Autoimmunity: Molecular Mechanisms and Therapeutic Potential. *Front. Immunol.* **2017**, *7*, 697. [CrossRef] [PubMed]
55. Shuwa, H.A.; Shaw, T.N.; Knight, S.B.; Wemyss, K.; McClure, F.A.; Pearmain, L.; Prise, I.; Jagger, C.; Morgan, D.J.; Khan, S.; et al. Alterations in T and B cell function persist in convalescent COVID-19 patients. *Med* **2021**, *2*, 720–735.e4. [CrossRef] [PubMed]
56. Kudryavtsev, I.V.; Arsentieva, N.A.; Batsunov, O.K.; Korobova, Z.R.; Khamitova, I.V.; Isakov, D.V.; Kuznetsova, R.N.; Rubinstein, A.A.; Stanevich, O.V.; Lebedeva, A.A.; et al. Alterations in B Cell and Follicular T-Helper Cell Subsets in Patients with Acute COVID-19 and COVID-19 Convalescents. *Curr. Issues Mol. Biol.* **2022**, *44*, 194–205. [CrossRef] [PubMed]

MDPI

St. Alban-Anlage 66

4052 Basel

Switzerland

Tel. +41 61 683 77 34

Fax +41 61 302 89 18

www.mdpi.com

Nutrients Editorial Office

E-mail: nutrients@mdpi.com

www.mdpi.com/journal/nutrients